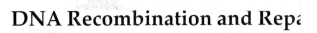

# DNA Recombination and Repa

# Frontiers in Molecular Biology

SERIES EDITORS

**B. D. Hames**

*Department of Biochemistry*
*and Molecular Biology*
*University of Leeds, Leeds LS2 9JT, UK*

**D. M. Glover**

*Department of Genetics,*
*University of Cambridge,*
*Cambridge, UK*

## TITLES IN THE SERIES

# DNA Recombination and Repair

EDITED BY

## Paul J. Smith and Christopher J. Jones

*Department of Pathology*
*University of Wales College of Medicine*
*Cardiff, Wales*

OXFORD

UNIVERSITY PRESS

# OXFORD

UNIVERSITY PRESS

Great Clarendon Street, Oxford OX2 6DP

Oxford University Press is a department of the University of Oxford
and furthers the University's aim of excellence in research, scholarship,
and education by publishing worldwide in

Oxford  New York

Athens  Auckland  Bangkok  Bogotá  Buenos Aires  Calcutta
Cape Town  Chennai  Dar es Salaam  Delhi  Florence  Hong Kong  Istanbul
Karachi  Kuala Lumpur  Madrid  Melbourne  Mexico City  Mumbai
Nairobi  Paris  São Paulo  Singapore  Taipei  Tokyo  Toronto  Warsaw

with associated companies in Berlin  Ibadan

Oxford is a registered trade mark of Oxford University Press
in the UK and in certain other countries

Published in the United States
by Oxford University Press Inc., New York

A catalogue record for this book is available from the British Library

Library of Congress Cataloging in Publication Data
DNA recombination and repair / edited by Paul J. Smith and Christopher
J. Jones
(Frontiers in molecular biology; 22)
Includes bibliographical references and index.
1. DNA repair.   2. Genetic recombination.   I. Smith, Paul J.
(Paul James), 1953–   .  II. Jones, Christopher J. (Christopher
John), 1965–   .  III. Series.
QH467.D156  1999     572.8'6459–dc21      99–32315

ISBN 0 19 963707 5 (Hbk)
ISBN 0 19 963706 7 (Pbk)

Typeset by Footnote Graphics, Warminster, Wiltshire

Printed by The Bath Press, Avon

# Preface

All organisms whether of single or multicellular origin invest in the preservation of the integrity of their genetic material. For a number of years there was a tacit acceptance by the wider scientific community that there must be mechanisms in place to maintain genomic stability. Furthermore, there was an expectation that studying cells under stress, particularly after the induction of DNA damage, would help to uncover the cellular tactics involved. However, many researchers were dismayed by the lack of information on the underlying processes which could support such mechanisms and the need to describe pathways in nebulous terms. The discovery of the significant overlap of DNA repair mechanisms and basic cellular transactions such as DNA replication and basal transcription gave the field new impetus. The journal *Science* recognized years of careful and patient research when awarding DNA repair the rather cumbersome title 'Molecule of the Year' in 1994. With the identification of repair genes and characterization of their biochemical activities it is becoming increasingly apparent that lesion handling strategies lie at the heart of the cell's integrated response to a variety of stresses.

No short book can address the myriad of issues that have projected the cellular processes of DNA repair and recombination to centre stage for so many researchers working in what were originally thought to be disparate fields. This publication has attempted to highlight a selection of some of the issues currently at the frontiers of our understanding of these processes.

Although repair was originally viewed as a process that prevented the 'fixation' of DNA lesions, limiting chromosome aberrations and mutation, it is now realistic to view DNA repair as only one facet of the cellular response to DNA damage. As such repair must be integrated with other pathways to effect physiologically and genetically satisfactory outcomes for the individual cell, affected tissue system, and organism. It is becoming increasingly clear that the fidelity of these integrated responses is affected by the accrual of some of the genetic lesions that constitute the multistep progression typical of neoplasia. The cancer predisposition disorders offer various insights into the consequences of the dysfunction of integration. Thus at many points the chapter authors have touched upon the relevance of DNA repair and recombination for human ill health, in particular cancer.

Integration of DNA repair with other cellular processes has occurred throughout evolution, linking pathways for stress signalling, adaptive responses, recombination, recovery mechanisms, cell-cycle delay, transcription and replication, and not least the molecular guillotine of programmed cell death, apoptosis. Although clearly important to the resolution of stress responses in normal and tumour tissues, the molecular pathways for apoptosis induction are not formally considered in this book but are clearly referenced where appropriate.

The book starts with the vital issue of DNA structure and folding dynamics, which eventually all processes involving protein–DNA interactions must address. David Leach's introductory chapter on molecular processing of DNA folding anomalies in *Escherichia coli* explores the consequences of different DNA structures and repeat sequence modification, leading to a discussion on the inhibition of DNA replication and initiation of homologous recombination by SbcCD protein. Developing the relationships between double-strand break repair and V(D)J recombination the chapter by Belinda Singleton and Penny Jeggo considers how the important features of recognition and cleavage of signal sequences impact upon double-strand break repair, with implications for homologous recombination and non-homologous end-joining.

The following two chapters consider different strategies for dealing with DNA lesions or mismatches. The ability of *E. coli* and yeast cells to enact translesion replication, 'usually a strategy of last resort' is examined in detail in the chapter by Christopher Lawrence and Roger Woodgate, with speculations on translesion repli-cation in humans. Bacterial mismatch repair is briefly discussed in the chapter by Peter Karran and Margherita Bignami. The dramatic role of mismatch repair in human cancers is dealt with in more detail together with an important overview of areas ripe for the future study, such as the linkages between mismatch repair, transcription-coupled excision repair and recombinational repair. Again this chapter echoes the need to view the responses to DNA damage as an integrated response with a discussion of mismatch repair, cell-cycle checkpoints, and apoptosis.

One of the great success stories in DNA repair research, resulting from the tenacious work of many groups, has been the elucidation of much of the enzymology of excision repair. An excellent chapter by Hanspeter Naegeli explores the complex enzymology of human nucleotide excision repair, highlighting the substrate dis-crimination problem and potential sensors. This discovery, now a paradigm for the selective repair of DNA lesions for biological advantage, is discussed in an overview of transcription-coupled repair and global genome repair in yeast and humans by Marcel Tijsterman, Richard Verhage, and Jaap Brouwer. The authors have also ex-plored the connection between nucleotide excision repair and other repair pathways such as mismatch repair. Along with xeroderma pigmentosum, the rare cancer-prone disorder ataxia–telangiectasia has acted to reveal the complexity of the re-sponses of human cells to DNA damage, and the chapter by Martin Lavin and Kum Kum Khanna details the nature of the ATM protein and gene family. The ATM story has, perhaps more than any other, served to reveal the multiple levels at which stress responses integrate with proliferation control.

The final chapter by Paul Smith and Chris Jones takes up the theme of the integra-tion of DNA repair into other cellular pathways by viewing the field primarily from the perspective of the tumour suppressor gene product p53. In doing so this chapter echoes themes developed in other chapters, including a perspective on cancer predisposition in the general population. In particular, the overview touches upon the induction, processing, and repair of DNA damage and the integration of these events with proliferation control. One area addressed is the spectrum of unusual forms of DNA damage induced by anticancer agents and the consequences for DNA repair and drug resistance, since an expanding area of anticancer research involves the identification of specific repair inhibitors. Importantly, the 'physiological' role of endogenous damage induction, as a programmed route for the development of senescence barriers to continuous cell proliferation, is highlighted through the events controlling telomere dynamics.

We thank the authors for responding to our initial suggestions with vigour, and expanding upon the themes in a manner that reflects the exciting nature of research in this field, and connects with subjects addressed in other publications in the 'Frontiers' series.

*Cardiff*                                                                                    P.J.S.
August 1999                                                                         C.J.J.

# Contents

## 5   Enzymology of human nucleotide excision repair    99

HANSPETER NAEGELI

## 6 Transcription-coupled and global genome repair in yeast and humans

MARCEL TIJSTERMAN, RICHARD A. VERHAGE, and JAAP BROUWER

## 7 The *ATM* gene and stress response

MARTIN F. LAVIN and KUM KUM KHANNA

## 8 p53 and the integrated response to DNA damage     202

PAUL J. SMITH and CHRISTOPHER J. JONES

# Contributors

M. BIGNAMI
Istituto Superiore di Sanitá, Viale Regina Elena, 299, 00161 Rome, Italy.

JAAP BROUWER
Laboratory of Molecular Genetics, Leiden Institute of Chemistry, Gorlaeus Laboratories, Leiden University, PO Box 9502, 2300 RA Leiden, The Netherlands.

PENNY A. JEGGO
MRC Cell Mutation Unit, University of Sussex, Falmer, Brighton BN1 9RR, UK.

CHRISTOPHER J. JONES
Department of Pathology, University of Wales College of Medicine, Heath Park, Cardiff CF4 4XN, Wales.

P. KARRAN
Imperial Cancer Research Fund, Clare Hall Laboratories, Blanche Lane, South Mimms, Potters Bar, Herts EN6 3LD, UK.

KUM KUM KHANNA
Queensland Institute of Medical Research, The Bancroft Centre, PO Royal Brisbane Hospital, Herston, Brisbane, 4029, Australia; and Dept Surgery, University of Queensland, St Lucia, Brisbane, Australia.

MARTIN F. LAVIN
Queensland Institute of Medical Research, The Bancroft Centre, PO Royal Brisbane Hospital, Herston, Brisbane, 4029, Australia; and Dept Surgery, University of Queensland, St Lucia, Brisbane, Australia.

CHRISTOPHER LAWRENCE
Department of Biochemistry and Biophysics, School of Medicine and Dentistry, University of Rochester, 601 Elmwood Ave, Rochester, NY 1462, USA.

DAVID R. F. LEACH
Institute of Cell and Molecular Biology, University of Edinburgh, King's Buildings, Edinburgh EH9 3JR, Scotland.

HANSPETER NAEGELI
Institute of Pharmacology and Toxicology, University of Zurich-Tierspital, Winterthurerstrasse 260, 8057 Zurich, Switzerland.

BELINDA K. SINGLETON
MRC Cell Mutation Unit, University of Sussex, Falmer, Brighton, BN1 9RR, UK.
Current address: Bristol Institute for Transfusion Sciences, Southmead Road, Bristol, BS10 5ND, UK.

PAUL J. SMITH

Department of Pathology, University of Wales College of Medicine, Heath Park, Cardiff CF4 4XN, Wales.

MARCEL TIJSTERMAN

Laboratory of Molecular Genetics, Leiden Institute of Chemistry, Gorlaeus Laboratories, Leiden University, PO Box 9502, 2300 RA Leiden, The Netherlands.

RICHARD A. VERHAGE

Laboratory of Molecular Genetics, Leiden Institute of Chemistry, Gorlaeus Laboratories, Leiden University, PO Box 9502, 2300 RA Leiden, The Netherlands.

ROGER WOODGATE

Section on DNA Replication, Repair and Mutagenesis, National Institute of Child Health and Human Development, National Institutes of Health, 9000 Rockville Pike, Bethesda, MD 20892-2725, USA.

# Abbreviations

| | |
|---|---|
| 6–4pp | pyrimidine(6–4)pyrimidone photoproduct |
| 6-meTG | 6-methylthioguanine |
| 6-TG | 6-thioguanine |
| 6MG | $O^6$-methylguanine |
| A–T | ataxia–telangiectasia |
| AAF | acetylaminofluorene |
| AP | apurinic–apyrimidinic sites |
| APC | adenomatous polyposis coli |
| AraC | 1-beta-D-arabinofuranosylcytosine |
| ATM | ataxia–telangiectasia mutated |
| Atr | *a*taxia–*t*elangiectasia and *r*ad3 |
| BER | base excision repair |
| BLM | bleomycin |
| bp | base pair |
| BPDE | benzo[*a*]pyrene diol-epoxide |
| CAK | cyclin-dependent kinase-activating kinase |
| CCNU | *N*-(2-chloroethyl)-*N*′-cyclohexyl-*N*-nitrosourea |
| cdk | cyclin-dependent kinase |
| CHL | chlorambucil |
| CPD | cyclobutane–pyrimidine dimer |
| CPT | camptothecin |
| CS | Cockayne syndrome |
| CTD | carboxy-terminal domain |
| D | diversity segment |
| DDB | damaged DNA binding |
| DIR | direct and inverted repeat |
| DNA pol ε | DNA polymerase ε |
| DNA-PK | DNA-dependent protein kinase |
| DNA-PKcs | DNA-PK catalytic subunit |
| dsb | double-strand break(s) |
| dsDNA | double-stranded DNA |
| DTIC | dacarbazine (dimethyl-triazenylimidazole carboxamide) |
| EBV | Epstein–Barr virus |
| EGFR | epidermal growth factor receptor |
| *ERCC* | *excision repair cross-complementing gene* |
| ERCC1 | excision repair cross-complementing group 1 protein |
| ERICs | enterobacterial repetitive intergenic consensus (also known as IRUs) |
| ES | embryonic stem cell |

FEN-1              flap endonuclease 1
FISH              fluorescent *in-situ* hybridization
FMTC              familial medullary thyroid cancer
GTF              general transcription factor
H-DNA              half of the purine-rich strand of DNA
*H-DNA              half of the pyrimidine-rich strand of DNA
HHR23B              human homologue of Rad23B
HMG              high-mobility group
h-mtTFA              human mitochondrial transcription factor
HNPCC              hereditary non-polyposis colorectal cancer (syndrome)
HPRT              hypoxanthine guanine phosphoribosyltransferase
hTERC              RNA component of telomerase
hTERT              human telomerase reverse transcriptase
hUBF              human upstream binding factor
ICL              interstrand cross-link
IGF-II              insulin-dependent growth factor-receptor II
IXR1              intrastrand cross-link recognition
J              joining segment
kb              kilobase
LEF-1              lymphoid enhancer binding factor
LFS              Li–Fraumeni syndrome
*m*-AMSA              acridinylamino-methanesulfonyl-*m*-anisidine
MAT              mating-type (locus)
MAT1              *ménage à trois* factor
MEN              multiple endocrine neoplasia
Mfd              mutation frequency decline
MGMT              $O^6$-methylguanine-DNA methyltransferase
MI              microsatellite instability
MMR              mismatch repair functions
MN              meiotic nodules
MNNG              $N$-methyl-$N'$-nitro-$N$-nitrosoguanidine
MNU              $N$-methyl-$N$-nitrosourea
MTX              methotrexate
NBCCS              nevoid basal-cell carcinoma syndrome
NBS              Nijmegen breakage syndrome
NER              nucleotide excision repair
NHEJ              non-homologous end-joining
NK              natural killer (cell)
Npf              *Nar*I processing factor
$O^6$-meGua              $O^6$-methylguanine
ORF              open-reading frame
PARP              poly(ADP-ribose) polymerase
PCNA              proliferating cell nuclear antigen
PCR              polymerase chain reaction

| PI | phosphoinositide |
|---|---|
| pRB | retinoblastoma protein |
| *RAD* | *rad*iation sensitive gene |
| *RAG* | recombination activating gene |
| RB | retinoblastoma |
| RDS | radioresistant DNA synthesis |
| RER$^+$ | replication *error* |
| RFC | replication factor C |
| RN | recombination nodules |
| RNAP | RNA polymerase |
| RNAPII | RNA polymerase II |
| RNase | ribonuclease |
| ROI | reactive oxygen intermediates |
| RPA | replication protein A |
| RSS | recombination signal sequences |
| S-DNA | slipped-strand DNA |
| SAPK | stress-activated protein kinase |
| SC | synaptonemal complex |
| scid | severe combined immune deficiency |
| SSA | single-strand annealing |
| SSB | single-strand binding |
| ssDNA | single-stranded DNA |
| TBP | TATA-binding protein |
| TCF-1α | T cell-specific transcription factor-1α |
| TCR | transcription-coupled repair |
| TdT | terminal deoxynucleotidyl transferase |
| TFIIH | transcription factor IIH |
| TGF-β | transforming growth factor-β |
| TR | translesion replication |
| TRAP | telomeric repeat amplification protocol |
| TRCF | transcription-repair coupling factor |
| UTR | untranslated regions |
| UV | ultraviolet |
| UVM | ultraviolet-induced modulation (of mutagenesis) |
| V | variable segment |
| XP | xeroderma pigmentosum |
| XPA–G | xeroderma pigmentosum complementation group A to G proteins |
| XP-V | xeroderma pigmentosum variant |
| XRCC | X-ray repair cross-complementing |

# 1 | Molecular processing of DNA folding anomalies in *Escherichia coli*

DAVID R. F. LEACH

## 1. Introduction

For evolution to proceed, there must be a balance between genetic stability and rearrangement. Chromosomal arrangements that prove beneficial to an organism are expected to be maintained while retaining the potential to generate radically new organizations over an evolutionary time-scale. In order to maintain this balance, cells have evolved mechanisms to reduce the frequency of chromosomal rearrangements. An important threat to DNA stability comes from sequences that have the potential for unusual folding. Unusually folded DNAs also pose problems for DNA replication, which may be particularly critical for organisms where there is an evolutionary advantage associated with rapid cell duplication. In principle, these problems can be reduced by lowering the genomic content of foldable DNA, reducing its frequency of folding, unfolding it when it does fold, and destroying folded structures if all else fails. It should be remembered, in this context, that 'random' single-stranded DNA sequences will fold upon themselves, so the structures studied within non-random sequences may best be considered as extreme examples of what is possible. Studies of the effects on DNA instability of sequences with the potential to form unusual secondary structures in the bacterium *Escherichia coli* have revealed a number of interesting rearrangements, which may have general relevance to living cells. *E. coli* has also provided a simple experimental system where questions concerning the formation and processing of unusual secondary structures can be addressed *in vivo*. Finally, the detailed knowledge of *E. coli* genetics is being used to identify genes and proteins that process unusually folded DNA.

## 2. Folding anomalies in DNA

### 2.1 Hairpin DNA

DNA is often thought of as a B-form helix. However, it is well known that deviations from the B-form average structure abound and that particular steps between base pairs can have consequences on structure and flexibility (see ref. 1). In addition to these deviations from the average conformation of B-DNA, there are radically different folded structures that can form within particular DNA sequences. Perhaps the simplest of these folded structures is the hairpin (Fig. 1.1(a)). Here the DNA helix still approximates B-DNA, but the hydrogen bonding is intramolecular. This type of structure can only form if there is intrastrand complementarity as occurs in an inverted repeat sequence. This type of sequence is palindromic within the context of double-stranded DNA if no unique DNA separates the inverted repeats. If both DNA strands form hairpins opposite each other, the structure generated is a cruciform (Fig. 1.1(b)). Hairpins can form within single-stranded DNA if that DNA is not protected from folding by the binding of proteins. Cruciforms, on the other hand, must form from double-stranded DNA and are only favoured energetically within negatively supercoiled substrates. This is because of the energy released as a consequence of relaxation of supercoiling that accompanies cruciform extrusion. Although cruciform extrusion is favoured by the energy available from negative supercoiling, the reaction can be very slow under physiological conditions because the initial central melting and protocruciform formation that initiates the reaction gives rise to a substantial kinetic barrier for most DNA sequences (2–7). Attempts have been made to assess whether cruciforms can be generated *in vivo* in *E. coli* cells. Palindromes with a normal base composition form cruciforms very rarely inside cells (2, 8, 9). This low frequency of cruciform formation can be increased for palindromes with A/T rich centres and G/C rich stems and in *topA* mutants that have an elevated level of negative supercoiling (10–12). Some cruciforms can also be detected in cells treated with chloramphenicol (13, 14). DNA sequences with $d(AT)_n \cdot d(AT)_n$ centres that do not have a kinetic barrier to extrusion *in vitro* can form cruciforms *in vivo* as judged by chemical cleavage of single-stranded loops, after elevation of intracellular supercoiling by high-salt shock or in *topA* mutants (15), or after induction of transcription (16). Indirect evidence for a centre-dependent pathway of cruciform formation *in vivo* has been obtained from studies of plaque-size inhibition of palindrome containing λ phage grown in an *sbcC* mutant (see later for a discussion of *sbcC*) (17–19). These studies suggest that transitions between a fully base-paired, double-stranded B-DNA structure and cruciform DNA do occur sufficiently frequently to have effects on the time-scale of the lytic cycle of a λ phage. This system has been used to show that DNA sequences capable of forming tight hairpin loops (*in vitro*) favour the transition to cruciform structures *in vivo* (19). Hairpin DNA, by contrast, can form easily within single-stranded DNA, and it is likely that any palindromic sequence (and even inverted repeats separated by short spacer regions) form hairpins much more frequently than cruciforms (see refs 20, 21). Hairpin formation is facilitated by enzymatic processes that generate single-stranded DNA (e.g. discontinuous DNA synthesis on the lagging-strand of the replication fork).

## 2.2  Pseudo-hairpin DNA and S-DNA

Pseudo-hairpins can form from DNA strands that have enough self-complementarity to fold back on themselves and form stable structures (Fig. 1.1(c)).

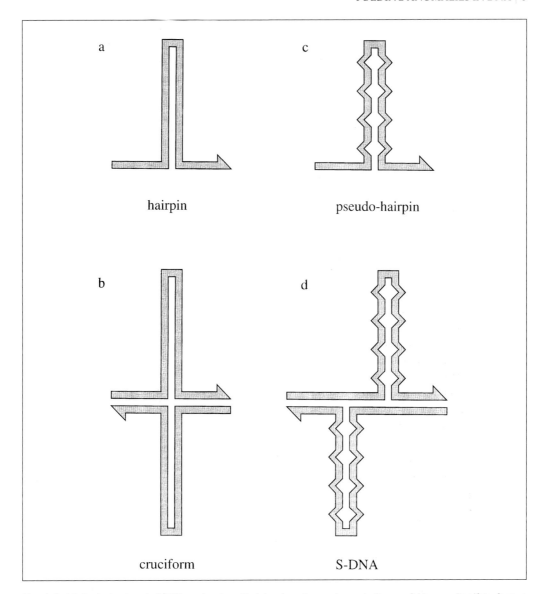

a

hairpin

c

pseudo-hairpin

b

cruciform

d

S-DNA

**Fig. 1.1** (a) A single-stranded DNA molecule with intrastrand complementarity can fold upon itself to form a hairpin structure. (b) If both complementary DNA strands form hairpins opposite each other a cruciform structure is generated. (c) A pseudo-hairpin can form if there is sufficient intrastrand complementarity to form a stable structure. This is the case for repetitive DNA sequences such as $d(CXG)_n$ triplet repeats. (d) In simple repetitive DNA sequences, such as $d(CAG)_n \cdot d(CTG)_n$, pseudo-hairpins may form on both strands and because of the repetitive nature of the DNA these structures may not lie directly opposite each other. This type of DNA has been called 'slipped structure DNA' or S-DNA.

Interest has recently focused on the triplet repeat sequences $d(CAG)_n \cdot d(CTG)_n$, $d(CGG)_n \cdot d(CCG)_n$, or $d(GAA)_n \cdot d(TTC)_n$ that have been implicated in a range of neurological diseases. See refs 22–25 for recent reviews that look particularly at aspects of these repeats relating to DNA structure. The pyrimidine-rich strands of the first two of these repeats can form pseudo-hairpins stabilized by C·G base pairs. In any triplet repeat sequence there are three potential frames of folding, and for each of these there is an odd and even loop configuration. *In vivo* experiments in *E. coli* cells have argued that $(CTG)_n$ prefers to fold in the frame $d(CTG) \cdot d(CTG)$ with an even number of bases in the loop (26) and the sequence $d(CCG)_n$ likes to fold as $d(CCG) \cdot d(CCG)$ with an even loop (26, 27). This same preferred fold also appears to be favoured *in vitro* when $n = 11$ or less (see ref. 23 for a detailed analysis of this argument). For $n = 15$ (or $d(CCG)_2$ at nuclear magnetic resonance (NMR) concentrations, where multimerization may occur), it has been suggested to prefer to fold as $d(GCC) \cdot d(GCC)$ with the unpaired C residues folded back 5′ into the minor groove (28, 29). For the $d(GCC) \cdot d(GCC)$ pairing-frame a preference for an odd number of bases in the loop has been observed *in vivo* (27). The purine-rich strands may also be able to form pseudo-hairpins but they (at least in the case of dCGG) can form quadruplex structures too (see refs 22–24). The two strands of $d(CAG)_n \cdot d(CTG)_n$ can fold back on themselves at different positions in the two strands to generate molecules with pseudo-hairpins extruded at different positions. These structures (Fig. 1.1(d)) have been shown to be stable when formed in fragments of linear DNA where they have been called S-DNA (30–32). Slipped structures were originally observed in minisatellite repeats, where the extruded structures on opposite strands can pair with each other to generate pseudo-knots (33).

## 2.3   H-DNA, *H-DNA, and nodule DNA

Bimolecular triplex structures can form in DNA sequences with a strong bias for purines in one strand and pyrimidines in the other. They can be generated in two different ways. Either the pyrimidine-rich strand folds back on itself and pairs with half of the purine-rich strand (H-DNA), or the purine-rich strand folds back on itself and pairs with half of the pyrimidine-rich strand (*H-DNA). In H-DNA the 3′ half of the pyrimidine-rich strand and in *H-DNA the 3′ half of the purine-rich strand are preferentially donated to the triplex. This means that the H-y3 isomer of H-DNA and the H-r3 isomer of *H-DNA are usually formed in preference to the H-y5 and H-r5 isomers (see Figs 1.2(a–d)). A combination of H-DNA and *H-DNA can be generated in what is known as nodule DNA (Fig. 1.2(e)). (See ref. 34 for a review of triplex DNA structures.) As is the case for cruciform DNA, H-DNA and *H-DNA formation *in vivo* is dependent on the release of supercoiling that accompanies the formation of the structure. Several studies using chemical probing techniques have indicated that H and *H-DNA can form in *E. coli* cells, and that formation is stimulated by conditions (such as treatment with chloramphenicol, induction of transcription, or the use of *topA* mutants) which cause elevation of supercoiling (35–37).

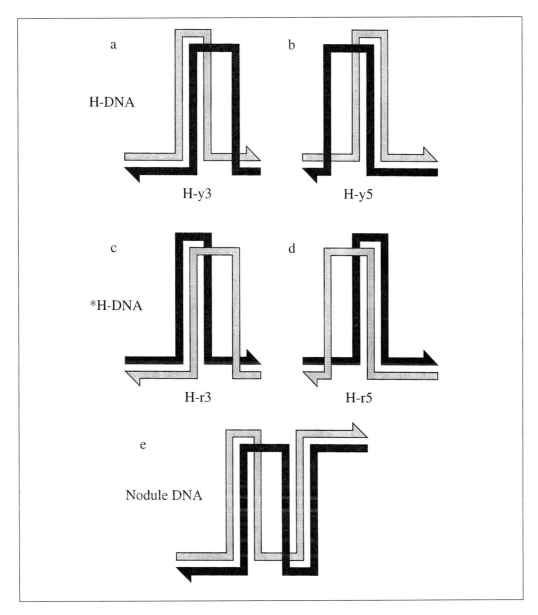

**Fig 1.2** Many different forms of triplex DNA can form between DNA strands that are purine-rich (shown in black) and pyrimidine-rich (shown shaded). In this figure three different forms of bimolecular triplexes known as H-DNA, *H-DNA, and nodule DNA are illustrated. Bimolecular triplexes either form between a folded pyrimidine-rich strand and half a purine-rich strand to give H-DNA (a, b), or between a folded purine-rich strand and half a pyrimidine-rich strand to give *H-DNA (c, d), or a combination of both to give nodule DNA (e). (a) H-y3 DNA, where the pyrimidine-rich strand is folded and paired to the 3′ half of the purine-rich strand. (b) H-y5 DNA, where the pyrimidine-rich strand is folded and paired to the 5′ half of the purine-rich strand. (c) H-r3 DNA, where the purine-rich strand is folded and paired to the 3′ half of the pyrimidine-rich strand. (d) H-r5 DNA, where the purine-rich strand is folded and paired to the 5′ half of the pyrimidine-rich strand. (e) Nodule DNA; the nodule DNA illustrated here is that formed with a combination of H-y3 and H-r3 triplexes.

## 2.4 Z-DNA

Left-handed Z-DNA (see ref. 38 for a recent review) can form in alternating stretches of purines and pyrimidines. $d(CG)_n \cdot d(CG)_n$ sequences form Z-DNA more easily than $d(TG)_n \cdot d(CA)_n$ which do so more easily than $d(TA)_n \cdot d(TA)_n$ sequences. Z-DNA has been shown to form *in vivo* in *E. coli* by a combination of chemical probing experiments and by monitoring protection from *Eco*RI methylation at GAATTC sites embedded in Z-forming sequence (9, 12, 39–44). Since the formation of left-handed DNA not only requires the unwinding of the two DNA strands (as is the case for cruciforms or H-type DNAs) but their winding in the opposite direction, the release of negative supercoiling that accompanies this event is greater. For this reason a sequence such as $d(CG)_n \cdot d(CG)_n$ which has the potential to adopt either Z- or cruciform conformations, prefers to adopt Z when extruded from a supercoiled plasmid substrate.

## 2.5 Quadruplex DNA

*In vitro*, four-stranded (quadruplex) DNAs can form in a number of parallel and antiparallel orientations of DNA strands and are stabilized by $G_4$ quartets. Recently some evidence in favour of $G_2C_2$ quartets has also been advanced (45), and it has been proposed that $d(CGG)_n$ oligonucleotides might form quadruplexes *in vitro* with either two $G_4$ quartets per triplet or one $G_4$ quartet and two $G_2C_2$ quartets per triplet (see refs 22–24 for discussions of this issue). It is difficult to see how parallel-stranded structures have much relevance to the quadruplexes that might potentially form in cloned sequences introduced into *E. coli* cells. On the other hand, it is possible to envisage how unimolecular antiparallel quadruplexes might form within a single strand of DNA (see Fig. 1.3). I am unaware of data relating to whether such structures can be detected in *E. coli* cells, but quadruplex DNA is extremely stable and it may be difficult (or indeed impossible) to replicate DNA sequences that have formed this structure. *In vitro*, unimolecular quadruplex DNA causes very strong inhibition of DNA synthesis (46–48).

# 3. Deletions at direct repeats stimulated by closely spaced inverted repeats

Directly repeated DNA sequences have the potential to cause instability due to mis-alignment during DNA replication, repair, or homologous genetic recombination. Misalignment implies that a strand of one copy of a repeated sequence becomes paired to the other strand of another copy. There are two classically defined ways of bringing this about—break–join and copy–choice. Reactions can either be intra-molecular or intermolecular, but often the nature of the reaction cannot be deter-mined by the structure of the product since several alternative pathways can lead to the same end. However, studies of material transfer from substrates to products, the influence of genes known to be implicated in particular reactions, and the

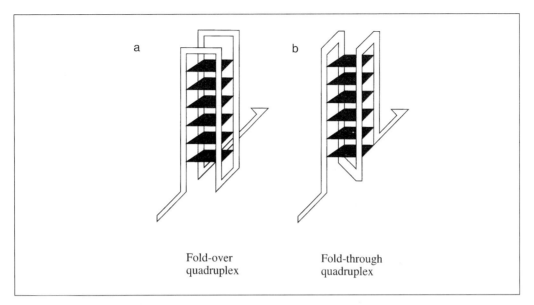

**Fig. 1.3** Two classes of quadruplex DNA. (a) Fold-over quadruplex DNA can be envisaged as a hairpin folding over on itself to give a four-stranded structure. (b) Fold-through quadruplex DNA has a central unpaired hairpin strand folded between the two outer strands of the four-stranded structure.

consequences of the rearrangements in relation to flanking DNA have led, in certain instances, to insight into the mechanisms involved.

Spontaneous deletions at short, directly repeated sequences are rare events (49). Therefore, in order to study the nature of the process, it has been necessary either to develop selective systems to isolate the rare deletion derivatives or to increase the frequency of the deletion events. One way to increase deletion frequencies is to use replicons that can be induced to replicate via a single-stranded intermediate (50). Phages (e.g. M13 and φX174) or plasmids containing an origin of single-strand replication can be induced to delete regions of DNA at a high frequency when copying a single-stranded template, particularly if the template is folded because of the presence of a closely spaced, inverted repeat sequence. In this way two pathways of deletion formation have been documented. In the first of these pathways strand-slippage during replication bypasses a folded DNA structure. Using induced, single-strand DNA formation in a substrate containing an inverted repeat sequence it has been possible to follow material transfer from parental to recombinant molecules and in this way prove that deletion occurred by coping (copy–choice recombination) (51). In the second pathway of deletion formation, a DNA break formed at the origin of replication is joined to a break generated at a DNA replication pause site (52–54). Palindromic DNA sequences are known to cause pausing of DNA replication (55) and have been shown to generate hot spots in this second deletion pathway (56).

Selective systems have been set up to study deletion events in plasmids. Here, it has often been possible to take advantage of insertions of inverted repeats within drug-resistance genes which can be deleted to restore antibiotic resistance. The direct repeats chosen to delete the inverted repeats have been shown to be affected by the length of the inverted repeat. As inverted repeats are made longer, deletion frequencies increase and the choice of the direct repeat used changes from sequences flanking the insertion to one copy within the inverted repeat and one located outside the repeat (57–59). This suggests a directionality of the event, one direct repeat within the palindrome acting as a donor and another located to one side acting as a target (see ref. 60 for a discussion of this). In most situations where the direction of DNA replication has been known, the orientation of the repeats suggests that deletion occurs primarily on the lagging strand (61–63). Recently it has been shown that the directionality of deletion of a 246-base pair (bp), palindromic DNA sequence inserted into the *E. coli* chromosome is also consistent with replication slippage on the lagging strand (63). Evidence for the occurrence of palindrome-stimulated strand-slippage in natural sequences comes from the sequencing of partial deletions of ERIC sequences. ERICs (also known of as IRUs) are imperfectly palindromic sequences of about 126 bp that are distributed in multiple copies in the chromosomes of *E. coli* and *Salmonella typhimurium* (64, 65). Sequencing has revealed the existence of direct repeats at the deletion endpoints, as expected if DNA loss had occurred via strand-slippage (66).

An analogous type of deletion to that stimulated by inverted repeats is that stimulated by triple helical DNA. It has been shown that a $d(G)_n \cdot d(C)_n$ sequence that can form *H-DNA after induction of transcription causes deletion events between directly repeated sequences flanking the structure (67). These events are RecA-independent (as are the palindrome-stimulated events) and may also be due to strand-slippage during replication.

# 4. Duplications and inversions stimulated by closely spaced inverted repeats

Closely spaced inverted repeats also have the potential to stimulate more complex rearrangements such as duplications and inversions. Inversions stimulated by inverted repeats have been studied in plasmid substrates. It has been shown that inversion of the DNA between inverted repeats is often accompanied by head-to-head dimerization of the plasmid carrying the repeats (68). In this situation, it is unclear whether DNA folding is implicated in the reaction, but a model that involves a dumb-bell intermediate has been proposed (69). Evidence that such an intermediate can lead to the formation of the observed products has been obtained by transformation of *E. coli* with dumb-bells constructed *in vitro* (69). Another rearrangement that appears to have been stimulated by adjacent inverted repeats and may be associated with plasmid dimerization is the formation of a direct and inverted repeat (DIR) (70). Here, a short sequence immediately adjacent to a long inverted repeat is found to be triplicated in a head-to-tail-to-head-to-tail fashion and this rearrangement has been accompanied by loss of the inverted repeat. The rearrangement is associated with plasmid dimerization, although this may not be intrinsic to the mechanism. The dimer formed is a head-to-tail dimer in contrast to the head-to-head dimers generated in the inversion pathway. DIR formation has been observed in 'normal' chromosomal DNA (without inverted repeats) where it is unknown whether DNA folding plays any part (71, 72). Similar complex re-arrangements have also been shown to occur in *E. coli dnaE173* mutants (73) and in eukaryotic cells (74). One situation in the *E. coli* genome where DNA folding may have contributed to insertions resembling DIRs is the enlargement of ERICs (75), although an alternative suggestion is that the enlargement is derived from copying another location on the *E. coli* chromosome (76).

# 5. Deletions and amplifications of triplet repeats

Several human genetic diseases, including Huntington's disease, myotonic dystrophy, Friedreich's ataxia, and the fragile X syndrome, are associated with unstable triplet repeats (see refs 24 and 25 for recent reviews focusing on the relation between DNA structure and instability). The sequences involved fall into two classes—the CXG repeats $d(CAG)_n \cdot d(CTG)_n$ and $d(CGG)_n \cdot d(CCG)_n$ that are responsible for all known diseases, except Friedreich's ataxia where the repeat is $d(GAA)_n \cdot d(TTC)_n$. As indicated above, these types of repeats have the potential to adopt unusual DNA secondary structures such as pseudo-hairpins, S-DNA, and quadruplexes for CXG repeats (see reviews listed above) and H-DNA for $d(GAA)_n \cdot d(TTC)_n$ repeats (77–80). Of all the triplet repeats, the instability of $d(CAG)_n \cdot d(CTG)_n$ has been most extensively studied in *E. coli*. Most instability at moderate repeat lengths (e.g. $n = 41$) appears to be replicational and caused by small changes (expansion or deletion) of a few triplets (Schmidt, Abbott, and Leach, unpublished). However, in longer repeat arrays (e.g. $n = 180$) large deletions appear to be stimulated by secondary structure formation of the CTG repeat, especially when it is present as the lagging-strand template (81). In these long repeat tracts, mismatch repair can stimulate folding of the d(CTG) repeat and the formation of deletions when it is the leading-strand template (81). With the shorter repeat tracts (e.g. $n = 43$) the stimulatory effect of mismatch repair on secondary structure formation and deletion can be seen when the CTG repeat is the lagging-strand template (Schmidt, Abbott, and Leach, unpublished). Deletion of these repeats is enhanced in mutants defective in single-strand binding protein, consistent with the idea that folding of single-stranded DNA is implicated in the events (82). Repeated sequences are rare in *E. coli*, but in other bacteria they are more common and have been implicated in the control of gene expression (e.g. the gene encoding the outer membrane protein PII of *Neisseria gonorrhoeae* (83)). Nevertheless, even in *E. coli* the strongest hot-spot for spontaneous mutagenesis in the well-studied *lacI* gene is a repeat of three copies of the sequence CTGG. Mutagenesis consists in the expansion or deletion of single copies of this tetranucleotide (84).

# 6. Inhibition of DNA replication and initiation of homologous recombination by SbcCD protein

A nuclease exists in *E. coli* that specifically targets and cleaves palindromic DNA sequences, thus preventing their replication if longer than about 150–200 bp. This nuclease is known as SbcCD because mutants were originally isolated as suppressors of the recombination deficiency of *recBC* mutants (85, 86). The action of *sbcCD* has been shown to be mediated via DNA replication (87, 88). The purified protein is an ATP-dependent double-strand exonuclease and ATP-independent single-strand endonuclease (89, 90) that can cleave DNA hairpins in the presence of ATP or the non-hydrolysable analogue ATPγS (91). Cleavage is just 5′ of the hairpin loop and leaves a 3′ hydroxyl and a 5′ phosphate at the site of cleavage. It is not yet clear what the natural role of the protein is, but its action on a palindrome cloned in the *E. coli* chromosome leads to homologous recombination with the sister 'chromatid' (92). This recombination has been shown to be conservative in that it does not result in palindrome deletion. A similar stimulation of homologous recombination by the presence of palindromic DNA has been observed in yeast (93) and in mammalian cells (94). Furthermore, a palindromic sequence in yeast meiotic cells has been shown to induce double-strand breaks, as expected if it can act as a recombination initiation site (95).

## 7. Conclusions

The bacterium *E. coli* has a streamlined genome that includes very little repetitive DNA. This may be due to selective pressures to minimize the time taken for DNA replication in the cell's bid to reproduce as fast as possible. Nevertheless, imperfect inverted repeats are commonplace. Some of these act as binding sites for regulatory proteins such as the *lac* repressor (LacI). Others (such as Rep sequences) are likely to participate in regulation at the level of RNA processing. And finally a third class (including the ERIC sequences) have unknown function. These naturally occurring sequences result in a significant potential for intrastrand folding in the *E. coli* genome if this were not controlled. It is likely that this control is exercised at several different levels. The first of these is the prevention of secondary structure formation. This is one of the important roles of single-strand binding protein (SSB). If structures do form it is likely that one of the roles of the replicative helicases is their removal. Finally, when structures persist, the cell encodes a nuclease (SbcCD) which can cleave the DNA that is improperly folded and channel the broken end into productive recombination with the sister 'chromatid' to restore a good replication fork.

Experiments performed to study the fate of unusual DNA sequences (artificially introduced into *E. coli*) have revealed significant aspects of the DNA processing in this organism. It is likely that these reflect how the organism deals with unusually folded DNA in its compromise to retain important sequence information in DNA that is intrinsically able to fold, to minimize folding that might interfere with DNA replication, and to remove DNA which has an intrinsically unacceptable frequency of mutagenesis. In so far as all organisms have to negotiate this three-way compromise and to reach a satisfactory balance in the light of their own biology, it is likely that significant aspects of the processing of unusually folded DNA in *E. coli* have universal relevance.

# Acknowledgements

I would like to thank the B.B.S.R.C., the M.R.C., and the Wellcome Trust for funding work carried out in my laboratory.

# References

1. Calladine, C. R. and Drew, H. R. (1997) *Understanding DNA: the molecule and how it works* (2nd edn). Academic Press, San Diego.
2. Courey, A. J. and Wang, J. C. (1983) Cruciform formation in a negatively supercoiled DNA may be kinetically forbidden under physiological conditions. *Cell*, **33**, 817–29.
3. Courey, A. J. and Wang, J. C. (1988) Influence of DNA sequence and supercoiling on the process of cruciform formation. *J. Mol. Biol.*, **202**, 35–43.
4. Gellert, M., O'Dea, M. H., and Mizuuchi, K. (1983) Slow cruciform transitions in palindromic DNA. *Proc. Natl. Acad. Sci. USA*, **80**, 5545–9.
5. Lilley, D. M. J. (1985) The kinetic properties of cruciform extrusion are determined by DNA base-sequence. *Nucleic Acids Res.*, **13**, 1443–65.
6. Murchie, A. I. H. and Lilley, D. M. J. (1987) The mechanism of cruciform formation in supercoiled DNA: initial opening of central basepairs in salt dependent reaction. *Nucleic Acids Res.*, **15**, 9641–54.
7. Sullivan, K. M. and Lilley, D. M. J. (1987) Influence of cation size and charge on the extrusion of a salt-dependent cruciform. *J. Mol. Biol.*, **193**, 397–404.
8. Sinden, R. R., Broyles, S. S., and Pettijohn, D. E. (1983) Perfect palindromic *lac* operator DNA sequence exists as a stable cruciform structure *in vitro* but not *in vivo*. *Proc. Natl Acad. Sci. USA*, **80**, 1797–801.
9. Jaworski, A., Hsieh, W.-T., Blaho, J. A., Larson, J. E., and Wells, R. D. (1987) Left-handed DNA *in vivo*. *Science*, **238**, 773–7.
10. Sinden, R. R., Zheng, G., Brankamp, R. G., and Allen, K. N. (1991) On the deletion of inverted repeated DNA in *Escherichia coli*: effects of length, thermal stability and cruciform formation *in vivo*. *Genetics*, **129**, 991–1005.
11. Zheng, G. and Sinden, R. R. (1988) Effects of base composition at the centre of inverted repeated DNA sequences on cruciform transitions in DNA. *J. Biol. Chem.*, **263**, 5356–61.
12. Zheng, G., Kochel, T., Hoepfner, R. W., Timmons, S. E., and Sinden, R. R. (1991) Torsionally tuned cruciform and Z-DNA probes for measuring unrestrained supercoiling at specific sites in DNA of living cells. *J. Mol. Biol.*, **221**, 107–29.
13. Haniford, D. B. and Pulleyblank, D. E. (1985) Transition of a cloned d(AT)n–d(TA)n tract to a cruciform *in vivo*. *Nucleic Acids Res.*, **13**, 4343–63.
14. Blaho, J. A., Larson, J. E., McLean, M. J., and Wells, R. D. (1988) Multiple DNA secondary structures in perfect inverted repeat inserts in plasmids. *J. Biol. Chem.*, **263**, 14446–55.
15. McClellan, J. A., Boublikova, P., Palecek, E., and Lilley, D. M. J. (1990) Superhelical torsion in cellular DNA responds directly to environmental and genetic factors. *Proc. Natl Acad. Sci. USA*, **87**, 8373–7.
16. Dayn, A., Malkhosyan, S., and Mirkin, S. M. (1992) Transcriptionally driven cruciform formation *in vivo*. *Nucleic Acids Res.*, **20**, 5991–7.
17. Chalker, A. F., Okely, E. A., Davison, A., and Leach, D. R. F. (1993) The effects of central asymmetry on the propagation of palindromic DNA in bacteriophage l are consistent with cruciform extrusion *in vivo*. *Genetics*, **133**, 143–8.

18. Davison, A. and Leach, D. R. F. (1994) The effects of nucleotide sequence changes on DNA secondary structure formation in *Escherichia coli* are consistent with cruciform extrusion *in vivo*. *Genetics*, **137**, 361–8.

19. Davison, A. and Leach, D. R. F. (1994) Two-base DNA hairpin-loop structures *in vivo*. *Nucleic Acids Res.*, **22**, 4361–3.

20. Allers, T. A. and Leach, D. R. F. (1995) DNA palindromes adopt a methylation-resistant conformation that is consistent with DNA cruciform or hairpin formation *in vivo*. *J. Mol. Biol.*, **252**, 70–85.

21. Leach, D. R. F. (1996) Cloning and characterisation of DNAs with palindromic sequences. *Genetic Engineering*, Vol. 18 (ed. J. Setlow), pp. 1–11. Plenum Press, New York.

22. Mitas, M. (1997) Trinucleotide repeats associated with human disease. *Nucleic Acids Res.*, **25**, 2245–53.

23. Darlow, J. M. and Leach, D. R. F. (1998) Secondary structures in d(CGG) and d(CCG) repeat tracts. *J. Mol. Biol.*, **275**, 3–16.

24. Pearson, C. E. and Sinden, R. R. (1998) Trinucleotide repeat DNA structures: dynamic mutations from dynamic DNA. *Curr. Opin. Struct. Biol.*, **8**, 321–30.

25. Wells, R. D. (1998) DNA structure, triplet repeats and hereditary neurological diseases. *J. Biochem. Mol. Biol.*, **31**, 2–19.

26. Darlow, J. M. and Leach, D. R. F. (1995) The effects of trinucleotide repeats found in human inherited disorders on palindrome inviability in *Escherichia coli* suggest folding preferences *in vivo*. *Genetics*, **141**, 825–32.

27. Darlow, J. M. and Leach, D. R. F. (1998) Evidence for two preferred hairpin folding patterns in d(CGG)·d(CCG) repeat tracts *in vivo*. *J. Mol. Biol.*, **275**, 17–23.

28. Gao, X., Huang, X., Kenneth, S. G., Zheng, M., and Liu, H. (1995) New antiparallel duplex motif of DNA CCG repeats that is stabilised by extrahelical bases symmetrically located in the minor groove. *J. Am. Chem. Soc.*, **117**, 8883–4.

29. Yu, A., Barron, M. D., Romero, R. M., Christy, M., Gold, B., Jianli, D., *et al.* (1997) At physiological pH d(CCG)15 forms a hairpin containing protonated cytosines and a distorted helix. *Biochemistry*, **36**, 3687–99.

30. Pearson, C. E. and Sinden, R. R. (1996) Alternative DNA structures within the trinucleotide repeats of the myotonic dystrophy and fragile X locus. *Biochemistry*, **35**, 5041–53.

31. Pearson, C. E., Wang, Y.-H., Griffith, J. D., and Sinden, R. R. (1998) Structural analysis of slipped strand DNA (S-DNA) formed in $(CTG)_n \cdot (CAG)_n$ repeats from the myotonic dystrophy locus. *Nucleic Acids Res.*, **26**, 816–23.

32. Pearson, C. E., Eichler, E. E., Lorenzetti, D., Kramer, S. F., Zoghbi, H. Y., Nelson, D. L., *et al.* (1998) Interruptions in the triplet repeats of SCA1 and FRAXA reduce the propensity and complexity of slipped strand DNA (S-DNA) formation. *Biochemistry*, **37**, 2701–8.

33. Coggins, L. W., O'Prey, M. O., and Akhter, S. (1992) Intrahelical pseudoknots and inter-helical associations mediated by mispaired human minisatellite DNA sequences *in vitro*. *Gene*, **121**, 279–85.

34. Frank Kamenetskii, M. D. and Mirkin, S., M. (1995) Triplex DNA structures. *Annu. Rev. Biochem.*, **64**, 65–95.

35. Karlovsky, P., Pecinka, P., Vojtiskova, M., Makatuova, E., and Palecek, E. (1990) Protonated triplex DNA in *E. coli* cells as detected by chemical probing. *FEBS Lett.*, **274**, 39–42.

36. Kohwi, Y., Malkhosyan, S. R., and Kohwi-Shigematsu, T. (1992) Intramolecular dG·dG·dC triplex detected in *Escherichia coli* cells. *J. Mol. Biol.*, **223**, 817–22.

37. Ussery, D. W. and Sinden, R. R. (1993) Environmental influences on the *in vivo* level of intramolecular triplex DNA in *Escherichia coli*. *Biochemistry*, **32**, 6206–13.

38. Rich, A. (1996) The biology of left-handed Z-DNA. *J. Biol. Chem.*, **271**, 11595–8.
39. Palecek, E., Boublikova, P., and Karlovsky, P. (1987) Osmium-tetroxide recognizes structural distortions at junctions between right-handed and left-handed DNA in a bacterial cell. *Gen. Physiol. Biophys.*, **6**, 593–608.
40. Jaworski, A., Zacharias, W., Hsieh, W. T., Blaho, J. A., Larson, J. E., and Wells, R. D. (1988) *In vivo* existence of left-handed DNA. *Gene*, **74**, 215–20.
41. Palecek, E., Rasouvska, E., and Boublikova, P. (1988) Probing of DNA polymorphic structure in the cell with osmium-tetroxide. *Biochem. Biophys. Res. Commun.*, **150**, 731–8.
42. Zacharias, W., Jaworski, A., Larson, J. E., and Wells, R. D. (1988) The B- to Z-DNA equilibrium *in vivo* is perturbed by biological processes. *Proc. Natl Acad. Sci. USA*, **85**, 7069–73.
43. Rahmouni, A. R. and Wells, R. D. (1989) Stabilization of Z-DNA *in vivo* by localized supercoiling. *Science*, **246**, 358–63.
44. Jiang, H., Zacharias, W., and Amirhaeri, S. (1991) Potassium permanganate as an *in situ* probe for B-Z and Z-Z junctions. *Nucleic Acids Res.*, **19**, 6943–8.
45. Ketani, A., Kumar, R. A., and Patel, D. J. (1995) Solution structure of a quadruplex containing the fragile X syndrome triplet repeat. *J. Mol. Biol.*, **254**, 638–56.
46. Woodford, K. J., Howell, R. M., and Usdin, K. (1994) A novel $K^+$-dependent DNA synthesis arrest site in a commonly occurring sequence motif in eukaryotes. *J. Biol. Chem.*, **269**, 27029–35.
47. Usdin, K. and Woodford, K. J. (1995) CGG repeats associated with DNA instability and chromosome fragility form structures that block DNA synthesis *in vitro*. *Nucleic Acids Res.*, **23**, 4202–9.
48. Weitzmann, M. N., Woodford, K. J., and Usdin, K. (1996) The development and use of a polymerase arrest assay for the evaluation of parameters affecting intrastrand quadruplex formation. *J. Biol. Chem.*, **271**, 20958–64.
49. Albertini, A. M., Hofer, M., Calos, M. P., and Miller, J. H. (1982) On the formation of spontaneous deletions: the importance of short sequence homologies in the generation of large deletions. *Cell*, **29**, 319–28.
50. Ehrlich, S. D., Bierne, H., d'Alençon, E., Villette, D., Petranovic, M., Noirot, P., *et al.* (1993) Mechanisms of illegitimate recombination. *Gene*, **135**, 161–6.
51. d'Alençon, E., Petranovic, M., Michel, B., Noirot, P., Aucouturier, A., Uzest, M., *et al.* (1994) Copy-choice illegitimate DNA recombination revisited. *EMBO J.*, **13**, 2725–34.
52. Michel, B. and Ehrlich, S. D. (1986) Illegitimate recombination at the replication origin of bacteriophage M13. *Proc. Natl Acad. Sci. USA*, **83**, 3386–90.
53. Michel, B. and Ehrlich, S. D. (1986) Illegitimate recombination occurs between the replication origin of the plasmid pC194 and a progressing replication fork. *EMBO J.*, **5**, 3691–6.
54. Bierne, H., Ehrlich, S. D., and Michel, B. (1991) The replication termination signal *TerB* of the *Escherichia coli* chromosome is a deletion hotspot. *EMBO J.*, **10**, 2699–705.
55. La Duca, R. J., Fay, P. J., Chuang, C., McHenry, C. S., and Banbara, R. A. (1983) Site-specific pausing of deoxyribonucleic acid synthesis catalysed by four forms of *Escherichia coli* DNA polymerase III. *Biochemistry*, **22**, 5177–88.
56. Michel, B., d'Alençon, E., and Ehrlich, S. D. (1989) Deletion hot spots in chimeric *Escherichia coli* plasmids. *J. Bacteriol.*, **171**, 1846–53.
57. Weston-Hafer, K. and Berg, D. E. (1989) Palindromy and the location of deletion end points in *Escherichia coli*. *Genetics*, **121**, 651–8.
58. Weston-Hafer, C. and Berg, D. E. (1991) Deletions in plasmid pBR322: replication slippage involving leading and lagging strands. *Genetics*, **127**, 649–55.

59. Weston-Hafer, K. and Berg, D. E. (1991) Limits to the role of palindromy in deletion formation. *J. Bacteriol.*, **173**, 315–18.
60. Leach, D. R. F. (1994) Long palindromes, cruciform structures, genetic instability and secondary structure repair. *BioEssays*, **16**, 893–900.
61. Trinh, T. Q. and Sinden, R. R. (1991) Preferential DNA secondary structure mutagenesis in the lagging strand of DNA replication in *E. coli*. *Nature*, **352**, 544–7.
62. Rosche, W. A., Trinh, T. Q., and Sinden, R. R. (1995) Differential DNA secondary structure-mediated deletion mutation in the leading and lagging strands. *J. Bacteriol.*, **177**, 4385–91.
63. Pinder, D. J., Blake, C. E., Lindsey, J. C., and Leach, D. R. F. (1998) Replication strand preference for deletions associated with DNA palindromes. *Mol. Microbiol.*, **28**, 719–27.
64. Sharples, G. J. and Lloyd, R. G. (1990) A novel repeated DNA sequence located in the intergenic regions of bacterial chromosomes. *Nucleic Acids Res.*, **18**, 6053–8.
65. Hulton, C. S. J., Higgins, C. F., and Sharp, P. M. (1991) ERIC sequences: a novel family of repetitive elements in the genomes of *Escherichia coli*, *Salmonella typhimurium* and other enterobacteria. *Mol. Microbiol.*, **5**, 825–34.
66. Sharp, P. M. and Leach, D. R. F. (1996) Palindrome-induced deletion in enterobacterial repetitive sequences. *Mol. Microbiol.*, **22**, 1055–6.
67. Kohwi, Y. and Panchenko, Y. (1993) Transcription-dependent recombination induced by triple-helix formation. *Genes Dev.*, **7**, 1766–78.
68. Bi, X. and Liu, L. F. (1996) DNA rearrangement mediated by inverted repeats. *Proc. Natl Acad. Sci. USA*, **93**, 819–23.
69. Lin, C.-T., Lyu, Y. L., and Liu, L. F. (1997) A cruciform-dumbbell model for inverted dimer formation mediated by inverted repeats. *Nucleic Acids Res.*, **25**, 3009–16.
70. Pinder, D. J., Blake, C. E., and Leach, D. R. F. (1997) DIR: a novel DNA rearrangement associated with inverted repeats. *Nucleic Acids Res.*, **25**, 523–9.
71. Charlier, D., Crabeel, M., Cunin, R., and Glansdorff, N. (1979) Tandem and inverted repeats of arginine genes in *Escherichia coli*. *Mol. Gen. Genet.*, **174**, 75–88.
72. Charlier, D., Severne, Y., Zafarullah, M., and Glansdorff, N. (1983) Turn-on of inactive genes by promoter recruitment in *Escherichia coli*: inverted repeats resulting in artificial divergent operons. *Genetics*, **105**, 469–88.
73. Mo, J.-Y., Maki, H., and Sekiguchi, M. (1991) Mutational specificity of the *dnaE173* mutator associated with a defect in the catalytic subunit of DNA polymerase III of *Escherichia coli*. *J. Mol. Biol.*, **222**, 925–36.
74. Gordon, A. J. E. and Halliday, J. A. (1995) Inversions with deletions and duplications. *Genetics*, **140**, 411–14.
75. Cromie, G., Collins, J., and Leach, D. R. F. (1997) Sequence interruptions in enterobacterial repeated elements retain their ability to encode well-folded RNA secondary structures. *Mol. Microbiol.*, **24**, 1311–14.
76. Sharp, P. M. (1997) Insertions within ERIC sequences. *Mol. Microbiol.*, **24**, 1314–15.
77. Shimizu, M., Hanvey, J. C., and Wells, R. D. (1989) Intramolecular DNA triplexes in supercoiled plasmids. I. Effect of loop size on formation and stability. *J. Biol. Chem.*, **264**, 5944–9.
78. Hanvey, J. C., Shimuzu, M., and Wells, R. D. (1989) Intramolecular DNA triplexes in supercoiled plasmids. II. Effect of base composition and noncentral interruptions on formation and stability. *J. Biol. Chem.*, **264**, 5950–6.
79. Bidichandani, S. I., Ashizawa, T., and Patel, P. I. (1998) The GAA triplet-repeat expansion in Friedreich's ataxia interferes with transcription and may be associated with an unusual DNA structure. *Am. J. Hum. Genet.*, **62**, 111–21.

80. Gacy, A. M., Goeller, G. M., Spiro, C., Chen, X., Gupta, G., Bradbury, E. M., *et al.* (1998) GAA instability in Friedreich's-ataxia shares a common, DNA-directed and intra-allelic mechanism with other trinucleotide diseases. *Mol. Cell*, **1**, 583–93.

81. Jaworski, A., Rosche, W. A., Gellibolian, R., Kang, S., Shimizu, M., Bowater, R. P., *et al.* (1995) Mismatch repair in *Escherichia coli* enhances instability of (CTG)$_n$ triplet repeats from human hereditary diseases. *Proc. Natl Acad. Sci. USA*, **92**, 11019–23.

82. Roshe, W. A., Jaworski, A., Kang, S., Kramer, S. F., Larson, J. E., Giedroc, D. P., *et al.* (1996) Single-stranded DNA binding protein enhances the stability of CTG triplet repeats in *Escherichia coli*. *J. Bacteriol.*, **178**, 5042–4.

83. Murphy, G. L., Connell, T. D., Barritt, D. S., Koomey, M., and Cannon, J. G. (1989) Phase variation of gonococcal protein II: regulation of gene expression by slipped-strand mispairing of a repetitive DNA sequence. *Cell*, **56**, 539–47.

84. Farabaugh, P. J., Schmeissner, U., Hofer, M., and Miller, J. H. (1978) Genetic studies of the *lac* repressor. VII. On the molecular nature of spontaneous hotspots in the *lacI* gene of *Escherichia coli*. *J. Mol. Biol.*, **126**, 847–57.

85. Lloyd, R. G. and Buckman, C. (1985) Identification and genetic analysis of *sbcC* mutations in commonly used *recBC sbcB* strains of *Escherichia coli* K-12. *J. Bacteriol.*, **164**, 836–44.

86. Gibson, F. P., Leach, D. R. F., and Lloyd, R. G. (1992) Identification of *sbcD* mutations as co-suppressers of *recBC* that allow propagation of DNA palindromes in *Escherichia coli* K-12. *J. Bacteriol.*, **174**, 1222–8.

87. Lindsey, J. C. and Leach, D. R. F. (1989) Slow replication of palindrome-containing DNA. *J. Mol. Biol.*, **206**, 7024–7.

88. Shurvinton, C. E., Stahl, M. M., and Stahl, F. W. (1987) Large palindromes in the λ phage genome are preserved in a *rec*$^+$ host by inhibiting λ DNA replication. *Proc. Natl Acad. Sci. USA*, **84**, 1624–8.

89. Connelly, J. C. and Leach, D. R. F. (1996) The *sbcC* and *sbcD* genes of *Escherichia coli* encode a nuclease involved in palindrome inviability and genetic recombination. *Genes Cells*, **1**, 285–91.

90. Connelly, J. C., de Leau, E. S., Okely, E. A., and Leach, D. R. F. (1997) Overexpression, purification, and characterisation of the SbcCD protein from *Escherichia coli*. *J. Biol. Chem.*, **272**, 19819–26.

91. Connelly, J. C., Kirkham, L. C., and Leach, D. R. F. (1998) The SbcCD nuclease of *Escherichia coli* is a structural maintenance of chromosomes (SMC) family protein that cleaves hairpin DNA. *Proc. Natl Acad. Sci. USA*, **95**, 7969–74.

92. Leach, D. R. F., Okely, E. A., and Pinder, D. J. (1997) Repair by recombination of DNA containing a palindromic sequence. *Mol. Microbiol.*, **26**, 597–606.

93. Lobachev, K. S., Shor, B. M., Tran, H. T., Taylor, W., Keen, J. D., Resnick, M. A., *et al.* (1998) Factors affecting inverted repeat stimulation of recombination and deletion in *Saccharomyces cerevisiae*. *Genetics*, **148**, 1507–24.

94. Akgun, E., Zahn, J., Baumes, S., Brown, G., Liang, F., Romanienko, P. J., *et al.* (1997) Palindrome resolution and recombination in the mammalian germ line. *Mol. Cell. Biol.*, **17**, 5559–70.

95. Nag, D. K. and Kurst, A. (1997) A 140-bp-long palindromic sequence induces double-strand breaks during meiosis in the yeast *Saccharomyces cerevisiae*. *Genetics*, **146**, 835–47.

# 2 | Double-strand break repair and V(D)J recombination

BELINDA K. SINGLETON and PENNY A. JEGGO

## 1. Introduction

DNA damaging agents are present both exogenously in the environment and also endogenously as by-products of metabolic processing. Such DNA-damaging agents can cause many lesions, the most crucial of which may be a double-strand break (dsb). Misrepaired dsbs can result in gene mutations or chromosome rearrangements, while unrepaired dsbs may lead to cell death. Therefore, cells have evolved a number of pathways to deal with dsbs. Different species utilize these pathways to different extents. For example, homologous recombination is the predominant dsb repair pathway in lower eukaryotes, whereas higher eukaryotes mainly repair dsbs by mechanisms that require little or no DNA sequence homology.

Dsbs also occur as intermediates in the rearrangement of gene segments during development of the vertebrate lymphoid system. This process, known as V(D)J recombination, generates the enormous diversity of T- and B-cell receptor loci required for an efficient immune response.

Several years ago, these two distinct areas of study were linked by the discovery that a group of proteins involved in dsb repair also play a role in the processing of dsbs introduced during V(D)J recombination. The components linking these two processes form the basis of this chapter.

## 2. V(D)J recombination

Component V (variable), D (diversity), and J (joining) segments in the germline are assembled into functional immunoglobulin and T-cell receptor genes during the development of lymphoid cells (reviewed in refs 1–3). The only known lymphoid-specific proteins required for V(D)J recombination are those encoded by the recombination activating genes, *RAG1* and *RAG2* (4, 5), and the non-essential terminal deoxynucleotidyl transferase, TdT (6). The RAG proteins are not only essential for the reaction but are sufficient to induce V(D)J recombination in non-lymphoid cells. Therefore, virtually any cell type can undergo V(D)J recombination following the co-transfection of *RAG1* and *RAG2* genes and an extrachromosomal V(D)J substrate.

The coding segments to be joined are flanked by recombination signal sequences (RSS) consisting of partially conserved heptamer and nonamer motifs separated by a spacer of 12 or 23 base pairs. Efficient joining occurs only between signals with different spacer lengths (the '12/23 rule'). Double-strand cleavage between the coding and signal sequences results in the generation of blunt signal ends and a covalently closed hairpin at each coding end. The signal ends are joined in a precise manner but the coding ends are processed before joining and may undergo deletion or addition of nucleotides, thereby increasing the diversity of antigen receptors. For ease, the reaction can be divided into two stages, as shown in Fig. 2.1.

## 2.1  Recognition and cleavage of the signal sequences

Substantial progress in understanding this stage of the process has come from the recent development of an *in vitro* cleavage system (7). Studies with this system have shown that both RAG proteins play a direct role in V(D)J recombination (reviewed in ref. 8). The nonamer motif is required for the initial sequence-specific binding of RAG1, whereas the heptamer appears to be essential for the correct targeting of cleavage. *In vitro*, nicking occurs at the border between the coding sequence and the heptamer, yielding a 3' hydroxyl group which attacks the complementary strand resulting in a closed hairpin at the coding end and a blunt 5' phosphorylated signal end. Whilst this cleavage can occur in the presence of either $Mg^{2+}$ or $Mn^{2+}$, it only follows the 12/23 rule if the divalent cation is $Mg^{2+}$. This reaction is facilitated by the presence of a crude cell extract, suggesting the involvement of other factors in the cleavage step, possibly in the synapsis of the two signal sequences.

## 2.2  Processing and joining the dsbs

Once cleavage has occurred, the synapsed signal ends (but not coding ends) are held together in a complex containing RAG1, RAG2, the high-mobility group protein HMG-1, and components of the DNA-dependent protein kinase, DNA-PK (9). Whilst the precise function of such a complex is not clear, its presence may inhibit signal-joint formation and, by protecting the ends from nucleolytic degradation, could explain why signal joints, once formed, are precise.

In contrast, the highly modified coding ends are rapidly joined. The first step in the processing of coding ends is the resolution of the hairpin. If this occurs off-centre, then palindromic overhangs are generated which may either be removed or result in the addition of so-called 'P nucleotides'. In addition, further nucleotides may be deleted or non-templated 'N' nucleotides may be added by TdT, if present.

Despite V(D)J recombination being such a specialized process, at this stage the signal and coding ends are, in effect, dsbs, and hence components of the general dsb repair machinery are recruited to rejoin them. The properties and possible functions of these repair proteins will be discussed in more detail later in this chapter.

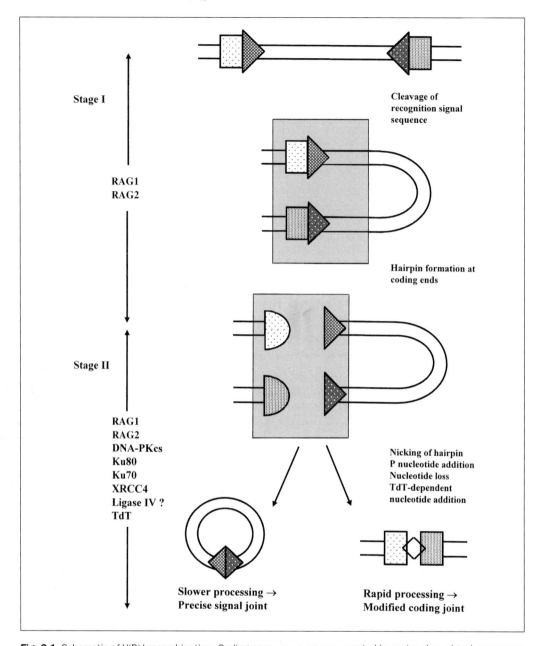

**Fig. 2.1** Schematic of V(D)J recombination. Coding sequences are represented by rectangles, signal sequences by triangles, and the synaptic complex by the large shaded box. Proteins possibly involved are listed on the left of the figure. Stage I: recognition of the signal sequences by the RAG proteins, synapsis of the signals, and cleavage between heptamer and coding sequences, resulting in hairpin formation at the coding ends. Stage II: Signal ends are processed slowly but protected by synaptic complex, resulting in a precise signal joint. Hairpin coding ends are released from synaptic complex and processed rapidly. Nucleotide additions and deletions may occur, resulting in a modified coding joint.

# 3. Mechanisms of DNA double-strand break repair

In lower eukaryotes such as *Saccharomyces cerevisiae*, repair of DNA dsbs occurs primarily by homologous recombination during the late $S/G_2$ phase of the cell cycle (see Fig. 2.2). However, at least two alternative repair pathways exist—single-strand annealing (SSA, Section 3.2) and non-homologous end-joining (NHEJ, Section 3.3). In contrast, mammalian cells predominantly repair dsbs by mechanisms that require little or no sequence homology (10), and which have been suggested to take place during the $G_1/S$ phase of the cell cycle (see, for example, ref. 11).

## 3.1 Homologous recombination

Homologous recombination is a mechanism with two apparently opposing functions. It can be used to increase genetic variation (by recombination between homologous chromosomes during meiosis) and to maintain genomic stability in mitotic cells (by repair of chromosome breaks using the sister chromatid as a template).

In yeast, the process is dependent on the *RAD52* epistasis group of genes (*RAD50—55, RAD57, RAD59, MRE11*, and *XRS2*) (reviewed in ref. 12). As expected, loss-of-function mutations in any of these genes result in extreme sensitivity to agents that cause DNA dsbs. Whilst members of the *RAD52* epistasis group have been implicated in mitotic and meiotic recombination, plasmid integration, and mating type switching, *RAD50, MRE11*, and *XRS2* are only involved in a subset of *RAD52*-dependent recombination functions and additionally play a role in a *RAD52*-independent non-homologous dsb repair process. Rad54p belongs to the family of Swi2/Snf2 helicase-like proteins that may have a role in chromatin remodelling. Although it has been reported that Rad54p interacts with Rad51p (13–15) and hence may be part of a recombination complex, its precise role remains unclear.

As shown in Fig. 2.3, a DNA dsb is processed to reveal single-stranded regions of DNA (ssDNA). Rad51p (the homologue of the bacterial RecA protein), together with the single-stranded binding protein RPA (replication protein A), coats the ssDNA to form a nucleoprotein filament. This filament searches for a homologous sequence in the genome and invades the donor duplex to form a joint molecule. Recent biochemical studies have shown that Rad52p binds to Rad51p/RPA and stimulates homologous pairing (16–18). The DNA gap is repaired by fill-in synthesis using the donor strand as template, followed by branch migration and resolution of the Holliday junctions to release the repaired DNA molecules.

Recently, mammalian homologues of *RAD50, RAD51, RAD52, RAD54*, and *MRE11* have been identified and cloned, underlining the evolutionary importance of homologous recombination (19–23). Consistent with this, expression of the human homologues hRad51p and hRad52p appears to be lowest during the $G_1$ phase of the cell cycle and upregulated during late $S/G_2$ (24). However, targeting of *mRAD51* results in early embryonic lethality, indicating an essential role in mouse development in addition to a role in dsb repair (25, 26).

A recent and intriguing finding is that both *BRCA1* and *BRCA2* (tumour

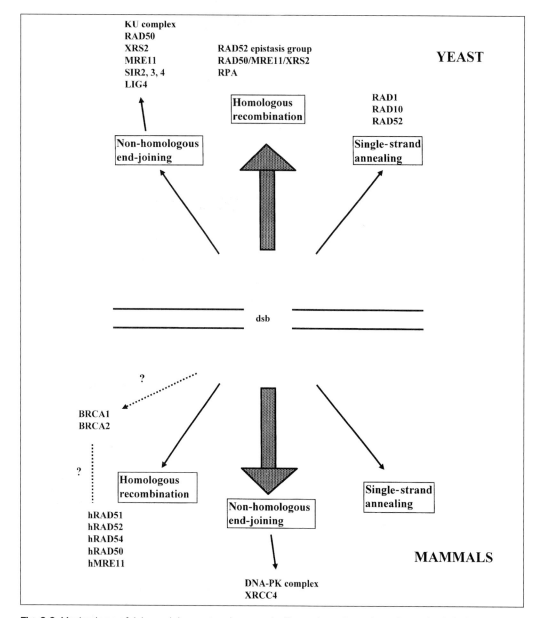

**Fig. 2.2** Mechanisms of dsb repair in yeast and mammals. The major pathway in each species is indicated by the large arrows. In yeast, dsbs are mainly repaired by the process of homologous recombination which utilizes the *RAD52* epistasis group of genes. In contrast, mammals predominantly utilize NHEJ mediated by the DNA-PK complex. In mammals, the tumour suppressor genes *BRCA1* and *BRCA2* may possibly perform a role in dsb repair, in conjunction with *hRAD51*.

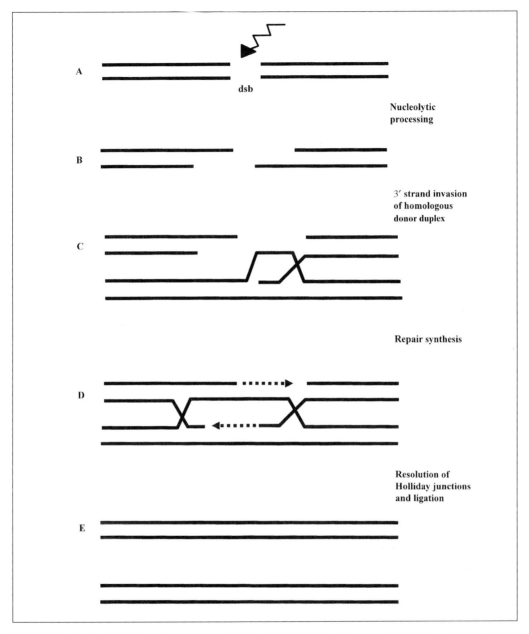

**Fig. 2.3** Repair of a dsb by homologous recombination. The DNA duplex is represented by the pair of black lines and the dsb is induced by ionizing radiation. Nucleolytic processing produces 3′ ssDNA ends that become bound by Rad51p and RPA. The 3′ strand searches for and invades a homologous donor duplex, forming a joint molecule. Pairing of molecules is promoted by binding of Rad52p to Rad51p. The donor strand is used as a template for repair synthesis and the process is completed by resolution of the Holliday junctions and ligation.

suppressor genes predisposing to familial cases of breast cancer) interact directly with hRad51p (27, 28). Furthermore, cells derived from both *RAD51* and *BRCA2* 'knock-out' embryos show hypersensitivity to ionizing radiation (25, 28). This raises the exciting prospect that the three proteins function as a complex in dsb repair, either together with Rad52p or in a separate repair pathway (27, 28).

## 3.2 Single-strand annealing

An additional mitotic recombination pathway exists in both mammals and yeast and is involved in dsb-induced recombination between long direct repeats. This mechanism is known as single-strand annealing (SSA) and was first proposed by Lin and co-workers in 1984 (29) to account for the non-conservative recombination products detected after the transfection of DNA into mouse cells (see Fig. 2.4).

Unlike the homologous recombination model outlined above (Section 3.1), SSA in yeast is dependent on *RAD52* but independent of other members of the epistasis group. Recombination is initiated by a dsb but, unlike homologous recombination, the dsb does not need to be within the region of homology. The DNA ends are degraded by a strand-specific exonuclease or unwound by a helicase to reveal single-stranded direct repeat sequences. The homologous regions realign to form heteroduplex DNA and the 3' ssDNA tails are removed. In *S. cerevisiae*, this is performed by the Rad1/Rad10 complex. Recombination is completed by repair synthesis and ligation.

The Rad1/Rad10 complex has been shown to have an endonuclease activity that cleaves at the junctions of single/double-stranded DNA (30) and is thought to be involved in both mitotic recombination and nucleotide excision repair (NER). The human homologues of *RAD1* and *RAD10* are the *ERCC4* and *ERCC1* genes that are defective in the NER syndrome, xeroderma pigmentosum (31, 32). These proteins also have a strand-specific endonuclease activity that is thought to perform the 5' incision step during NER (31). The hypersensitivity of *ERCC4*/*ERCC1* rodent mutants to DNA cross-linking agents has substantiated the suggestion that the proteins may have a role in the recombinational repair of interstrand cross-links in mammalian cells (31).

An important feature of the SSA model is that sequences are always lost during the process, in contrast to the conservative nature of homologous recombination. Therefore, the mechanism is intrinsically error-prone.

## 3.3 Non-homologous end-joining (NHEJ)

Whilst the repair of DNA dsbs in yeast occurs mainly by homologous recombination (see Fig. 2.2 and Section 3.1), studies involving the integration of DNA into mammalian genomes have shown that homologous recombination occurs between 100 to 10 000 times less frequently than non-homologous mechanisms (33). The genes involved in this process will be discussed in the ensuing section.

Potential non-homologous end-joining mechanisms for a range of DNA end-

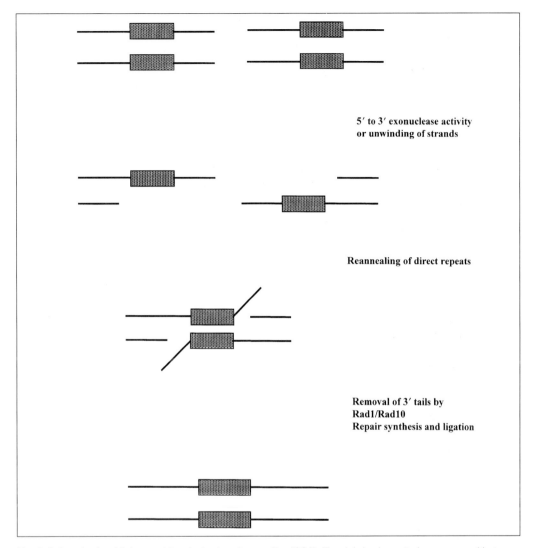

5' to 3' exonuclease activity
or unwinding of strands

Reannealing of direct repeats

Removal of 3' tails by
Rad1/Rad10
Repair synthesis and ligation

**Fig. 2.4** Repair of a dsb in yeast by single-strand annealing (SSA). The dsb is shown to have occurred between two regions of homology (indicated by shaded boxes). Homologous sequence is exposed by a 5' to 3' exonuclease activity or unwinding of the strands. The DNA reanneals at the repeated sequences and the overhanging 3' ssDNA ends are removed by Rad1/Rad10 endonuclease activity. The process is completed by repair synthesis and ligation.

structures are illustrated in Fig. 2.5. Structure 'A' involves direct blunt-end to blunt-end ligation, whereas repair of 'B' and 'C' could be mediated by single-strand ligation followed by fill-in synthesis and ligation (34). Direct end-joining of linear DNA molecules is commonly seen in mammalian cells (34) and in extracts of *Xenopus laevis* eggs (35). No sequence homology is required to generate these types of junction. In contrast, repair of structure 'D' frequently utilizes 1–6 base pairs (bp) of

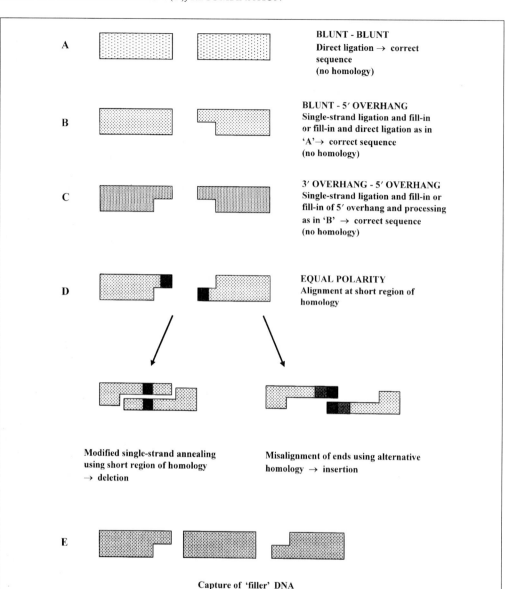

**Fig. 2.5** Potential mechanisms of NHEJ. The shaded boxes represent DNA ends and may be blunt or overhanging ss ends. (A) Repair does not require homology and may be by direct ligation of blunt ends. (B) Repair does not require homology. Overhang may be filled in by 5′ to 3′ synthesis, resulting in two blunt ends and processing as in A, or repair could involve single-strand ligation, followed by fill-in synthesis. (C) Repair does not require homology. The 5′ overhang may be filled in by 5′ to 3′ synthesis, followed by processing as in B, or repair could involve single-strand ligation and fill-in synthesis. (D) Ends share a short region of homology (1–6 bp). Three possible outcomes are shown, depending on the location of the homologies. Complementary ends may align, followed by repair synthesis and ligation. Ends with internal homologies may be repaired by a modified SSA mechanism, resulting in the deletion of sequences. Homologous ends may misalign using alternative homology, resulting in the insertion of sequences. (E) A variety of end-structures may be repaired by the capture of 'filler' DNA such as retrotransposons or oligonucleotides.

microhomology at or close to the ends of the DNA (34) and, as shown in Fig. 2.5, may result in either deletion or a short insertion. Microhomology-dependent deletion junctions are thought to occur by a modified SSA mechanism and have been observed in a number of different systems including: the repair of linearized plasmids transfected into various cell types (e.g. refs 34, 36, 37) *in vitro* by human cell extracts (e.g. refs 38, 39) and *Xenopus laevis* egg extracts (e.g. ref. 35); and *in vivo* at sites of chromosome rearrangements and deletions in human genetic diseases (e.g. ref. 40). In *S. cerevisiae*, both deletion and insertion junctions, as illustrated in Fig. 2.5 (arising from structure 'D'), have been observed during the repair of an HO endonuclease-induced dsb at the chromosomal mating-type (MAT) locus (41, 42). The lack of a requirement for substantial homology and thus a stable heteroduplex DNA molecule has led to the postulation of an alignment protein in NHEJ (35).

An alternative, non-homologous end-joining mechanism involves the 'capture' of DNA at the site of joining. 'Filler DNA' at dsb junctions in non-lymphoid cells may consist of extra nucleotides (similar to the addition of nucleotides by terminal deoxynucleotidyl transferase, TdT, during V(D)J recombination in lymphoid cells (see Section 2.2)), or possibly longer blocks of preformed oligonucleotides (33). Furthermore, a rare non-homologous event at the MAT locus in *S. cerevisiae* has shown captured DNA from the Ty1 retrotransposon at the HO cleavage site, providing a possible mechanism for the insertion of pseudogenes and repetitive sequences (SINES and LINES) into the genome of mammalian cells (42).

# 4. Identification of the genes involved in NHEJ

A number of different approaches have been utilized to identify the genes involved in NHEJ in yeast and mammals (see Table 2.1 and references therein). The importance of NHEJ was first established by studies on mammalian cells and was subsequently found to be conserved in yeast. In both organisms, the DNA-PK pathway appears to be the most common form of NHEJ.

## 4.1 DNA-dependent protein kinase, DNA-PK

The characterization of a group of ionizing radiation-sensitive hamster mutants has provided valuable information on the genes involved in the repair of DNA breaks. Cell-fusion studies have defined at least 11 complementation groups (44) and the human genes correcting the defects have been designated XRCC (X-ray repair cross-complementing) genes (43). Mutants in three of these complementation groups (groups 5, 6, and 7) are defective in components of the DNA-PK complex. DNA-PK consists of a 460-kDa catalytic subunit (DNA-PKcs) which is recruited by DNA-bound Ku protein, itself a heterodimer of Ku70 and Ku80 (65, 66). The binding of DNA-PKcs to DNA-bound Ku induces its kinase activity (66, 67). DNA-PK can phosphorylate many substrates *in vitro*, including Ku, p53, RPA, and several transcription factors, but the relevant *in vivo* substrates are still to be identified (see, for example, refs 68, 69).

**Table 2.1** Approaches used to identify genes involved in NHEJ

| Species | Gene | Identification of gene (refs) |
|---------|------|-------------------------------|
| Mammal | XRCC5/Ku80 | Characterization of defective cell lines; gene cloning using defective cell lines (reviewed in 43, 44) |
| | XRCC6/Ku70 | Inferred from Ku80 involvement; generation of knock-out ES cell lines (reviewed in 43, 44) |
| | XRCC7/DNA-PKcs | Inferred from Ku80 data; characterization of defective cell lines (reviewed in 43, 44) |
| | XRCC4 | Gene cloning using defective cell line (45) |
| | Ligase IV | Interaction with XRCC4 (by yeast two-hybrid analysis and co-immunoprecipitation (46, 47) |
| | hRAD50 | Homology to yeast RAD50 (21) |
| | hMRE11 | Homology to yeast MRE11 (20) |
| Yeast | HDF2/YKU80 | Purification of DNA end-binding activity; characterization of yeast mutants for γ-ray survival and NHEJ (48–50) |
| | HDF1/YKU70 | Purification of DNA end-binding activity; characterization of yeast mutants for γ-ray survival and NHEJ (49, 51–55) |
| | LIG4/DNL4 | Homology to human DNA ligase IV; characterization of yeast mutants for γ-ray survival and NHEJ (56–58) |
| | RAD50 | Characterisation of yeast mutants |
| | MRE11 | for γ-ray survival |
| | XRS2 | and NHEJ (49, 59–62) |
| | SIR2 | Interaction of Sir4 with YKU70 (by yeast two-hybrid |
| | SIR3 | analysis); characterization of yeast mutants |
| | SIR4 | for γ-ray survival and NHEJ (63, 64) |

The human Ku protein was first identified more than 15 years ago, as an auto-antigen present in the sera of autoimmune patients (70), and was found to have double-stranded (ds) DNA end-binding activity *in vitro* (e.g. refs 71, 72). Members of complementation group 5 were found to lack the DNA end-binding activity and were subsequently shown to be defective in *Ku80* (see Table 2.2 and refs 73, 74). Although no mutants have yet been identified with defects in *Ku70*, this gene has been assigned to complementation group 6. Several rodent mutants have been identified with defects in *XRCC7* and both complementation and mapping studies have confirmed that the defective gene is *DNA-PKcs*, the large catalytic subunit of DNA-PK (75–80). The phenotypes of these mutants are listed in Table 2.2 and will be discussed in detail in Section 5.

## 4.2 XRCC4 and DNA ligase IV

The *XRCC4* gene defines a fourth complementation group with defective DNA dsb rejoining and complements the defects observed in the *XR-1* hamster mutant (see Table 2.2). The phenotypic similarities between *XR-1* and members of complement-ation groups 5, 6, and 7 implicate *XRCC4* in the DNA-PK pathway (see Section 5). The XRCC4 protein interacts with and stimulates the activity of human DNA ligase IV (46, 47, 81), suggesting that this enzyme may perform the final ligation step of

NHEJ and V(D)J recombination. Furthermore, a homologue of DNA ligase IV has been identified in *S. cerevisiae* (*LIG4* (57, 58) or *DNL4* (56)), and null mutants generated. Plasmid rejoining assays have confirmed the involvement of *lig4* in Ku-dependent DNA repair.

## 4.3 Yeast as a model system

The contribution of NHEJ to dsb repair in yeast is only detected in the absence of homologous recombination, i.e. in *rad52* mutant strains or where there are no homologous sequences (e.g. refs 49, 53). The use of plasmid rejoining assays in *rad52* mutants of *S. cerevisiae* has helped to establish the involvement of a number of genes in NHEJ, as shown in Table 2.1. Yeast *Ku* homologues (*YKU80* and *YKU70*) have been found and characterized (48, 50, 51, 54, 55), although neither a *DNA-PKcs* nor an *XRCC4* homologue has been detected despite searching the entire yeast database. Yeast *Ku* mutants (*yku80* and *yku70*), like their mammalian counterparts, lack DNA end-binding activity, have reduced ability to rejoin DNA ends, and (in a *rad52* background) are sensitive to ionizing radiation (49–51, 53, 54, 63). In addition, yeast *Ku* mutants show a temperature-sensitive growth defect and have shortened telomeres, suggesting that Ku is crucial for telomere maintenance (50, 51, 82). Consistent with this proposed role is the finding that yku70p interacts with Sir4, a protein involved in transcriptional silencing at telomeres and mating type loci (63). Furthermore, mutations in the *SIR2*, *SIR3*, and *SIR4* genes result in reduced NHEJ using the plasmid assay and, when in a *rad52* background, increased sensitivity to ionizing radiation (63, 64). The *sir yku70* double mutants are no more defective than the *yku70* single mutant, suggesting that the Sir proteins are involved in the Ku pathway of NHEJ. Furthermore, yku70 may recruit the Sir proteins to a dsb, leading to a heterochromatin-like state that may be essential for the rejoining of DNA ends (63, 83). Figure 2.6 suggests a model for Ku-mediated repair of a DNA dsb.

The Rad50/Mre11/Xrs2 complex was first identified in yeast and was suggested to be involved in meiotic recombination based on the inability of defective mutants to undergo meiotic recombination (88). The complex was subsequently found to function in a second, distinct pathway: dsb repair in the absence of *RAD52*, since *rad50*, *mre11*, and *xrs2* mutants show a similar phenotype to *yku70*, *yku80*, and *lig4* mutants in both plasmid and chromosome NHEJ assays (49, 60, 62, 63). The recent identification of human homologues of *RAD50* and *MRE11* and their protein interaction suggests that these proteins have a conserved function in mammals, although it is not clear how they function in relation to the other known NHEJ proteins (20, 21).

## 5. The phenotypes of mammalian mutants defective in NHEJ

Mutants in four of the eleven complementation groups described in Section 4.1 have defects in dsb repair (reviewed in refs 43, 44). The phenotypes of these mutants

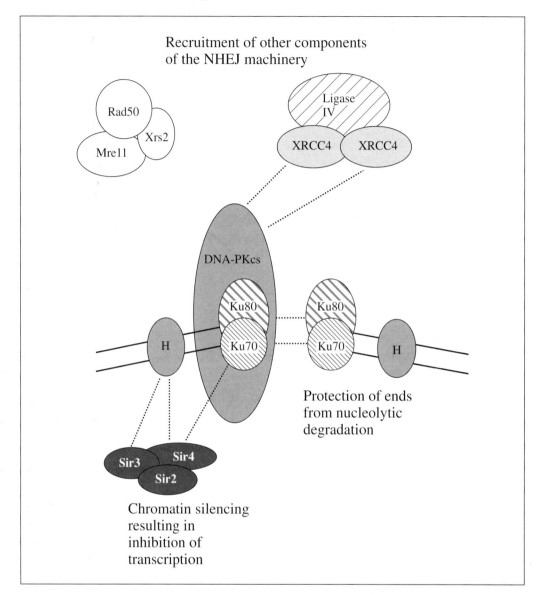

**Fig. 2.6** Model of Ku-mediated dsb repair. The DNA duplex is represented by the two solid black lines. Ovals marked 'H' represent the histone octamers. Potential protein interactions are depicted by dashed lines. Ku protein is shown binding to both DNA ends, but only one Ku has recruited DNA-PKcs. Ku has a direct role in the repair (84), but repair may involve either interaction between two DNA-bound Ku molecules (85), as shown, or one Ku bound simultaneously to two DNA molecules (86). In both cases, Ku could protect the DNA ends from nucleolytic degradation. In yeast, Ku70 may recruit the Sir proteins to the dsb, possibly inhibiting transcription by chromatin silencing. In mammals, interaction of XRCC4 with DNA-PKcs may recruit ligase IV to the dsb. XRCC4 is shown functioning as a dimer (47, 87). It is unclear how the Rad50/Mrell/Xrs2 complex functions.

**Table 2.2** Phenotypes of rodent mutants defective in NHEJ

| Properties | Group 4 | Group 5 | Group 6 | Group 7 |
|---|---|---|---|---|
| Gene | XRCC4 | XRCC5/Ku80 | XRCC6/Ku70 | XRCC7/DNA-PKcs |
| Chromosomal locus | 5q | 2q33–35 | 22q13 | 8q11 |
| Defective cell lines | XR-1 | xrs1, 2, 4, 5, 6, 7 XR-V15B, XR-V9B sxi-2, 3 | ES cell line | scid mouse V3 irs20 |
| Repair defect (dsb) | 2–4-fold | 2–4-fold | 4–5-fold | 2–4-fold |
| γ-Ray sensitivity | 6-fold | 6-fold | 3–6-fold | 6-fold |
| Ku DNA-binding activity | Normal | Absent | Absent | Normal |
| DNA-PKcs activity | Normal | Undetectable | Undetectable | Undetectable |
| V(D)J recombination: – Signal joints | $10^3$-fold decrease; deletions | $10^3$-fold decrease; deletions | $10^2$-fold decrease; deletions | 2–5-fold decrease |
| – Coding joints | $10^3$-fold decrease | $10^3$-fold decrease | $10–10^2$-fold decrease | $10^3$-fold decrease |

(groups 4, 5, 6, and 7) are summarized in Table 2.2. These mutants are extremely sensitive to ionizing radiation, and in addition to a defect in the rejoining of DNA dsbs, are unable to perform V(D)J recombination correctly. This link was first established by the observation that cells derived from the *scid* (severe combined immune deficiency) mouse (89) were radiosensitive and dsb repair-deficient (see, for example, ref. 90). This discovery prompted the analysis of the hamster dsb repair-deficient mutants for their ability to carry out V(D)J recombination (e.g. ref. 91). The observed defects in both dsb repair and the specific process of V(D)J recombination suggested that these two processes were mechanistically linked, and a DNA dsb was identified as the common factor. As described above, the genes defective in these hamster mutants have now been identified (see Sections 4.1 and 4.2 and Table 2.2).

The *XR-1* mutant is the only member of complementation group 4 and appears to be deleted for *XRCC4* (45, 92). The XRCC4 protein has no obvious domains or motifs and has an unknown function.

A number of group 5 mutants have been characterized and, where examined, can be complemented by transfection with *Ku80* cDNA (see Table 2.2 and references 93–99). In addition, causative mutations have recently been identified in several group 5 mutants (97, 99). *Ku80*-deficient mice have been generated, allowing further characterization of the function of Ku80 *in vivo* (100, 101). These mice show the predicted immune deficiency resulting from impaired V(D)J recombination, but they also exhibit growth retardation (100), an unexpected phenotype, suggesting an important role for Ku80 in growth and development. *Ku80* mutants are severely impaired in the production of both signal joints and coding joints (e.g. ref. 91). Based on the end-binding properties of Ku it has been suggested that Ku protects DNA ends from nucleolytic degradation (see, for example, refs 102, 103). However, blunt-ended and full-length signal ends are detected in *xrs6* cells and in *Ku80* 'knock-out' mice, suggesting that *Ku80* is not required for either the protection or stabilization of signal ends (101, 104).

Although several *Ku80* hamster mutants have been independently isolated, genetic screens based on radiosensitivity have not produced mutants defective in *Ku70*. It has been postulated that the rodent cell lines used contain two active copies of *Ku70*, whereas one copy of *Ku80* appears to be 'silenced' by methylation (97, 99, 105). Nevertheless, embryonic stem (ES) cells deficient in Ku70 have now been constructed and display the defects expected of an *XRCC6* mutant (see Table 2.2 and reference 106). In addition, *Ku70* 'knock-out' mice show a similar growth retardation phenotype to *Ku80* deficient mice and have impaired rejoining of V(D)J recombination signal sequences. However, unlike *Ku80* deficient mice that lack mature T and B cells, *Ku70* mice are capable of developing mature T cells, suggesting a *Ku70*-independent rescue pathway of T-cell receptor V(D)J recombination (107, 108).

DNA-PKcs belongs to a subfamily of phosphatidylinositol (PI) 3-kinases that includes a number of other large proteins (> 200 kDa) involved in DNA damage surveillance (109). Despite sharing homology with PI-3 kinases, DNA-PKcs is a serine/threonine protein kinase and does not have detectable lipid phosphorylation activity (109). A nonsense mutation in the highly conserved carboxy-terminal region of DNA-PKcs has recently been identified as the causative mutation in the *scid* mouse (110, 111), suggesting that this region is important for protein kinase activity. This is supported by the finding that the *irs20* cell line (another group 7 mutant) has a mutation resulting in an amino acid substitution of the fourth residue from the carboxy-terminus, although since the full 14-kb cDNA has not been sequenced this may not necessarily be the causative mutation (112).

Although mutants of the complementation groups 4, 5, 6, and 7 all have defects in components of the DNA-PK pathway, and share similar dsb repair defects and ionizing radiation sensitivities, they differ in their defects in V(D)J recombination. The most severe defects are observed in group 4 and 5 mutants, in which both coding- and signal-junction formation are dramatically decreased (see Table 2.2). In contrast, group 7 mutants display a minor defect in signal-joint formation, although coding-joint formation is significantly reduced. These findings indicate an important mechanistic difference between signal- and coding-joint formation and suggest that while Ku and XRCC4 proteins are essential for both processes, DNA-PKcs is dispensable for the joining of signal ends.

# 6. Conclusions

A combination of approaches, involving both yeast and mammalian cells, has provided valuable insights into the mechanisms of DNA dsb repair and its link with the process of V(D)J recombination. The continued characterization of the genes and proteins involved should facilitate the determination of their precise roles and result in a greater understanding of the repair of radiation-induced damage *in vivo*.

The overlap between the ionizing radiation sensitivity and defects in immune response that has been identified in rodent mutants has implications for human health. An understanding of the proteins involved in these responses may aid the identification of human individuals who are sensitive to ionizing radiation. For example, the inappropriate use of radiotherapy could be avoided if individual patients with breast cancer could be predicted to be ionizing radiation-sensitive. Furthermore, the isolation of inhibitors of DNA repair (e.g. dominant-negative mutants) to radiosensitize cells could greatly benefit radiotherapeutic treatments.

Hopefully, ongoing biochemical analyses will identify additional interacting proteins and help to provide further information on a range of cellular processes in addition to dsb repair.

# References

1. Alt, F. W., Oltz, E. M., Young, F., Gorman, J., Taccioli, G., and Chen, J. (1992) VDJ recombination. *Immunol. Today*, **13**, 306.
2. Gellert, M. (1992) Molecular analysis of V(D)J recombination. *Annu. Rev. Genet.*, **26**, 425.
3. Lewis, S. M. (1994) The mechanism of V(D)J joining: lessons from molecular, immunological, and comparative analyses. *Adv. Immunol.*, **56**, 27.
4. Schatz, D. G., Oettinger, M. A., and Baltimore, D. (1989) The V(D)J recombination activating gene, RAG-1. *Cell*, **59**, 1035.
5. Oettinger, M. A., Schatz, D. G., Gorka, C., and Baltimore, D. (1990) RAG-1 and RAG-2, adjacent genes that synergistically activate V(D)J recombination. *Science*, **248**, 1517.
6. Alt, F. W. and Baltimore, D. (1982) Joining of immunoglobulin heavy chain gene segments: implications from a chromosome with evidence three D–JH fusions. *Proc. Natl Acad. Sci. USA*, **79**, 4118.
7. van Gent, D. C., McBlane, J. F., Ramsden, D. A., Sadofsky, M. J., Hesse, J. E., and Gellert, M. (1995) Initiation of V(D)J recombination in a cell-free system. *Cell*, **81**, 925.
8. Schatz, D. G. (1997) V(D)J recombination moves *in vitro*. *Semin. Immunol.*, **9**, 149.
9. Agrawal, A. and Schatz, D. G. (1997) RAG1 and RAG2 form a stable postcleavage synaptic complex with DNA containing signal ends in V(D)J recombination. *Cell*, **89**, 43.
10. Roth, D. B., Porter, T. N., and Wilson, J. H. (1985) Mechanisms of nonhomologous recombination in mammalian cells. *Mol. Cell Biol.*, **5**, 2599.
11. Lee, S. E., Mitchell, R. A., Cheng, A., and Hendrickson, E. A. (1997) Evidence for DNA-PK-dependent and -independent DNA double-strand break repair pathways in mammalian cells as a function of the cell cycle. *Mol. Cell Biol.*, **17**, 1425.
12. Petes, T. D., Malone, R. E., and Symington, L. S. (1991) Recombination in yeast. In *The molecular and cellular biology of the yeast* Saccharomyces (ed. J. R. Broach, J. R. Pringle, and E. W. Jones), p. 407. Cold Spring Harbor Laboratory Press, Cold Spring Harbor, New York.
13. Jiang, H., Xie, Y. Q., Houston, P., Stemkehale, K., Mortensen, U. H., Rothstein, R., *et al.* (1996) Direct association between the yeast Rad51 and Rad54 recombination proteins. *J. Biol. Chem.*, **271**, 33181.
14. Golub, E. I., Kovalenko, O. V., Gupta, R. C., Ward, D. C., and Radding, C. M. (1997) Interaction of human recombination proteins Rad51 and Rad54. *Nucl. Acids Res.*, **25**, 4106.
15. Clever, B., Interthal, H., Schmuckli-Maurer, J., King, J., Sigrist, M., and Heyer, W. D. (1997)

Recombinational repair in yeast: functional interactions between Rad51 and Rad54 proteins. *EMBO J.*, **16**, 2535.

16. Benson, F. E., Baumann, P., and West, S. C. (1998) Synergistic actions of Rad51 and Rad52 in recombination and repair. *Nature,* **391**, 401.

17. New, J. H., Sugiyama, T., Zaitseva, E., and Kowalczykowski, S. C. (1998) Rad52 protein stimulates DNA strand exchange by Rad51 and replication protein A. *Nature,* **391**, 407.

18. Shinohara, A. and Ogawa, T. (1998) Stimulation by Rad52 of yeast Rad51-mediated recombination. *Nature,* **391**, 404.

19. Muris, D. F. R., Bezzubova, O., Buerstedde, J. M., Vreeken, K., Balajee, A. S., Osgood, C. J., *et al.* (1994) Cloning of human and mouse genes homologous to *RAD52*, a yeast gene involved in DNA repair and recombination. *Mutat. Res.*, **315**, 295.

20. Petrini, J. H., Walsh, M. E., DiMare, C., Chen, X. N., Korenberg, J. R., and Weaver, D. T. (1995) Isolation and characterization of the human MRE11 homologue. *Genomics,* **29**, 80.

21. Dolganov, G. M., Maser, R. S., Novikov, A., Tosto, L., Chong, S., Bressan, D. A., *et al.* (1996) Human RAD50 is physically associated with human Mre11: identification of a conserved multiprotein complex implicated in recombinational DNA repair. *Mol. Cell Biol.*, **16**, 4832.

22. Shinohara, A., Ogawa, H., Matsuda, Y., Ushio, N., Ikeo, K., and Ogawa, T. (1993) Cloning of human, mouse and fission yeast recombination genes homologous to *RAD51* and *recA*. *Nature Genetics*, **4**, 239.

23. Kanaar, R., Troelstra, C., Swagemakers, S. M. A., Essers, J., Smit, B., Franssen, J.-H., *et al.* (1996) Human and mouse homologs of the *Saccharomyces cerevisiae RAD54* DNA repair gene: evidence for functional conservation. *Curr. Biol.*, **6**, 828.

24. Chen, F., Nastasi, A., Shen, Z., Brenneman, M., Crissman, H., and Chen, D. J. (1997) Cell cycle-dependent protein expression of mammalian homologues of yeast DNA double-strand break repair genes Rad51 and Rad52. *Mutat. Res.*, **384**, 205.

25. Lim, D.-S. and Hasty, P. (1996) A mutation in mouse *rad51* results in an early embryonic lethal that is suppressed by a mutation in *p53*. *Mol. Cell Biol.*, **16**, 7133.

26. Tsuzuki, T., Fujii, Y., Sakumi, K., Tominaga, Y., Nakao, K., Sekiguchi, M., *et al.* (1996) Targeted disruption of the Rad51 gene leads to lethality in embryonic mice. *Proc. Natl Acad. Sci. USA*, **93**, 6236.

27. Scully, R., Chen, J., Plug, A., Xiao, Y., Weaver, D., Feunteun, J., *et al.* (1997) Association of BRCA1 with Rad51 in mitotic and meiotic cells. *Cell*, **88**, 265.

28. Sharan, S. K., Morimatsu, M., Albrecht, U., Lim, D.-S., Regel, E., Dinh, C., *et al.* (1997) Embryonic lethality and radiation hypersensitivity mediated by Rad51 in mice lacking *Brca2*. *Nature*, **386**, 804.

29. Lin, F.-L., Sperle, K., and Sternberg, N. (1984) Model for homologous recombination during transfer of DNA into mouse L cells: role for DNA ends in the recombination process. *Mol. Cell Biol.*, **4**, 1020.

30. Bardwell, A. J., Bardwell, L., Tomkinson, A. E., and Friedberg, E. C. (1994) Specific cleavage of model recombination and repair intermediates by the yeast Rad1-Rad10 DNA endonuclease. *Science*, **265**, 2082.

31. Sijbers, A. M., de Laat, W. L., Ariza, R. R., Biggerstaff, M., Wei, Y.-F., Moggs, J. G., *et al.* (1996) Xeroderma pigmentosum group F caused by a defect in a structure-specific DNA repair endonuclease. *Cell*, **86**, 811.

32. van Duin, M., de Wit, J., Odijk, H., Westerveld, A., Yasui, A., Koken, M. H. M., *et al.* (1986) Molecular characterization of the human excision repair gene ERCC-1: cDNA cloning and amino acid homology with the yeast DNA repair gene RAD10. *Cell*, **44**, 913.

33. Roth, D. and Wilson, J. (1988) Illegitimate recombination in mammalian cells. In *Genetic recombination* (ed. R. Kucherlapati and G. R. Smith), p. 621. American Society of Microbiology, Washington, DC.

34. Roth, D. B. and Wilson, J. H. (1986) Nonhomologous recombination in mammalian cells: role for short sequence homologies in the joining reaction. *Mol. Cell Biol.*, **6**, 4295.

35. Pfeiffer, P. and Vielmetter, W. (1988) Joining of nonhomologous DNA double strand breaks *in vitro*. *Nucl. Acids Res.*, **16**, 907.

36. King, J. S., Valcarcel, E. R., Rufer, J. T., Phillips, J. W., and Morgan, W. F. (1993) Noncomplementary DNA double-strand-break rejoining in bacterial and human cells. *Nucl. Acids Res.*, **21**, 1055.

37. Goedecke, W., Pfeiffer, P., and Vielmetter, W. (1994) Nonhomologous DNA end joining in *Schizosaccharomyces pombe* efficiently eliminates DNA double-strand breaks from haploid sequences. *Nucl. Acids Res.*, **22**, 2094.

38. Thacker, J., Chalk, J., Ganesh, A., and North, P. (1992) A mechanism for deletion formation in DNA by human cell extracts: the involvement of short sequence repeats. *Nucl. Acids Res.*, **20**, 6183.

39. Nicolas, A. L., Munz, P. L., and Young, C. S. H. (1995) A modified single-strand annealing model best explains the joining of DNA double-strand breaks in mammalian cells and cell extracts. *Nucl. Acids Res.*, **23**, 1036.

40. Krawczak, M. and Cooper, D. N. (1991) Gene deletions causing human genetic disease: mechanisims of mutagenesis and the role of the local DNA sequence environment. *Hum. Genet.*, **86**, 425.

41. Kramer, K. M., Brock, J. A., Bloom, K., Moore, J. K., and Haber, J. E. (1994) Two different types of double-strand breaks in *Saccharomyces cerevisiae* are repaired by similar *RAD52*-independent, nonhomologous recombination events. *Mol. Cell Biol.*, **14**, 1293.

42. Moore, J. K. and Haber, J. E. (1996) Capture of retrotransposon DNA at the sites of chromosomal double-strand breaks. *Nature*, **383**, 644.

43. Thompson, L. H. and Jeggo, P. A. (1995) Nomenclature of human genes involved in ionizing radiation sensitivity. *Mutat. Res.*, **337**, 131.

44. Zdzienicka, M. Z. (1995) Mammalian mutants defective in the response to ionizing radiation-induced DNA damage. *Mutat. Res.*, **336**, 203.

45. Li, Z., Otevrel, T., Gao, Y., Cheng, H.-L., Seed, B., Stamato, T. D., *et al.* (1995) The *XRCC4* gene encodes a novel protein involved in DNA double-strand break repair and V(D)J recombination. *Cell*, **83**, 1079.

46. Grawunder, U., Wilm, M., Wu, X., Kulesza, P., Wilson, T. E., Mann, M., *et al.* (1997) Activity of DNA ligase IV stimulated by complex formation with XRCC4 protein in mammalian cells. *Nature*, **388**, 492.

47. Critchlow, S. E., Bowater, R. P., and Jackson, S. P. (1997) Mammalian DNA double-strand break repair protein XRCC4 interacts with DNA ligase IV. *Curr. Biol.*, **7**, 588.

48. Feldmann, H., Driller, L., Meier, B., Mages, G., Kellermann, J., and Winnacker, E. L. (1996) HDF2, the second subunit of the Ku homologue from *Saccharomyces cerevisiae*. *J. Biol. Chem.*, **271**, 27765.

49. Milne, G. T., Jin, S., Shannon, K. B., and Weaver, D. T. (1996) Mutations in two Ku homologs define a DNA end-joining repair pathway in *Saccharomyces cerevisiae*. *Mol. Cell Biol.*, **16**, 4189.

50. Boulton, S. J. and Jackson, S. P. (1996) Identification of a *Saccharomyces cerevisiae* Ku80 homologue: roles in DNA double-strand break repair and in telomeric maintenance. *Nucl. Acids Res.*, **24**, 4639.

51. Feldmann, H. and Winnacker, E. L. (1993) A putative homologue of the human autoantigen Ku from *Saccharomyces cerevisiae*. *J. Biol. Chem.*, **268**, 12895.

52. Tsukamoto, Y., Kato, J. I., and Ikeda, H. (1996) Hdf1, a yeast Ku-protein homologue, is involved in illegitimate recombination, but not in homologous recombination. *Nucl. Acids Res.*, **24**, 2067.

53. Siede, W., Friedl, A. A., Dianova, I., Eckardt-Schupp, F., and Friedberg, E. C. (1996) The *Saccharomyces cerevisiae* Ku autoantigen homologue affects radiosensitivity only in the absence of homologous recombination. *Genetics*, **142**, 91.

54. Boulton, S. J. and Jackson, S. P. (1996) *Saccharomyces cerevisiae* Ku70 potentiates illegitimate DNA double-strand break repair and serves as a barrier to error-prone DNA repair pathways. *EMBO J.*, **15**, 5093.

55. Mages, G. J., Feldmann, H. M., and Winnacker, E. L. (1996) Involvement of the *Saccharomyces cerevisiae* HDF1 gene in DNA double-strand break repair and recombination. *J. Biol. Chem.*, **271**, 7910.

56. Wilson, T. E., Grawunder, U., and Lieber, M. R. (1997) Yeast DNA ligase IV mediates non-homologous DNA end joining. *Nature*, **388**, 495.

57. Schar, P., Herrmann, G., Daly, G., and Lindahl, T. (1997) A newly identified DNA ligase of *Saccharomyces cerevisiae* involved in RAD52-independent repair of DNA double-strand breaks. *Genes Dev.*, **11**, 1912.

58. Teo, S.-H. and Jackson, S. P. (1997) Identification of *Saccharomyces cerevisiae* DNA ligase IV: involvement in DNA double-strand break repair. *EMBO J.*, **16**, 4788.

59. Schiestl, R. H., Zhu, J., and Petes, T. D. (1994) Effect of mutations in genes affecting homologous recombination on restriction enzyme-mediated and illegitimate recombination in *Saccharomyces cerevisiae*. *Mol. Cell Biol.*, **14**, 4493.

60. Tsukamoto, Y., Kato, J., and Ikeda, H. (1996) Effects of mutations of *RAD50*, *RAD51*, *RAD52*, and related genes on illegitimate recombination in *Saccharomyces cerevisiae*. *Genetics*, **142**, 383.

61. Johzuka, K. and Ogawa, H. (1995) Interaction of Mre11 and Rad50: two proteins required for DNA repair and meiosis-specific double-strand break formation in *Saccharomyces cerevisiae*. *Genetics*, **139**, 1521.

62. Moore, J. K. and Haber, J. E. (1996) Cell cycle and genetic requirements of two pathways of nonhomologous end-joining repair of double-strand breaks in *Saccharomyces cerevisiae*. *Mol. Cell Biol.*, **16**, 2164.

63. Tsukamoto, Y., Kato, J., and Ikeda, H. (1997) Silencing factors participate in DNA repair and recombination in *Saccharomyces cerevisiae*. *Nature*, **388**, 900.

64. Boulton, S. J. and Jackson, S. P. (1998) Components of the Ku-dependent non-homologous end-joining pathway are involved in telomeric length maintenance and telomeric silencing. *EMBO J.*, **17**, 1819.

65. Dvir, A., Peterson, S. R., Knuth, M. W., Lu, H., and Dynan, W. S. (1992) Ku autoantigen is the regulatory component of a template-associated protein kinase that phosphorylates RNA polymerase II. *Proc. Natl Acad. Sci. USA*, **89**, 11920.

66. Gottlieb, T. M. and Jackson, S. P. (1993) The DNA-dependent protein kinase: requirement of DNA ends and association with Ku antigen. *Cell*, **72**, 131.

67. Suwa, A., Hirakata, M., Takeda, Y., Jesch, S. A., Mimori, T., and Hardin, J. A. (1994) DNA-dependent protein kinase (Ku protein—p350 complex) assembles on double-stranded DNA. *Proc. Natl Acad. Sci. USA*, **91**, 6904.

68. LeesMiller, S. P., Chen, Y. R., and Anderson, C. W. (1990) Human cells contain a DNA-

activated protein kinase that phosphorylates simian virus 40 T antigen, mouse p53, and the human Ku autoantigen. *Mol. Cell Biol.*, **10**, 6472.

69. Anderson, C. W. and Lees-Miller, S. P. (1992) The nuclear serine/threonine protein kinase DNA-PK. *Crit. Rev. Eukaryot. Gene Expr.*, **2**, 283.

70. Mimori, T., Akizuki, M., Yamagata, H., Inada, S., Yoshida, S., and Homma, M. (1981) Characterization of a high molecular weight acidic nuclear protein recognized by autoantibodies in sera from patients with polymyositis-scleroderma overlap syndrome. *J. Clin. Invest.*, **68**, 611.

71. Mimori, T. and Hardin, J. A. (1986) Mechanism of interaction between Ku protein and DNA. *J. Cell. Biochem.*, **261**, 375.

72. Paillard, S. and Strauss, F. (1991) Analysis of the mechanism of interaction of simian Ku protein with DNA. *Nucl. Acids Res.*, **19**, 5619.

73. Rathmell, W. K. and Chu, G. (1994) A DNA end-binding factor involved in double-strand break repair and V(D)J recombination. *Mol. Cell Biol.*, **14**, 4741.

74. Getts, R. C. and Stamato, T. D. (1994) Absence of a Ku-like DNA end binding-activity in the *xrs* double-strand DNA repair-deficient mutant. *J. Biol. Chem.*, **269**, 15981.

75. Miller, R. D., Hogg, J., Ozaki, J. H., Gell, D., Jackson, S. P., and Riblet, R. (1995) Gene for the catalytic subunit of mouse DNA-dependent protein kinase maps to the *scid* locus. *Proc. Natl Acad. Sci. USA*, **92**, 10792.

76. Sipley, J. D., Menninger, J. C., Hartley, K. O., Ward, D. C., Jackson, S. P., and Anderson, C. W. (1995) Gene for the catalytic subunit of the human DNA-activated protein kinase maps to the site of the *XRCC7* gene on chromosome 8. *Proc. Natl Acad. Sci. USA*, **92**, 7515.

77. Blunt, T., Finnie, N. J., Taccioli, G. E., Smith, G. C. M., Demengeot, J., Gottlieb, T. M., *et al.* (1995) Defective DNA-dependent protein kinase activity is linked to V(D)J recombination and DNA repair defects associated with the murine *scid* mutation. *Cell*, **80**, 813.

78. Kirchgessner, C. U., Patil, C. K., Evans, J. W., Cuomo, C. A., Fried, L. M., Carter, T., *et al.* (1995) DNA-dependent kinase (p350) as a candidate gene for murine SCID defect. *Science*, **267**, 1178.

79. Peterson, S. R., Kurimasa, A., Oshimura, M., Dynan, W. S., Bradbury, E. M., and Chen, D. J. (1995) Loss of the catalytic subunit of the DNA-dependent protein kinase in DNA double-strand-break-repair mutant mammalian cells. *Proc. Natl Acad. Sci. USA*, **92**, 3171.

80. Zdzienicka, M. Z., Jongmans, W., Oshimura, M., Priestley, A., Whitmore, G. F., and Jeggo, P. A. (1995) Complementation analysis of the murine *scid* cell line. *Radiat. Res.*, **143**, 238.

81. Wei, Y.-F., Robins, P., Carter, K., Caldecott, K., Pappin, D. J. C., Yu, G.-L., *et al.* (1995) Molecular cloning and expression of human cDNAs encoding a novel DNA ligase IV and DNA ligase III, an enzyme active in DNA repair and recombination. *Mol. Cell Biol.*, **15**, 3206.

82. Porter, S. E., Greenwell, P. W., Ritchie, K. B., and Petes, T. D. (1996) The DNA-binding protein Hdf1p (a putative Ku homologue) is required for maintaining normal telomere length in *Saccharomyces cerevisiae*. *Nucl. Acids Res.*, **24**, 582.

83. Jackson, S. P. (1997) Silencing and DNA repair connect. *Nature*, **388**, 829.

84. Ramsden, D. A. and Gellert, M. (1998) Ku protein stimulates DNA end joining by mammalian DNA ligases: a direct role for Ku in repair of DNA double-strand breaks. *EMBO J.*, **17**, 609.

85. Cary, R. B., Peterson, S. R., Wang, J., Bear, D. G., Bradbury, E. M., and Chen, D. J. (1997) DNA looping by Ku and the DNA-dependent protein kinase. *Proc. Natl Acad. Sci. USA*, **94**, 4267.

86. Bliss, T. M. and Lane, D. P. (1997) Ku selectively transfers between DNA molecules with homologous ends. *J. Biol. Chem.*, **272**, 5765.

87. Leber, R., Wise, T. W., Mizuta, R., and Meek, K. (1998) The XRCC4 gene product is a target for and interacts with the DNA-dependent protein kinase. *J. Biol. Chem.*, **273**, 1794.

88. Ogawa, H., Johzuka, K., Nakagawa, T., Leem, S. H., and Hagihara, A. H. (1995) Functions of the yeast meiotic recombination genes, MRE11 and MRE2. *Adv. Biophys.*, **31**, 67.

89. Bosma, G. C., Custer, R. P., and Bosma, M. J. (1983) A severe combined immuno-deficiency mutation in the mouse. *Nature,* **301**, 527.

90. Fulop, G. M. and Phillips, R. A. (1990) The *scid* mutation in mice causes a general defect in DNA repair. *Nature*, **374**, 479.

91. Taccioli, G. E., Rathbun, G., Oltz, E., Stamato, T., Jeggo, P. A., and Alt, F. W. (1993) Impairment of V(D)J recombination in double-strand break repair mutants. *Science*, **260**, 207.

92. Giaccia, A., Weinstein, R., Hu, J., and Stamato, T. D. (1985) Cell cycle-dependent repair of double-strand DNA breaks in a gamma-ray sensitive Chinese hamster cell. *Somat. Cell Mol. Genet.*, **11**, 485.

93. Taccioli, G. E., Gottlieb, T. M., Blunt, T., Priestley, A., Demengeot, J., Mizuta, R., *et al.* (1994) Ku80: product of the *XRCC5* gene. Role in DNA repair and V(D)J recombination. *Science*, **265**, 1442.

94. Smider, V., Rathmell, W. K., Lieber, M. R., and Chu, G. (1994) Restoration of X-ray resistance and V(D)J recombination in mutant-cells by Ku cDNA. *Science*, **266**, 288.

95. Boubnov, N. V., Hall, K. T., Wills, Z., Sang, E. L., Dong, M. H., Benjamin, D. M., *et al.* (1995) Complementation of the ionizing radiation sensitivity, DNA end binding, and V(D)J recombination defects of double-strand break repair mutants by the p86 Ku autoantigen. *Proc. Natl Acad. Sci. USA*, **92**, 890.

96. Ross, G. M., Eady, J. J., Mithal, N. P., Bush, C., Steel, G. G., and Jeggo, P. A. (1995) DNA strand break rejoining defect in *xrs-6* is complemented by transfection with the human *Ku80* gene. *Cancer Res.*, **55**, 1235.

97. Errami, A., Smider, V., Rathmeli, W. K., He, D. M., Hendrickson, E. A., Zdzienicka, M. Z., *et al.* (1996) Ku86 defines the genetic defect and restores X-ray resistance and V(D)J recombination to complementation group 5 hamster cell mutants. *Mol. Cell Biol.*, **16**, 1519.

98. He, D. M., Lee, S. E., and Hendrickson, E. A. (1996) Restoration of X-ray and etoposide resistance, Ku-end binding activity and V(D)J recombination to the Chinese hamster *sxi-3* mutant by a hamster Ku86 cDNA. *Mutat. Res.*, **363**, 43.

99. Singleton, B. K., Priestley, A., Gell, D., Blunt, T., Jackson, S. P., Lehmann, A. R., *et al.* (1997) Molecular and biochemical characterisation of mutants defective in Ku80. *Mol. Cell Biol.*, **17**, 1264.

100. Nussenzweig, A., Chen, C., da Costa Soares, V., Sanchez, M., Sokol, K., Nussenzweig, M. C., *et al.* (1996) Requirement for Ku80 in growth and immunoglobulin V(D)J recombination. *Nature*, **382**, 551.

101. Zhu, C. M., Bogue, M. A., Lim, D. S., Hasty, P., and Roth, D. B. (1996) Ku86-deficient mice exhibit severe combined immunodeficiency and defective processing of V(D)J recombination intermediates. *Cell*, **86**, 379.

102. Roth, D. B., Lindahl, T., and Gellert, M. (1995) How to make ends meet. *Curr. Biol.*, **5**, 496.

103. Jeggo, P. A., Taccioli, G. E., and Jackson, S. P. (1995) Menage à trois: double strand break repair, V(D)J recombination and DNA-PK. *Bioessays*, **17**, 949.

104. Han, J. O., Steen, S. B., and Roth, D. B. (1997) Ku86 is not required for protection of signal ends or for formation of nonstandard V(D)J recombination products. *Mol. Cell Biol.*, **17**, 2226.

105. Jeggo, P. A. (1997) DNA-PK: at the cross-roads of biochemistry and genetics. *Mutat. Res.*, **384**, 1.
106. Gu, Y., Jin, S., Gao, Y., Weaver, D. T., and Alt, F. W. (1997) Ku70-deficient ES cells have increased ionizing radiosensitivity, defective DNA end binding activity, and inability to support V(D)J recombination. *Proc. Natl Acad. Sci. USA*, **94**, 8076.
107. Gu, Y. S., Seidl, K. J., Rathbun, G. A., Zhu, C. M., Manis, J. P., van der Stoep, N., *et al.* (1997) Growth retardation and leaky SCID phenotype of Ku70-deficient mice. *Immunity*, **7**, 653.
108. Ouyang, H., Nussenzweig, A., Kurimasa, A., da Costa Soares, V., Li, X., Cordon-Cardo, C., *et al.* (1997) Ku70 is required for DNA repair but not for T cell antigen receptor gene recombination *in vivo*. *J. Exp. Med.*, **186**, 921.
109. Hartley, K. O., Gell, D., Smith, G. C. M., Zhang, H., Divecha, N., Connelly, M. A., *et al.* (1995) DNA-dependent protein kinase catalytic subunit: a relative of phospha-tidylinositol 3-kinase and the ataxia telangiectasia gene product. *Cell,* **82**, 849.
110. Blunt, T., Gell, D., Fox, M., Taccioli, G. E., Jackson, S. P., Lehmann, A. R., *et al.* (1996) Identification of a nonsense mutation in the carboxy-terminal region of DNA-dependent protein kinase catalytic subunit in the *scid* mouse. *Proc. Natl Acad. Sci. USA*, **93**, 10285.
111. Danska, J. S., Holland, D. P., Mariathasan, S., Williams, K. M., and Guidos, C. J. (1996) Biochemical and genetic defects in the DNA-dependent protein kinase in murine *scid* lymphocytes. *Mol. Cell Biol.*, **16**, 5507.
112. Priestley, A., Beamish, H. J., Gell, D., Amatucci, A. G., Muhlmann-Diaz, M. C., Singleton, B. K., Smith, G. C., Blunt, T., Schalkwyk, L. C., Bedford, J. S., Jackson, S. P., Jeggo, P. A., Taccioli, G. E. (1998) Molecular and biochemical characterisation of DNA-dependent protein kinase-defective rodent mutant *irs-20*. *Nucl. Acids Res.* **15**, 1965.

# 3 | Translesion replication

CHRISTOPHER LAWRENCE and ROGER WOODGATE

## 1. Introduction

Translesion replication (TR) is a specialized form of DNA replication that is used to extend nascent chains past sites of unrepaired template damage that would otherwise impede or block elongation. The progress of replication forks can be inhibited by a wide variety of DNA lesions, including UV photoproducts, chemical adducts, and abasic sites. They are sometimes collectively called 'bulky' lesions because each appreciably alters DNA structure. Translesion replication therefore enhances the ability of cells to tolerate DNA damage, and increases survival when their genomes contain unrepaired damage. On occasion, it is called 'error-prone repair', but this is a less suitable name because translesion replication does not itself repair DNA, though it does produce molecules that are substrates for repair by other processes. Although TR increases tolerance to DNA damage, it is also a major source of both spontaneous and induced mutations, which are rarely beneficial. Mutations arise principally because the accuracy with which DNA polymerases insert the correct number or type of nucleotide opposite a site of template damage is usually much lower than is typically found for replication on undamaged templates, though some mutations may also arise because of the lower inherent accuracy of the DNA polymerase employed in TR. Nevertheless, TR still favours survival, since its error frequency rarely approaches 100%, and moreover many mutations are essentially neutral.

The principal goal of this chapter is to discuss the mechanisms of translesion replication and DNA-damage induced mutagenesis in *Escherichia coli* and budding yeast, *Saccharomyces cerevisiae*, the two best-studied organisms. To understand these processes we need to examine two sets of issues. First, we need to identify the proteins employed, the nature of their enzymatic functions, and the way their expression is regulated. Second, we need to investigate the mechanisms by which DNA lesions elicit polymerase errors, including the extent to which their mutagenic properties depend on the lesion itself or on the particular replication proteins concerned, and the role that the structure of the lesion-containing DNA plays in this process.

Whether a particular lesion is a block to continued DNA replication is partly dependent on the chemical structure of the lesion, but this is much influenced by the enzyme replicating the lesion-containing DNA. For example, the very same lesion might negligibly impede one enzyme, but pose a considerable impediment to TR by another. An illustration of this is shown in Fig. 3.1. In this case, each of the three *E. coli* DNA polymerases was asked to replicate a synthetic oligonucleotide template containing a single abasic lesion (located at position 'X') *in vitro*. All three polymerases have difficulty replicating up to the lesion and there are strong 'pause' sites 1, 2, and 3 bases before the lesion itself. Interestingly, DNA polymerase I (pol I) is able to insert a nucleotide efficiently opposite the abasic site and is able to catalyse a small amount of TR unassisted by other proteins usually employed for this process *in vivo* (see below). By comparison, DNA polymerase II (pol II) catalyses limited insertion opposite the lesion, and no insertion is observed with DNA polymerase III (pol III). By changing the *in vitro* conditions however, a very different pattern of TR is observed. Simply substituting $Mn^{2+}$ for $Mg^{2+}$ ions in the reaction enables pol I and

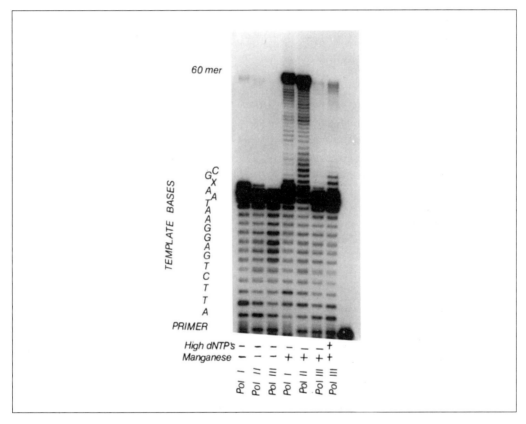

**Fig. 3.1** Ability of *E. coli* pol I, pol II, and pol III to perform TR of a single abasic lesion *in vitro*. Autoradiogram of a primer extension assay with a 60-bp template containing an abasic site (X). Reactions utilized 0.35 U pol I (Klenow fragment), 8 U pol II, or 1 U pol III holoenzyme. The normal assay includes dATP at 800 μM; dCTP, dGTP, and dTTP at 50 μM; and $Mg^{2+}$ at 8 mM. Where noted, manganese (5 mM) was substituted for $Mg^{2+}$ and the high dNTP reactions contained all dNTPs at 800 μM. Note that all three enzymes have difficulty in replicating up to the lesion and that all three polymerases exhibit differing abilities to insert a nucleotide opposite the lesion and perform unassisted TR. (We thank Sandra Randall for kindly supplying this figure.)

pol II to catalyse significant TR and pol III can be coerced into inserting a nucleotide opposite the lesion, though it is still unable to perform TR. In fact, under these assay conditions, TR by pol III is only seen in the presence of high levels of deoxy-nucleoside triphosphates. Such conditions are clearly non-physiological, but they nevertheless allow enzymologists to dissect the molecular mechanisms of TR *in vitro*. A recent and comprehensive description of the ability of various DNA polymerases to facilitate unassisted TR *in vitro* has recently been published (1), and is recommended reading for those interested in comparing various polymerases and their ability to perform TR past a variety of DNA lesions.

While some polymerases can clearly perform unassisted TR *in vitro*, many, like pol III, do so only under very non-physiological conditions and, more particularly, genetic evidence indicates that TR *in vivo* almost always requires the presence of specialized accessory factors (2). In *E. coli*, these are the well-characterized accessory proteins, RecA, UmuC, and UmuD, which assist the main replicative enzyme, pol III to perform TR. However, the enzymological basis for the promotion of TR by the UmuD′$_2$C complex is still unknown, and the subject of current research. Finally, although DNA polymerase III assisted by RecA and the UmuD′$_2$C complex appears to be responsible for most TR, for some lesions, or in a small proportion of cases with several lesions, TR may be carried out by pol II (3, 4), which does not require the accessory proteins. Much remains to be learnt about the regulation, as well as the enzymological mechanisms, of TR in budding yeast, *Saccharomyces cerevisiae*. It is known that at least one accessory factor, the Rev1 protein (5), is required for TR, and others may await discovery. But in this organism a completely new enzyme, DNA polymerase ζ, is usually utilized for TR rather than one of the major replicases (6).

A major reason for studying TR is that it is part of the seamless web of interrelated DNA transactions that include normal replication, the various types of recombination and repair, and transcription. This chapter summarizes recent studies of the mechanisms of TR in *E. coli* and *S. cerevisiae* within this context, and compares the two systems as a means of trying to discover general principles. But TR also has significance as an important, and perhaps the chief, source of nucleotide substitutions and frameshift mutations. Understanding how these arise is also an important aspect of TR. Because of this, the study of TR has the potential to help us understand genotoxic hazards, cancer, and other genetic diseases in humans, and perhaps ultimately to generate new ways of combating these conditions. An additional goal of this chapter is therefore to discuss the extension of information from the model systems to humans.

## 2. Translesion replication in *E. coli*

### 2.1 Translesion replication is usually a strategy of last resort

*E. coli*, like other organisms, employs a variety of repair and tolerance pathways (described in other chapters of this book) to avoid the immediate and deleterious consequences of DNA damage, principally the inability to carry out gene transcription and complete chromosomal replication. While some of the repair and tolerance proteins are constitutively expressed, many are damage-inducible and are expressed as part of the SOS response to the presence of unrepaired DNA damage. Elegant genetic and biochemical studies in the late 1970s and early 1980s (7–10) revealed that the response is controlled by two proteins: LexA, a transcriptional repressor that binds to sequences in the promoter/operator region of SOS genes so as to reduce their transcription; and RecA, which acts as a positive regulator (2, 11). Regulation of the SOS response is now well understood, and it is believed that the regions of single-stranded DNA generated during the futile attempts by pol III to replicate damaged DNA (12) constitutes the very signal which induces the SOS response (13). RecA protein binds to these single-stranded regions to form a spectacular spiral nucleo-protein filament. LexA molecules that are free in solution subsequently bind within the deep helical groove of the filament (14) and undergo a self-cleavage reaction (15), which inactivates its ability to function as a transcriptional repressor and as a consequence leads to the derepression of genes under LexA control. The key to the induction of a particular SOS gene is the affinity with which LexA binds to the site within the promoter/operator region; those genes with weak LexA binding sites will be expressed earlier than those which bind LexA more avidly. The prevailing hypothesis is that the SOS response has evolved in such a way that the cell tries to avoid error-prone TR (16). Error-free repair pathways such as nucleotide excision repair (NER) and postreplication recombination repair are, in general, induced very early in the SOS response (2) with the goal of repairing the damaged DNA before there is a need to utilize TR. By comparison, the *umu* operon binds LexA tightly (see below) and is probably one of the last in the SOS regulon to achieve full derepression.

## 2.2 Most translesion replication is dependent upon the Umu proteins

Although, as discussed in the previous section, *E. coli* tends to cope with DNA damage by first employing excision repair and postreplication tolerance mechanisms based on recombination before resorting to TR, this process is called into play when templates containing unrepaired damage are eventually replicated (17, 18). Both genetic evidence (19–21) and biochemical data (22–24) indicate that the key players in this process are the chromosomally encoded UmuD and UmuC proteins, or their plasmid-encoded homologues such as MucA and MucB (2, 25–27). The *umu* genes are arranged in an operon with the smaller *umuD* gene upstream of the larger *umuC* gene (28, 29). Mutations in either *umuD* or *umuC* virtually abolish cellular and phage mutagenesis induced by UV light and many chemicals (21, 30–32 ). The *umuDC* genes are not essential for viability, however, and their complete absence has only a modest effect on the survival of wild-type *E. coli* (33). Such phenotypes can be readily explained by the fact that the Umu proteins provide a backup to other error-free repair pathways. Indeed, in the experiments using a double-stranded plasmid carrying a single lesion described by Fuchs and colleagues (34), abolishing TR by deletion of *umuDC* (Δ*umuDC*) had no apparent effect on plasmid survival, indicating that many, if not all, lesions can in fact be repaired via error-free pathways, given enough time. In contrast, where no error-free pathways are available, as in single-stranded DNA phages on which NER or recombinational repair cannot operate, the Umu proteins have dramatic effects on TR and phage survival. TR past a single *cis-syn* T–T cyclobutane pyrimidine dimer located in such a phage is completely absent in Δ*umuDC* cells that are otherwise wild-type (21). In the presence of Umu proteins, or their functional homologues, TR increases dramatically so that close to 100% TR can be achieved under certain conditions (21). Although the Umu proteins clearly play a central role in pol III-dependent TR, there are instances, described in more detail in Section 2.6 below, where TR can in fact occur in the absence of the Umu proteins.

## 2.3 Umu-dependent TR is regulated with exquisite precision

Because, by promoting TR, the Umu proteins offer both the advantage of survival and the disadvantage of mutagenesis, it is hardly surprising that they are regulated with exquisite precision. Although it has been speculated that the mutagenesis arising from TR may be advantageous, and increase the overall fitness of the strain during periods of environmental stress (35), the tight temporal limits placed on TR suggest that such benefits may at best be limited.

Precise regulation is achieved by means of multiple mechanisms, rather than a single process (26). Like other genes in the SOS regulon, the *umu* genes are negatively regulated at the transcriptional level by the LexA protein. Studies have shown that LexA binds most efficiently to sites that more closely resemble the consensus binding site (28, 29). The LexA binding site in the *umu* promoter only deviates from the consensus at two positions and has a dissociation constant of 0.2 nM (28). As a result, the *umu* operon would be expected to be one of the last in the SOS regulon to become fully derepressed. By comparing the steady-state levels of the Umu proteins when uninduced to those expressed in a strain lacking a functional LexA repressor, Woodgate and Ennis (36) estimated that when fully derepressed, the Umu proteins would be induced approximately 12-fold and give rise to 2400 molecules of UmuD and 200 molecules of UmuC per cell. In a wild-type cell, however, full induction rarely occurs and, as a consequence, the level of Umu proteins in a cell exposed to transient DNA damage might be significantly lower (16).

Despite the fact that the *umu* operon is tightly regulated at the transcriptional level, limited damage-independent expression nevertheless occurs. The intracellular level of the UmuD and UmuC proteins is kept to a minimum, however, because of their rapid proteolysis by the Lon serine protease (37). Moreover, UmuD molecules that escape degradation remain functionally inactive until they undergo a RecA-mediated cleavage reaction which generates the shorter, but mutagenically active, UmuD' protein (38–40). This reaction is thought to be mechanistically similar to that which inactivates LexA protein (15, 41). The two reactions differ, however, in that LexA cleavage is thought to occur predominantly via intramolecular self-cleavage (15), whereas UmuD cleavage is thought to occur predominantly via an intermolecular cleavage reaction, in which one molecule of UmuD acts as an enzyme to facilitate cleavage of another substrate UmuD molecule (42). UmuD cleavage most likely occurs when it is in a dimeric state and the N-terminal tail (containing the cleavage site) of one monomer is brought into close proximity with the catalytic active site of another monomer (42). At the present time, it is not known if both UmuD molecules undergo cleavage simultaneously, or if there is preferential cleavage of one monomer over another. If the latter does occur, which might especially be the case under conditions of limited cellular DNA damage, a UmuD–UmuD' heterodimer could be formed by continued association of the resultant UmuD' and intact UmuD molecules. Interestingly, *E. coli* seems to have taken advantage of heterodimer formation as UmuD', which is normally quite stable in a homodimer (43), is rapidly degraded by the serine protease ClpXP when in a heterodimer (37).

Previous experiments investigating the *in vivo* cleavage of UmuD suggested that the reaction is intrinsically inefficient (36). Recent studies indicate, however, that it is not as inefficient as previously thought, but appears so because ClpXP proteolysis slows down the accumulation of UmuD' (44). Degradation of UmuD' in a heterodimer by ClpXP therefore allows the accumulation of mutagenically active UmuD' homodimers to be postponed (44). It also provides an opportunity to return the cell to a non SOS-induced state once DNA repair has occurred (45). Another factor that appears to effect conversion of UmuD to UmuD' is DinI (46). Like the Umu proteins, DinI is induced as part of the SOS response. Recent studies suggest that DinI (and its plasmid-encoded homologue, ImpC, see Fig. 3.2) interferes with the RecA-mediated

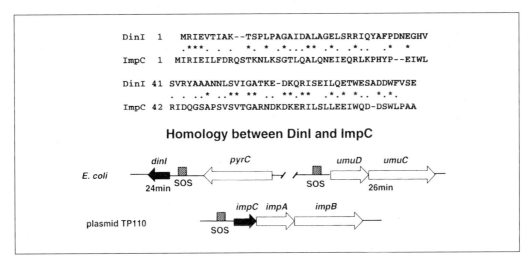

**Fig. 3.2** Homology between the chromosomally encoded DinI and plasmid-encoded ImpC proteins. Upper panel: alignment of DinI and ImpC proteins. Those residues that are identical are indicated with an asterisk (*), while those that are highly conserved are indicted with a full point (.). Lower panel: localization of DinI relative to UmuDC and of ImpC relative to ImpAB. *dinI* is located approximately 2 minutes away from the *E. coli* *umuDC* operon and is transcribed in the opposite direction (46). By comparison, *impC* is located immediately upstream of, and translated in the same direction as *impAB* (47). (We thank Haruo Ohmori and Takeshi Yasuda for kindly supplying this figure.)

cleavage of LexA and UmuD (46). Inhibition of the former reaction helps to switch off the SOS response, while inhibition of the latter slows UmuD cleavage and thereby channels UmuD' into the ClpXP degradation pathway.

An excellent example of how the various levels of transcriptional and post-translational regulation keep the mutagenically active UmuD'C proteins to a minimum is shown in Fig. 3.3. In this experiment, Sommer and colleagues (16) followed the UV-induced appearance of the Umu proteins in a wild-type cell. No detectable UmuD or UmuC protein was observed until 10–15 minutes after UV-irradiation. This is, in part, the consequence of the tight transcriptional regulation, but it also results from the rapid degradation by the Lon protease of the initial low level of UmuD and UmuC proteins. After about 15 minutes, synthesis of UmuD and UmuC outweighs their degradation and the intracellular concentration of the proteins begins to accumulate. At the same time, DinI is operating to inhibit UmuD cleavage and ClpXP operates to remove any UmuD' that is generated. As a consequence, the appearance of mutagenically active UmuD' is further delayed. Note that at about 20 min after UV irradiation there are approximately 200 molecules of UmuD and no molecules of UmuD', and after 30 min approximately 600 molecules of UmuD and only 30 molecules of UmuD'. Somewhere between 30 and 50 min after UV irradiation, the ClpXP degradation pathway becomes saturated and UmuD' begins to accumulate, peaking at around 50 min. Although some of the UmuD' molecules are likely to be trapped in a mutagenically inactive UmuD/UmuD'

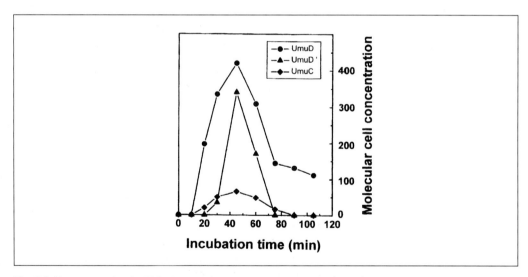

**Fig. 3.3** Time-course for the UV-induction of the Umu proteins. A *recA+ lexA+ ΔumuDC* strain harbouring a low-copy number plasmid expressing UmuDC was irradiated with 12 J/m$^2$ of UV light. At the times noted, whole-cell extracts were taken and the cellular concentration of Umu proteins determined by immunoblotting with antibodies raised against UmuD and UmuC as previously described (45). The concentration of Umu protein on the ordinate is that expected to be produced from the chromosome. In deriving this number, it is assumed that the low copy number Umu plasmid used in these experiments replicates with a copy number of five (16). (Adapted from ref. 16, with permission.)

heterodimer (37, 48), others will form UmuD′ homodimers that can associate with UmuC to form the mutagenically active UmuD′$_2$C complex (23, 49–50). As the inducing signal declines, because of the repair of the replication-blocking lesions, there is less production of UmuD′, and the little produced is degraded by ClpXP. Although 75 minutes after the initial UV dose there are still about 100 molecules of UmuD per cell, there are no detectable UmuD′ or UmuC molecules, and the cell has returned to a non SOS-induced state. Despite the fact that no molecules of UmuD′$_2$C are detected in a resting state, it is clear, however, that some must exist since the basal spontaneous mutation rate in wild-type *umu*$^+$ strains is 2–3-fold higher than that of otherwise isogenic *umu*$^-$ cells (51, 52). Such observations imply that only a few molecules of UmuD′$_2$C are required for error-prone replication (16, 53) and hence the need for their tight regulation to avoid excess gratuitous mutagenesis.

## 2.4  Formation of the mutasome and translesion replication

The existence of a 'mutasome' (Fig. 3.4), a term coined by the late Harrison Echols (49, 54), was inferred from a combination of data about the generation of the SOS inducing signal, the cleavage of LexA, and the regulation of UmuD′ activation, and it represents the set of associated proteins that are required for efficient TR. While we understand much about the steps that lead to its formation, the actual biochemical mechanism by which the mutasome facilitates TR still eludes discovery. Genetic experiments strongly indicate that TR occurs through UmuD′$_2$C–RecA–pol III protein–protein interactions (62, 63), and an appealing model for TR is that it occurs, at least in part, by inhibiting the 3′–5′ proofreading exonuclease activity of the ε subunit of pol III, encoded by *dnaQ* (64). Indeed, it has been hypothesized that 3′–5′ exonucleolytic functions are inactive during TR (65, 66). Furthermore, the level of UV-induced mutagenesis is significantly reduced when ε is overexpressed (67–69). While such a model is attractive, it is also clear that inhibition of proofreading alone cannot explain the activities of the Umu proteins, as in their presence there is a dramatic increase in TR even in mutant strains of *E. coli* and *Salmonella typhimurium* that are completely devoid of 3′–5′ proofreading (70, 71).

The major problem in elucidating the molecular mechanisms of mutagenesis has been the lack of a soluble purified form of UmuC. Earlier *in vitro* experiments (22, 49) utilized a denatured/renatured form of UmuC. In 1996, Bruck and colleagues (50) reported the purification of a soluble UmuD′$_2$C complex and demonstrated that it bound cooperatively to single-stranded DNA. Very recently, Tang and colleagues (23) have used the soluble UmuD′$_2$C complex to reconstitute a lesion-bypass assay *in vitro*. Using a different approach, Bacher Reuven and colleagues (24) have purified a soluble maltose-binding protein–UmuC chimeric protein which is also able to catalyse TR *in vitro*.

Based upon genetic models (19, 20), it was hypothesized that the Umu proteins only participated in the elongation from a mispair, but not in the actual mis-incorporation of an incorrect nucleotide. Tang and colleagues (23) however, observe a very different pattern of nucleotide insertion in the presence of the UmuD′$_2$C

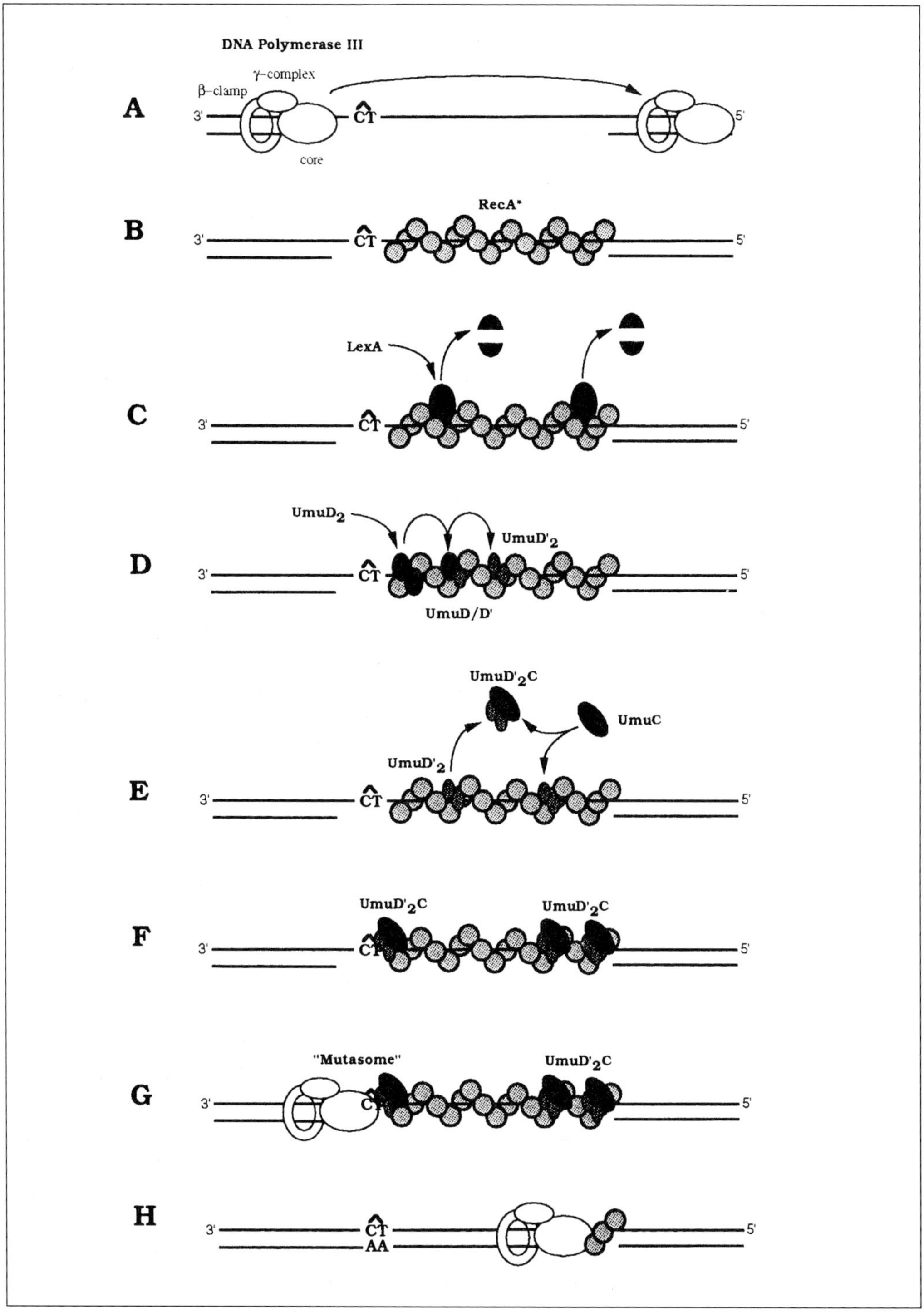

DNA Polymerase III
γ-complex
β-clamp
A
3'
5'
core
CT
B
RecA*
C
LexA
D
UmuD2
UmuD'2
UmuD/D'
E
UmuD'2C
UmuC
UmuD'2
F
UmuD'2C
UmuD'2C
G
"Mutasome"
UmuD'2C
H
CT
AA

**Fig. 3.4** Steps that lead to the formation of the mutasome and translesion replication. (A) DNA polymerase III holoenzyme (consisting of core, γ-complex, and β-sliding clamp (55)) encounters a replication-inhibiting lesion, in this case a 5′ TC 3′ pyrimidine dimer. The polymerase usually pauses 1–3 bases before the lesions but occasionally inserts a base opposite the lesion (56). In general, however, pol III is unable to replicate past the lesion and instead reinitiates synthesis approximately 1 kb downstream (12). (B) RecA protein binds avidly to the ssDNA region and forms a spectacular spiral nucleoprotein filament (57). (C) The LexA repressor recognizes the RecA nucleoprotein structure as a signal of environmental stress (2, 58) and binds within the deep helical groove of the filament (14). Such protein–protein interactions lead to the autocatalysis of LexA and its inactivation for repressor functions (15). (D) Like LexA, one of these damage-inducible proteins, UmuD also binds to the RecA nucleoprotein filament (59) and undergoes a mechanistically similar cleavage reaction to that of LexA (41). However, unlike LexA the UmuD reaction is predominately intermolecular in nature (42) and eventually leads to the generation of mutagenically active UmuD′$_2$ molecules (40). (E) The mutagenically active UmuD′$_2$ protein either remains bound to the RecA nucleoprotein filament and awaits the arrival of UmuC (59), or dissociates from the filament (to find UmuC in solution) to form a UmuD′$_2$C complex (50). (F) The UmuD′$_2$C complex binds to various sites along the RecA nucleoprotein filament (26). (G) Those UmuD′$_2$C molecules that bind to the very tip of the filament are positioned for an encounter with pol III (16, 43, 60) and form the so-called mutasome (49, 54), while those that bind within the filament itself lead to an inhibition of error-free recombination (43, 60, 61). (H) In the presence of UmuD′$_2$C-RecA, DNA polymerase III is able to replicate past the blocking lesion, but with a concomitant decrease in replication fidelity. In this case, an A has been misinserted opposite the C leading to a C→T transition. Although the lesion has not been repaired, the cell survives and the lesion is now a substrate for error-free repair pathways such as NER or recombinational repair.

proteins compared to that observed in their absence (Fig. 3.5). This fact, together with the observation that various Umu-like homologues exhibit differing mutational spectra (21, 72) imply that they may also act at the misincorporation step. Indeed, such phenotypes may be potentially explained by the differing abilities of the various Umu-like proteins to interact with specific subunits of pol III. Obvious candidate subunits that might be the target of Umu activity would be α (the polymerizing subunit), ε (the proofreading subunit), and β (the sliding clamp). With soluble UmuC now at hand, it is likely that the answers to these questions will be forthcoming and that the mechanism by which the mutasome promotes TR in *E. coli* will be solved in the near future.

## 2.5  The Umu proteins are generalized elongation factors

Originally, it was believed that the Umu proteins might function only at sites of DNA damage, but it now seems likely that the proteins act at any site that is kinetically unfavourable for extension by pol III (66). This hypothesis is based upon the elegant studies of Fijalkowska and colleagues (66) who investigated the nature of the modest Umu-dependent spontaneous mutator activity exhibited in certain *recA* mutant strains (73). It had previously been assumed that such mutagenesis resulted from TR proteins acting on spontaneously occurring cryptic lesions, such as abasic sites. Fijalkowska and colleagues demonstrated, however, that mutagenesis did not increase in strains defective in repairing abasic lesions; rather, they most likely occurred as a result of normal replication errors. Analysis of the types of mutations generated during Umu-dependent spontaneous mutagenesis revealed a preponderance of

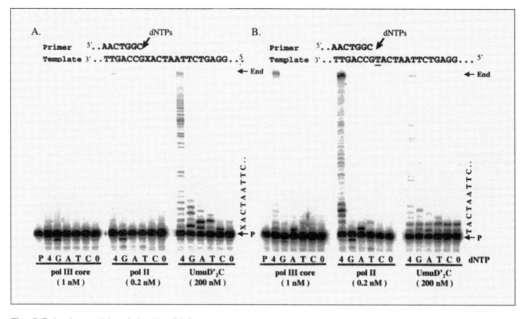

**Fig. 3.5** *In vitro* activity of the UmuD′$_2$C proteins on damaged and undamaged DNA templates. Standing-start replication assays were performed using pol III core, pol II, or UmuD′$_2$C. (A) Reactions carried out using a DNA template containing a single abasic lesion located at position, X. (B) Reactions carried out using a natural DNA template in which X is replaced by T. The lanes labelled as G, A, T, and C denote reactions carried out with a single dNTP substrate, dGTP, dATP, dTTP, and dCTP, respectively. The lanes labelled as 4 and 0 denote reactions carried out in the presence and absence of four dNTPs, respectively. Lane P contains the [32]P-labelled primer in the absence of proteins. The abasic lesion-containing and natural DNA templates are shown above each gel. The UmuD′$_2$C preparation used in this assay has intrinsic polymerase activity which may result from contamination with small quantities of pol II or pol III. Despite this fact, it clear that TR is absolutely UmuD′$_2$C-dependent as none is seen with comparable concentrations of pol II or pol II alone. (Reproduced from ref. 23, with permission.)

transversions. Fijalkowska and colleagues speculated that such spectra might result from the efficient methyl-directed mismatch repair of transition mutations, and the production of transversion mutations by Umu-facilitated extension of purine–purine base pairs, which are normally poorly extended by pol III (Fig. 3.6).

## 2.6 Translesion replication in the absence of Umu proteins

Although most pol III-dependent TR requires the Umu proteins, there are several notable instances where TR has been observed in its absence. For example, TR of a single acetylaminofluorene (AAF) lesion in single-stranded DNA occurs via two different pathways (34). One occurs as a result of extension from a non-slipped intermediate (Fig. 3.7) and requires the Umu proteins (32). The other is generated from a slipped intermediate and occurs independently of Umu activity. Instead, it apparently requires another SOS-inducible protein called Npf for *Nar*I processing factor (74). Very recent studies show that the need for the Npf protein in the slipped

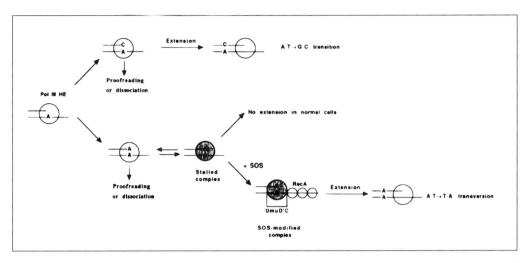

**Fig. 3.6** Model for the action of Umu proteins on undamaged DNA. This model, proposed by Fijalkowska *et al.* (66) depicts the processing of 3′ terminal transition and transversion mismatches resulting from polymerase misinsertion errors, and assumes that the former can be extended but that the latter cannot. Transition mutations are, however, kept to a minimum as they are efficiently corrected by the postreplication mismatch repair system. Under SOS-induced conditions, UmuD′$_2$C and RecA interact with pol III allowing extension from the mispair so that it becomes fixed as a transversion mutation. Such a model accounts for the preponderance of transitions generated during normal replication and for transversions under SOS-induced conditions. (Reproduced from ref. 66 with permission.)

intermediate pathway is obviated if the 3′–5′ proofreading activity of pol III is diminished (75). Such observations raise the possibility that the normal function of the Npf protein is to inhibit proofreading in an otherwise wild-type cell (75).

Along similar lines, Vandewiele and colleagues (71) have recently discovered that the inactivation of proofreading functions in a Δ*umuDC* strain leads to a dramatic increase in TR past a single T–T *cis-syn* cyclobutane dimer. Two processes, one LexA-dependent and the other LexA-independent, contribute to this Umu and proof-reading independent pathway of TR, which together give about half the amount of TR observed in a proofreading-deficient strain that carries a chromosomal copy of the *umuDC* operon. The basis for the SOS-dependent process is not known, although it does not appear to depend on the SOS-inducible pol II enzyme. Similarly, the mechanism for the SOS-independent process is also unknown. However, an interesting possibility is that it is related to the UVM response characterized by Humayun and colleagues (76–78). UVM, coined for the 'ultraviolet-induced modu-lation' of mutagenesis, was discovered by its ability to alter the mutation spectrum resulting from TR past an ethenocytosine lesion, but it was also found to enhance the frequency of TR. It has now been shown to occur in response to a variety of DNA damaging agents in addition to UV light, and is independent of the *recA*, *lexA*, *umuDC*, *polA*, *polB*, and *mutHLS* genes (76–78). Recent studies suggest, however, that this Umu-independent TR is affected by mutations in *mutA* (*glyV*) and *mutC* (*glyW*) (79), which themselves have been implicated in the inactivation of 3′–5′ proofreading

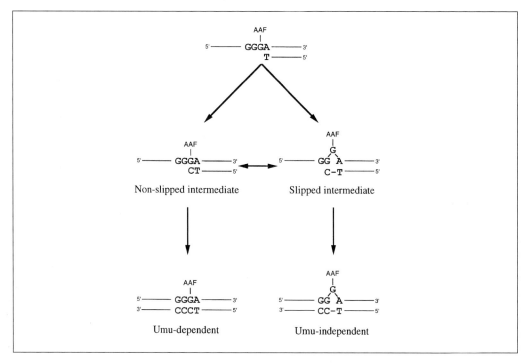

**Fig. 3.7** Structures of non-slipped and slipped mutagenic intermediates opposite an AAF adduct. AAF binds to guanines and in a monotonous run of Gs can lead to TR in two possible pathways (34). One, in which the incoming cytosine correctly pairs with the adducted guanine, and extension from which requires the Umu proteins (32, 34). The other, involving a slipped intermediate, in which the correctly incorporated cytosine pairs with the downstream guanine so that a –1 frameshift mutation occurs. This pathway is Umu-independent (32), but is, instead, dependent upon 3′–5′ proofreading as it apparently occurs constitutively in its absence (74). (Adapted from ref. 34, with permission.)

functions (80, 81). It is not yet known whether the same molecular mechanisms are employed in the three Umu-independent, proofreading-independent, pathways described above.

# 3. Translesion replication in yeast

## 3.1 Genes important for TR and their mutant phenotypes

In yeast, translesion replication and DNA-damage induced mutagenesis requires the products of the *REV1*, *REV3* (= *PSO1*), and *REV7* genes (5, 6, 82–84). Induced mutagenesis is substantially deficient in strains carrying null mutations in any one of these genes, and the strains also exhibit slightly increased sensitivity to killing by DNA damaging agents. Spontaneous mutation frequencies are also reduced in *REV3* mutants (85, 86), presumably because of the mutations caused by spontaneously occurring DNA damage. Spontaneous mutation associated with transcription (87), and double-strand break stimulated recombination (88) is also REV3-dependent, though the gene itself does not appear to be involved in either transcription or recombination. Based on their mutant phenotype, the products of the less well-studied *REV6* (89) and *NGM2* (90) genes may also be employed in translesion replication, and there is no reason to believe that all relevant genes have been identified.

Translesion replication in yeast appears to produce mostly base-pair substitutions and frameshift mutations, and is the major source of these events (84). The extent to which larger deletions or rearrangements are produced by this process has not been well investigated, but such events seem to be much less abundant in yeast than in the higher eukaryotes. On average, about 98% of base-pair substitutions, and more than 90% of frameshift mutations, induced by UV light are probably dependent on *REV3* function, based on the relative reversion frequencies of a set of nine defined alleles comprising amber, ochre, initiation, and proline missense mutations, and a set of eight contrasting frameshift alleles, induced in $REV3^+$ and null *rev3* mutant strains (91, 92). Since similar results were observed in γ-irradiated strains (93), these average values are probably typical for a variety of types of DNA damage. In keeping with this, other less extensive data suggest that *REV3* function is responsible for a substantial fraction of the mutations induced by methyl methane sulfonate, psoralen, *N*-methyl-*N*-nitrosoguanidine, and AAF, though not by ethyl methane sulfonate (94–96). However, for unknown reasons, at some specific genetic sites the induction of base-pair substitutions and in particular of frameshift events can be much less dependent on *REV3* function (91, 92). In addition to its role in the generation of mutations induced by overt mutagen damage to DNA, *REV3* function is probably also responsible for somewhere between 50 and 80% of spontaneous mutations, though again estimates of the *REV3*-dependent fraction vary widely according to the test system used, presumably because the test systems detect different kinds of events at different genetic sites (85, 86). Although much less studied, the *REV7* mutant phenotype appears to be very similar to that of the *REV3* gene (97). Investigation of the *REV1* mutant phenotype, based on the UV-induced reversion of the same set of alleles used in the *REV3* study, suggests that about 95% of base substitution events depend on this gene. However, the *REV1* gene appears to play a much smaller role in the production of frame-shift mutations at a majority of the genetic sites tested (92, 98).

## 3.2  Regulation of translesion replication

Translesion replication is one of the functions performed by the *RAD6* pathway in yeast, which is concerned with promoting tolerance of DNA damage (84). The *REV* genes have been assigned to this pathway because *rad6 rev* double-mutant strains are no more sensitive to killing by UV (or any one of a variety of other DNA damaging agents) than single mutant strains carrying a *rad6* mutation alone. The UV sensitivity of strains carrying mutant alleles of genes active in different repair pathways is at least the sum of the sensitivities of the two single mutants, and may be greater than this. The *RAD6* pathway is regulated by the *RAD6* and *RAD18* genes, each of which have very similar mutant phenotypes that encompass all the mutant phenotypes of other genes assigned to the pathway. Strains with mutations in either the *RAD6* or *RAD18* genes are deficient in DNA-damage induced mutagenesis (99), though interestingly not in spontaneous mutagenesis, and are highly sensitive to killing by UV and other such agents. The fact that they are much more sensitive than the *REV* mutants indicates that the *RAD6* pathway performs another tolerance function. This employs proliferating cell nuclear antigen (PCNA) (100), the sliding clamp protein that enhances the processivity of DNA polymerases $\delta$ and $\varepsilon$, and also the Rad5 (101) and Rad 30 proteins (102). The mechanism underlying this non-mutagenic process in not known, but may entail transient template strand switching (18) as a means of allowing error-free replication past sites of template damage that could potentially impede the replication fork.

The *RAD6* pathway is regulated by the ubiquitination of certain target proteins, a process that is therefore quite different from SOS-regulation in *E. coli*. The key factor in this regulation is a stable heterodimer of Rad6p, which possesses the ubiquitinating activity, and Rad18p which is a DNA binding protein (103–105). The 20-kDa Rad6 protein (106) is one of a set of 10 or more ubiquitin-conjugating (E2) enzymes that attach the 8.5-kDa ubiquitin protein covalently to lysine residues in certain target proteins, the identity of which is determined by the particular E2 enzyme used. In performing this function, a ubiquitin molecule is covalently linked by means of a thiolester bond to cysteine 88 in Rad6p, and then transferred to the target. Ubiquitination usually consigns the target molecules to rapid proteolysis by the proteosome, though it may possibly activate the protein in some cases.

Each of the *RAD6* pathway DNA-damage tolerance functions depend specifically on Rad6p-dependent ubiquitination, because the phenotype of *rad6* mutants in which cysteine 88 has been converted to alanine, valine, or serine is identical to that of a deletion mutant (107, 108). The serine-88 mutant is particularly informative because serine 88 can be ubiquitinated, but the stable ester conjugate formed prevents the transfer of ubiquitin to the target protein. Unlike the other E2 enzymes, Rad6p is required for N-end-recognizing ubiquitination, in which it is targeted to proteins with certain amino-terminal residues by the *UBR1*-encoded E3 ligase protein (109). N-end-recognizing ubiquitination is not involved in translesion replication and mutagenesis, however, because a *rad6*-deletion mutant producing an amino-terminally truncated protein is fully proficient for mutagenesis but deficient in the *UBR1*-dependent process (110); the amino-terminus of targets important for the control of translesion replication presumably do not contain the residues recognized by the *UBR1* E3 ligase. The DNA binding capability of the 55-kDa Rad18p presumably directs the E2 conjugase to DNA-bound proteins, such as subunits in the replication complex, though the identities of the Rad6p target proteins and the way they regulate translesion replication are not yet known. Possible molecular mechanisms for Rad6p–Rad18p control might include ubiquitination-dependent disassembly of replication complexes into which the Revp need to be incorporated or which they need to replace, or the release of Revp from complexes in which they are sequestered. Alternatively, proteins important for TR might be directly or indirectly activated by ubiquitination.

The levels of *RAD6* and *RAD18* mRNA are relatively unchanged during the yeast mitotic cycle, but are both elevated in UV-irradiated cells, by about fivefold and fourfold, respectively (111, 112). *REV3* mRNA levels are also constant throughout the mitotic cycle, but in this case are at best only marginally increased by UV-irradiation (113). In keeping with this, Rev1p levels appear to be unchanged in response to UV-irradiation, as judged by the amounts of a *REV1::LacZ* fusion protein (114). In contrast to these events in *E. coli*, translesion replication and mutagenesis in yeast do not therefore appear to be subject to transcriptional regulation.

## 3.3 Molecular and enzymatic analysis of *REV* gene function

Sequence analyses of the cloned *REV* genes, together with enzymatic studies of their purified products, are beginning to uncover the mechanisms of translesion replication and mutagenesis in yeast and perhaps, since yeast is often a good eukaryotic model, for mammalian systems as well. Unlike *E. coli*, which principally uses its major replicase, pol III, for TR, yeast possesses a new class of DNA polymerase that appears to be dedicated exclusively to this process, and which seems to perform no other function in DNA replication, repair, or recombination (6). This enzyme, called DNA polymerase ζ, consists of a 173-kDa catalytic subunit encoded by the *REV3* gene (115), and at least one other subunit, the 29-kDa protein encoded by the *REV7* gene (116), which stabilizes the catalytic subunit but as yet possesses no known enzymatic activity. Since it has not yet been possible to isolate the native enzyme, it is possible that other subunits also occur. Pol ζ lacks 3′–5′ proofreading exonuclease activity and is a very non-processive enzyme; in half of the template binding/ dissociation cycles, three or fewer nucleotides are added to the primer (6).

The existence of a specialized DNA polymerase suggests that it may have unique properties suited to its purpose. As might be expected, one of these appears to be a superior ability to replicate past lesions that block the progress of other enzymes. *In vitro* studies indicate that pol ζ replicates past a T–T *cis-syn* cyclobutane dimer in about 10% of binding events, and past an abasic site in about 7% of these events, whereas yeast pol α is essentially incapable of such bypass (6). As discussed below, pol ζ is probably not required for replication past T–T dimers *in vivo*, but this lesion nevertheless serves to reveal the properties of the enzyme *in vitro*. One reason for the superior ability of pol ζ to perform TR is likely to be the enzyme's capacity to add nucleotides to the primer even when the primer/template structure at the primer terminus does not conform to that found in normal duplex DNA. *In vitro*, pol ζ can extend with modest efficiency a primer terminating in an adenine placed opposite the 3′ thymine of a T–T cyclobutane dimer, or in which adenines are placed opposite both the 3′ and the 5′ thymines, and extend with relatively high- efficiency primers that additionally form a correct base pair 5′ to the dimer. Yeast pol α, by contrast, is virtually incapable of extending any of these primers, even that ending with a correct base-pair beyond the dimer (Nelson *et al.*, in preparation). Pol ζ is also relatively efficient at adding nucleotides to a terminally mismatched primer, using a lesion-free template, again indicating the enzyme's tolerance of non-normal structure at the primer terminus. Compared to a correctly matched primer, pol $\zeta$ extended the various mismatched primers with an efficiency ranging from around 1% to 20%, as measured by the method of Goodman and colleagues (117). By contrast, pol $\alpha$ extended the mismatches with an efficiency that was usually at least 10-fold lower. Since neither of these enzymes possesses a 3'–5' exonuclease proofreading activity, the results only concern nucleotide addition (Nelson et al., in preparation).

Genetic evidence indicates that TR also requires the participation of the *REV1* gene product. The *REV1* gene encodes a 112-kDa protein (114) that possesses an activity necessary for elongation past any lesion that impedes the replication fork. It also encodes a deoxycytidyl activity that inserts deoxycytosine opposite abasic sites, which is probably concerned exclusively with TR past this lesion (5). Although the enzymatic basis for the general elongation function is as yet poorly defined, the existence of this activity is clearly indicated by the observation that Rev1p is needed for mutagenesis by UV photoproducts and other lesions in which cytosine insertion does not feature. The two Rev1p activities can be separated mutationally; the *rev1–1* mutant, originally isolated because of its deficiency in UV-induced mutagenesis (82), produces protein exhibiting an approximately wild-type level of deoxycytidyl activity, even though its general TR function is substantially impaired (Nelson et al., in preparation). The two activities may therefore be encoded in different domains within the Rev1 protein. *In vitro* studies show that the addition of Rev1p increases TR past an abasic site by pol $\zeta$ from 7% per binding/disassociation cycle to 30–40% (5), which is similar to the TR frequency *in vivo* (118). Both of the Rev1p activities seem to be required for TR past an abasic lesion *in vivo* (Nelson et al., in preparation). TR efficiencies *in vivo* were measured by normalizing the number of colonies resulting from transformation with a gapped duplex DNA vector, that carried an abasic site within a 28-nucleotide single-stranded region, to the number of colonies resulting from transformation with equal amounts of a similar, but lesion-free, vector; transformants occur only if gap-filling, and hence TR, take place. Although about 25% TR occurred in a wild-type strain, TR efficiency was reduced to < 1% in strains in which the *REV1* and *REV3* genes had been deleted. Interestingly, TR efficiency was also much reduced in a *rev1–1* strain in which, as noted above, the deoxycytidyl transferase—but not the elongation—activity is present. Sequence analysis of vectors propagated in the wild-type strain, in keeping with previous results (118), showed that TR involved 89% cytosine and 11% adenine insertions opposite the abasic site, and a similar result was also seen in the *rev1–1* mutant. As expected, only adenine insertion occurred in the few TR events observed in the *rev1* deletion strain. However, the same was found in the *rev3* deletion mutant, even though it carries a wild-type *REV1* gene, implying that pol $\zeta$ is mainly responsible for elongation from a terminus created by cytosine insertion opposite an abasic site. Moreover, pol $\zeta$ is also responsible for extension of that minority of primers in which adenine is inserted opposite the abasic site, an insertion probably performed by this enzyme.

Since there can be no assurance that all relevant genes and proteins have been identified, further progress in understanding TR in yeast may well depend on the discovery of such loci and factors. The *in vivo* levels of pol $\zeta$ appear to be very low, and perhaps as a consequence it has not yet been possible to isolate the native enzyme, so other subunits may remain to be detected. Low amounts of Rev3p are predicted from the presence of an out-of-frame ATG, 10 nucleotides 5′ to the open reading frame ATG in the *REV3* gene (115). Experiments with model systems suggest that the translation efficiency of *REV3* mRNA may be reduced by a 100-fold or more as a consequence of this feature (119).

# 4. The mutagenic properties of DNA damage

A prominent and important feature of translesion replication is that it leads to the production of mutations. Almost all those recovered result from errors in the type or number of nucleotides inserted opposite the template lesion, that is *targeted* mutations, though some lesions produce a small proportion of untargeted sequence errors and the number of untargeted mutations increases in strains lacking mismatch repair. Some of the mutations may occur because TR is carried out by DNA polymerases that lack 3′–5′ exonuclease proofreading activity, but in most cases the chemical alteration in the template produced by the mutagen itself appears, usually in some unknown way, to cause the DNA polymerase to make errors. An early hypothesis suggested that mutagenic bulky lesions are non-instructional and incapable of influencing nucleotide selection (120), with the particular nucleotide inserted depending on polymerase preference rather than on lesion structure, but this model is now known to be insufficient to explain the diversity of mutagenic properties. A better understanding of the problem has been gained from experiments in which cells are transformed with vectors carrying a single specific DNA lesion at a unique location (76, 96, 121–128). These observations show that individual lesions have characteristic mutational phenotypes, indicating that, in many instances, lesion structure is likely to influence nucleotide selection. They also show that TR can often be accurate and that the error rate can vary widely. In *E. coli*, for example, TR past a T–T *cis-syn* cyclobutane dimer is generally accurate, with errors occurring in only about 6% of events (121), but at the other extreme the T–T pyrimidine (6–4) pyrimidinone adduct, a different UV photoproduct, is nearly 100% mutagenic (124). Examination of the properties of the same mutagenic lesion introduced into different species, or into the same organism expressing different TR proteins, shows that a lesion's mutagenic properties are also strongly influenced by the particular proteins responsible for TR. Although the error rate of TR past the T–T (6–4) adduct is >91% in *E. coli* (124), it is <40% in yeast (127), and the TR error rate for a *cis-syn* T–T cyclobutane dimer in yeast is <1% (126). However, the error rate for the *trans-syn* isomer is higher in yeast than *E. coli*, indicating that TR in yeast is not uniformly more accurate (122, 126). A particularly telling example of the influence of TR proteins on the mutagenic properties of a *cis-syn* T–T dimer is observed when MucAB proteins are substituted for UmuDC proteins in *E. coli* (21). With MucAB proteins facilitating

TR, the error rate is decreased nearly twofold, and the ratio of the two major classes of mutations, 3′ T→A and 3′ T→C, is reversed; with Umu proteins it is around 5:1, but with Muc proteins it is about 1:5. For the most part, TR proteins appear to influence TR frequency and error rate to a greater extent than the *types* of mutations generated, suggesting that the kinds of errors induced are largely determined by the altered properties of the template alone. AAF adducts on guanine within a *Nar*I sequence produce −1 frameshifts by a slippage mechanism in both yeast and *E. coli* (96, 125) because the slipped intermediate is a relatively stable structure. Although the cause is not known in these cases, similar types of mutations are also induced in yeast and *E. coli* by a *trans-syn* T–T dimer and a T–T (6–4) adduct (122, 124, 126, 127). The types of nucleotide insertion opposite an abasic site are different in yeast and *E. coli*, but this is probably a special case. Cytosine insertion is the major event in yeast (128) because of the deoxycytidyl transferase activity of the Rev1p, but almost non-existent in the bacterium (123), which lacks a comparable enzyme.

# 5. Comparisons, speculations, and TR in humans

Investigation into the mechanisms of TR remains a fascinating and potentially rewarding field for research. Although TR is regulated in very different ways in *E. coli* and yeast, it is possible that similarities between the two species with respect to the enzymological mechanisms of nucleotide insertion and chain elongation may emerge as our relative ignorance of these issues is remedied. Similarly, comparisons between the enzymes performing TR and the major replicases, that are responsible for replication on damage-free templates but which are incapable of lesion bypass, are likely to be informative. Replication on undamaged templates is characterized by high accuracy, which is achieved by a combination of postreplication mismatch repair (see Chapter 4), 3′–5′ exonucleolytic proofreading, and the inherent stringency of nucleotide selection displayed by the replicases. This stringency is thought to reflect the tight limits imposed by the DNA polymerase on the geometry and dimensions of any potential base-pair, a process facilitated by the virtual identity of normal A·T and G·C pairs in these respects. The dimensional restraints are a virtue in replicases but may contribute towards their inability to perform TR. As discussed in Section 3.3, preliminary evidence suggests that nucleotide addition by yeast pol ζ is much less dependent on normal duplex DNA structure at the primer terminus, implying that dimensional stringency is probably relaxed. If so, one of the functions of the UmuD′$_2$C complex in *E. coli* may be to achieve a similar end by inducing an allosteric change in the α catalytic subunit of pol III. A variety of indirect evidence suggests that the complex may interact transiently with several pol III subunits. In addition to permitting the formation of base-pairs with non-normal dimensions, changes in the structure of the polymerase performing TR may be needed to accommodate distorted templates, and allow their rotation within the polymerase to maximize the occurrence of favourable interactions with incoming nucleotides. Definitive answers on these questions are likely to come from X-ray crystallographic analyses of DNA polymerases co-crystallized with templates, a field in which great progress is currently being made.

A comparison between the two model systems highlights the absence of information about the mechanism for lesion targeting in yeast. In *E. coli* this is achieved by a combination of RecA protein and the UmuD'$_2$C complex. The yeast *RAD51* gene encodes a RecA homologue (129) that is important for recombination, but it does not appear to be involved in mutagenesis (130). Perhaps lesion targeting in yeast depends on the stalled replication complex itself, and pol ζ may be recruited to the lesion by interactions with some of the replication proteins. A comparison between the two model systems also highlights the possibility that further yeast proteins remain to be discovered. A region of 150 residues in the Rev1p shows some similarity to the *E. coli* UmuC protein, and both are involved in promoting TR, but no homologue of the UmuD polypeptide has yet been found. Possibly UmuD function is also encoded in the Rev1p, which is much larger than both the UmuC and UmuD proteins combined, but no sequence similarity with *umuD* can be seen.

Finally, because TR is an important source of mutations, the study of the mechanisms underlying this process has the potential to help us understand, and perhaps combat, genotoxic hazards and genetic diseases such as cancer in humans. To this end, we need to identify human TR genes and proteins, and, as a start in this direction, human homologues of the *REV1* and *REV3* genes have been isolated by using sequence information from yeast (131 and Gibbs *et al.*, in preparation). The human homologue of *REV3* (called *REV3L*) and the yeast gene are 29% identical in an amino-terminal region of approximately 340 residues, 39% identical in a carboxy-terminal region of 850 residues, and 29% identical in a 55-residue region in the middle of the gene, but *REV3L* is twice the size of the yeast gene (3130 versus 1504 amino acids (aa)) as a result of two large internal non-homologous regions. Observations from human cells producing high levels of a *REV3L* antisense fragment, which presumably much reduces the cellular levels of pol ζ, suggest that the human gene performs a similar function to its yeast counterpart. Such cells show 10-fold lower frequencies of UV-induced mutations to thioguanine resistance and a slightly greater sensitivity to killing by UV light, a phenotype similar to yeast *rev3* mutants. They also grow normally, indicating that the human gene, like yeast *REV3*, is probably not essential for viability. The human homologue of Rev1p is also larger than the yeast protein (1251 versus 985 aa), largely as a consequence of a longer non-homologous carboxy-terminal region. Compared with yeast Rev1p, the human homologue is 40% identical within an amino-terminal region of 110 residues, 20% identical within an adjacent region of 93 residues, and 31% identical within an internal region of 330 residues. In addition, the 13 residue sequence motif I/VXHI/VDXDCFFXXV (where X = any residue), which is conserved in yeast, *Caenorhabditis elegans*, and humans, occurs between the second and third regions of identity. The function of this peptide is not yet known. Although much clearly remains to be done in the investigation of mechanisms of TR in humans, the existence of genes homologous to those in yeast, and the evidence suggesting that *REV3L* has a function similar to its yeast counterpart, encourage the belief that the molecular biology of TR in yeast and humans may be quite similar.

# References

1. Hatahet, Z. and Wallace, S. S. (1998) Translesion DNA synthesis. In *DNA damage and repair: DNA repair in prokaryotes and lower eukaryotes* (ed. J. A. Nickoloff and M. F. Hoekstra), p. 229. Humana Press, Totowa, NJ.

2. Friedberg, E. C., Walker, G. C., and Siede, W. (1995) *DNA repair and mutagenesis* American Society for Microbiology, Washington, DC.

3. Tessman, I. and Kennedy, M. A. (1993) DNA polymerase II in bypass of abasic sites *in vivo. Genetics*, **136**, 439.

4. Rangarajan, S., Gudmundsson, G., Qiu, Z., Foster, P. L., and Goodman, M. F. (1997) *Escherichia coli* DNA polymerase II catalyzes chromosomal and episomal DNA synthesis in vivo. *Proc. Natl Acad. Sci. USA*, **94**, 946.

5. Nelson, J. R., Lawrence, C. W., and Hinkle, D. C. (1996) Deoxycytidyl transferase activity of yeast *REV1* protein. *Nature*, **382**, 729.

6. Nelson, J. R., Lawrence, C. W., and Hinkle, D. C. (1996) Thymine–thymine dimer bypass by yeast DNA polymerase ζ. *Science*, **272**, 1646.

7. Mount, D. W. (1977) A mutant of *Escherichia coli* showing constitutive expression of the lysogenic induction and error-prone DNA repair pathways. *Proc. Natl Acad. Sci. USA*, **74**, 300.

8. Markham, B. E., Little, J. W., and Mount, D. W. (1981) Nucleotide sequence of the *lexA* gene of *Escherichia coli* K-12. *Nucl. Acids Res.*, **9**, 4149.

9. Little, J. W. and Mount, D. W. (1982) The SOS regulatory system of *Escherichia coli. Cell*, **29**, 11.

10. Little, J. W. (1984) Autodigestion of LexA and phage repressors. *Proc. Natl Acad. Sci. USA*, **81**, 1375.

11. Koch, W. H. and Woodgate, R. (1998) The SOS response. In *DNA damage and repair: DNA repair in prokaryotes and lower eukaryotes* (ed. J. A. Nickoloff and M. F. Hoekstra), p. 107. Humana Press, Totowa, NJ.

12. Rupp, W. D. and Howard-Flanders, P. (1968) Discontinuities in the DNA synthesized in an excision-defective strain of *Escherichia coli* following ultraviolet radiation. *J. Mol. Biol.*, **31**, 291.

13. Sassanfar, M. and Roberts, J. W. (1990) Nature of the SOS-inducing signal in *Escherichia coli*: the involvement of DNA replication. *J. Mol. Biol.*, **212**, 79.

14. Yu, X. and Egelman, E. H. (1993) The LexA repressor binds within the deep helical groove of the activated RecA filament. *J. Mol. Biol.*, **231**, 29.

15. Little, J. W. (1993) LexA cleavage and other self-processing reactions. *J. Bacteriol.*, **175**, 4943.

16. Sommer, S., Boudsocq, F., Devoret, R., and Bailone, A. (1998) Specific RecA amino acid changes affect RecA-UmuD′C interaction. *Mol. Microbiol.*, **28**, 281.

17. Rupp, W. D., Wilde, C. E., Reno, D. L., and Howard-Flanders, P. (1971) Exchanges between DNA strands in ultraviolet-irradiated *Escherichia coli. J. Mol. Biol.*, **61**, 25.

18. Higgins, N. P., Kato, K., and Strauss, B. (1976) A model for replication repair in mammalian cells. *J. Mol. Biol.*, **101**, 417.

19. Bridges, B. A. and Woodgate, R. (1984) Mutagenic repair in *Escherichia coli*, X. The *umuC* gene product may be required for replication past pyrimidine dimers but not for the coding error in UV mutagenesis. *Mol. Gen. Genet.*, **196**, 364.

20. Bridges, B. A. and Woodgate, R. (1985) Mutagenic repair in *Escherichia coli*: products of the *recA* gene and of the *umuD* and *umuC* genes act at different steps in UV-induced mutagenesis. *Proc. Natl Acad. Sci. USA*, **82**, 4193.

21. Szekeres, E. S. Jr., Woodgate, R., and Lawrence, C. W. (1996) Substitution of *mucAB* or *rumAB* for *umuDC* alters the relative frequencies of the two classes of mutations induced by a site-specific T–T cyclobutane dimer and the efficiency of translesion DNA synthesis. *J. Bacteriol.*, **178**, 2559.

22. Rajagopalan, M., Lu, C., Woodgate, R., O'Donnell, M., Goodman, M. F., and Echols, H. (1992) Activity of the purified mutagenesis proteins UmuC, UmuD' and RecA in replicative bypass of an abasic DNA lesion by DNA polymerase III. *Proc. Natl Acad. Sci. USA*, **89**, 10777.

23. Tang, M., Bruck, I., Eritja, R., Turner, J., Frank, E. G., Woodgate, R., *et al.* (1998) Biochemical basis of SOS-induced mutagenesis in *Escherichia coli*: reconstitution of *in vitro* lesion bypass dependent on the UmuD'$_2$C mutagenic complex and RecA. *Proc. Natl Acad. Sci. USA*, **95**, 9755.

24. Reuven, B. N., Tomer, G., and Livneh, Z. (1998) The mutagenesis proteins UmuD' and UmuC prevent lethal frameshifts while increasing base-substitution mutations. *Mol. Cell.*, **2**, 191.

25. Woodgate, R. and Sedgwick, S. G. (1992) Mutagenesis induced by bacterial UmuDC proteins and their plasmid homologues. *Mol. Microbiol.*, **6**, 2213.

26. Woodgate, R. and Levine, A. S. (1996) Damage inducible mutagenesis: recent insights into the activities of the Umu family of mutagenesis proteins. In *Cancer surveys: genetic instability in cancer*, Vol. 28 (ed. T. Lindahl), p. 117. Cold Spring Harbor Laboratory Press, Cold Spring Harbor, NY.

27. Smith, B. T. and Walker, G. C. (1998) Mutagenesis and more: *umuDC* and the *Escherichia coli* SOS response. *Genetics*, **148**, 1599.

28. Kitagawa, Y., Akaboshi, E., Shinagawa, H., Horii, T., Ogawa, H., and Kato, T. (1985) Structural analysis of the *umu* operon required for inducible mutagenesis in *Escherichia coli*. *Proc. Natl Acad. Sci. USA*, **82**, 4336.

29. Perry, K. L., Elledge, S. J., Mitchell, B., Marsh, L., and Walker, G. C. (1985) *umuDC* and *mucAB* operons whose products are required for UV light and chemical-induced mutagenesis: UmuD, MucA, and LexA products share homology. *Proc. Natl Acad. Sci. USA*, **82**, 4331.

30. Kato, T. and Shinoura, Y. (1977) Isolation and characterization of mutants of *Escherichia coli* deficient in induction of mutations by ultraviolet light. *Mol. Gen. Genet.*, **156**, 121.

31. Steinborn, G. (1978) *Uvm* mutants of *Escherichia coli* K12 deficient in UV mutagenesis. I. Isolation of *uvm* mutants and their phenotypical characterization in DNA repair and mutagenesis. *Mol. Gen. Genet.*, **165**, 87.

32. Napolitano, R. L., Lambert, I. B., and Fuchs, R. P. P. (1997) SOS factors involved in translesion synthesis. *Proc. Natl Acad. Sci. USA*, **94**, 5733.

33. Woodgate, R. (1992) Construction of a *umuDC* operon substitution mutation in *Escherichia coli*. *Mutat. Res.*, **281**, 221.

34. Koffel-Schwartz, N., Coin, F., Veaute, X., and Fuchs, R. P. P. (1996). Cellular strategies for accommodating replication-hindering adducts in DNA: control by the SOS response in *Escherichia coli*. *Proc. Natl Acad. Sci. USA*, **93**, 7805.

35. Echols, H. (1981) SOS functions, cancer and inducible evolution. *Cell*, **25**, 1.

36. Woodgate, R. and Ennis, D. G. (1991) Levels of chromosomally encoded Umu proteins and requirements for *in vivo* UmuD cleavage. *Mol. Gen. Genet.*, **229**, 10.

37. Frank, E. G., Ennis, D. G., Gonzalez, M., Levine, A. S., and Woodgate, R. (1996) Regulation of SOS mutagenesis by proteolysis. *Proc. Natl Acad. Sci. USA*, **93**, 10291.

38. Shinagawa, H., Iwasaki, H., Kato, T., and Nakata, A. (1988) RecA protein-dependent cleavage of UmuD protein and SOS mutagenesis. *Proc. Natl Acad. Sci. USA*, **85**, 1806.

39. Burckhardt, S. E., Woodgate, R., Scheuermann, R. H., and Echols, H. (1988) UmuD mutagenesis protein of *Escherichia coli*: overproduction, purification and cleavage by RecA. *Proc. Natl Acad. Sci. USA*, **85**, 1811.

40. Nohmi, T., Battista, J. R., Dodson, L. A., and Walker, G. C. (1988) RecA-mediated cleavage activates UmuD for mutagenesis: mechanistic relationship between transcriptional derepression and posttranslational activation. *Proc. Natl Acad. Sci. USA*, **85**, 1816.

41. Peat, T., Frank, E. G., McDonald, J. P., Levine, A. S., Woodgate, R., and Hendrickson, W. A. (1996) Structure of the UmuD' protein and its regulation in response to DNA damage. *Nature*, **380**, 727.

42. McDonald, J. P., Frank, E. G., Levine, A. S., and Woodgate, R. (1998) Intermolecular cleavage of the UmuD-like mutagenesis proteins. *Proc. Natl Acad. Sci. USA*, **95**, 1478.

43. Frank, E. G., Gonzalez, M., Ennis, D. G., Levine, A. S., and Woodgate, R. (1996) *In vivo* stability of the Umu mutagenesis proteins: a major role for RecA. *J. Bacteriol.*, **178**, 3550.

44. Gonzalez, M., Frank, E. G., McDonald, J. P., Levine, A. S., and Woodgate, R. (1998) Structural insights into the regulation of SOS mutagenesis. *Acta Biochemica Polonica*, **45**, 163.

45. Boudsocq, F., Campbell, M., Devoret, R., and Bailone, A. (1997) Quantitation of the inhibition of Hfr x F⁻recombination by the mutagenesis complex UmuD'C. *J. Mol. Biol.*, **270**, 201.

46. Yasuda, T., Morimatsu, K., Horii, T., Nagata, T., and Ohmori, H. (1998) Inhibition of *Escherichia coli* RecA coprotease activities by DinI. *EMBO. J.*, **17**, 3207.

47. Lodowick, D., Owen, D., and Strike, P. (1990) DNA sequence analysis of the *imp* UV protection and mutation operon of the plasmid TP110: identification of a third gene. *Nucl. Acids Res.*, **18**, 5045.

48. Battista, J. R., Ohta, T., Nohmi, T., Sun, W., and Walker, G. C. (1990) Dominant negative *umuD* mutations decreasing RecA-mediated cleavage suggest roles for intact UmuD in modulation of SOS mutagenesis. *Proc. Natl Acad. Sci. USA*, **87**, 7190.

49. Woodgate, R., Rajagopalan, M., Lu, C., and Echols, H. (1989) UmuC mutagenesis protein of *Escherichia coli*: purification and interaction with UmuD and UmuD'. *Proc. Natl Acad. Sci. USA*, **86**, 7301.

50. Bruck, I., Woodgate, R., McEntee, K., and Goodman, M. F. (1996) Purification of a soluble UmuD'C complex from *Escherichia coli*: cooperative binding of UmuD'C to single-stranded DNA. *J. Biol. Chem.*, **271**, 10767.

51. Sargentini, N. J. and Smith, K. C. (1981) Much of spontaneous mutagenesis in *Escherichia coli* is due to error-prone DNA repair: implications for spontaneous carcinogenesis. *Carcinogenesis*, **2**, 863.

52. Woodgate, R. (1992) Construction of a *umuDC* operon substitution mutation in *Escherichia coli*. *Mutat. Res.*, **281**, 221.

53. Woodgate, R. and Ennis, D. G. (1991) Levels of chromosomally encoded Umu proteins and requirements for *in vivo* UmuD cleavage. *Mol. Gen. Genet.*, **229**, 10.

54. Echols, H. and Goodman, M. F. (1990) Mutation induced by DNA damage: a many protein affair. *Mutat. Res.*, **236**, 301.

55. Kelman, Z. and O'Donnell, M. (1995) DNA polymerase III holoenzyme: structure and function of a chromosomal replicating machine. *Annu. Rev. Biochem.*, **64**, 171.

56. Bridges, B. A. (1988) Mutagenic DNA repair in *Escherichia coli*. XVI. Mutagenesis by ultraviolet light plus delayed photoreversal in *recA* strains. *Mutat. Res.*, **198**, 343.

57. Heuser, J. and Griffith, J. (1989) Visualization of RecA protein and its complexes with DNA by quick-freeze/deep-etch electron microscopy. *J. Mol. Biol.*, **210**, 473.

58. Woodgate, R. and Sedgwick, S. G. (1992) Mutagenesis induced by bacterial UmuDC proteins and their plasmid homologues. *Mol. Microbiol.*, **6**, 2213.

59. Frank, E. G., Hauser, J., Levine, A. S., and Woodgate, R. (1993) Targeting of the UmuD, UmuD' and MucA' mutagenesis proteins to DNA by RecA protein. *Proc. Natl Acad. Sci. USA*, **90**, 8169.

60. Sommer, S., Bailone, A., and Devoret, R. (1993) The appearance of the UmuD'C protein complex in *Escherichia coli* switches repair from homologous recombination to SOS mutagenesis. *Mol. Microbiol.*, **10**, 963.

61. Rehrauer, W. M., Bruck, I., Woodgate, R., Goodman, M. F., and Kowalczykowski, S. C. (1998) Modulation of recombination function by the mutagenic UmuD'C protein complex. *J. Biol. Chem.* 273, 32384

62. Hagensee, M. E., Timme, T., Bryan, S. K., and Moses, R. E. (1987) DNA polymerase III of *Escherichia coli* is required for UV and ethyl methanesulfonate mutagenesis. *Proc. Natl Acad. Sci. USA*, **84**, 4195.

63. Foster, P. L. and Sullivan, A. D. (1988) Interactions between epsilon, the proofreading subunit of DNA polymerase III, and proteins involved in the SOS response of *Escherichia coli*. *Mol. Gen. Genet.*, **214**, 467.

64. Villani, G., Boiteux, S., and Radman, M. (1978) Mechanism of ultraviolet-induced mutagenesis: extent and fidelity of *in vitro* DNA synthesis on irradiated templates. *Proc. Natl Acad. Sci. USA*, **75**, 3037.

65. Woodgate, R., Bridges, B. A., Herrera, G., and Blanco, M. (1987) Mutagenic repair in *Escherichia coli*, XIII. Proofreading exonuclease of DNA polymerase III holoenzyme is not operational during UV mutagenesis. *Mutat. Res.*, **183**, 31.

66. Fijalkowska, I. J., Dunn, R. L., and Schaaper, R. M. (1997) Genetic requirements and mutational specificity of the *Escherichia coli* SOS mutator activity. *J. Bacteriol.*, **179**, 7435.

67. Jonczyk, P., Fijalkowska, I., and Ciesla, Z. (1988) Overproduction of the ε subunit of DNA polymerase III counteracts the SOS mutagenic response of *Escherichia coli*. *Proc. Natl Acad. Sci. USA*, **85**, 9124.

68. Foster, P. L., Sullivan, A. D., and Franklin, S. B. (1989) Presence of the *dnaQ-rnh* divergent transcriptional unit on a multicopy plasmid inhibits induced mutagenesis in *Escherichia coli*. *J. Bacteriol.*, **171**, 3144.

69. Kanabus, M., Nowicka, A., Sledziewska-Gojska, E., Jonczyk, P., and Ciesla, Z. (1995) The antimutagenic effect of a truncated ε subunit of DNA polymerase III in *Escherichia coli* cells irradiated with UV light. *Mol. Gen. Genet.*, **247**, 216.

70. Slater, S. C. and Maurer, R. (1991) Requirements for bypass of UV-induced lesions in single-stranded DNA of bacteriophage φX174 in *Salmonella typhimurium*. *Proc. Natl Acad. Sci. USA*, **88**, 1251.

71. Vandewiele, D., Borden, A., O'Grady, P. I., Woodgate, R. and Lawrence, C. W. (1998) Efficient translesion replication in the absence of *Escherichia coli* Umu proteins and 3'–5' exonuclease proofreading function. *Proc. Natl Acad. Sci. USA*, 95, 15519.

72. Doyle, N. and Strike, P. (1995) The spectra of base substitutions induced by the *impCAB*, *mucAB* and *umuDC* error-prone DNA repair operons differ following exposure to methyl methansulfonate. *Mol. Gen. Genet.*, **247**, 735.

73. Sweasy, J. B., Witkin, E. M., Sinha, N., and Roegner-Maniscalco, V. (1990) RecA protein of *Escherichia coli* has a third essential role in SOS mutator activity. *J. Bacteriol.*, **172**, 3030.

74. Maenhaut-Michel, G., Janel-Bintz, R., and Fuchs, R. P. P. (1992) A *umuDC*-independent SOS pathway for frameshift mutagenesis. *Mol. Gen. Genet.*, **235**, 373.

75. Fuchs, R. P. P. and Napolitano, R. L. (1998) Inactivation of DNA proofreading obviates the need for SOS induction in frameshift mutagenesis. *Proc. Natl Acad. Sci. USA*, 95, 13114.

76. Palejwala, V. A., Pandya, G. A., Bhanot, O. S., Solomon, J. J., Murphy, H. S., Dunman, P. M., *et al.* (1994) UVM, an ultraviolet-inducible RecA-independent mutagenic phenomenon in *Escherichia coli*. *J. Biol. Chem.*, **269**, 27433.

77. Palejwala, V. A., Wang, G. E., Murphy, H. S., and Humayun, M. Z. (1995) Functional *recA*, *lexA*, *umuD*, *umuC*, *polA*, and *polB* genes are not required for the *Escherichia coli* UVM response. *J. Bacteriol.*, **177**, 6041.

78. Murphy, H. S., Palejwala, V. A., Rahman, M. S., Dunman, P. M., Wang, G., and Humayun, M. Z. (1996) Role of mismatch repair in the *Escherichia coli* UVM response. *J. Bacteriol.*, **178**, 6651.

79. Murphy, H. S. and Humayun, M. Z. (1997) *Escherichia coli* cells expressing a mutant *glyV* (glycine tRNA) gene have a UVM-constitutive phenotype: implications for mechanisms underlying the *mutA* or *mutC* mutator effect. *J. Bacteriol.*, **179**, 7507.

80. Slupska, M. M., Baikalov, C., Lloyd, R., and Miller, J. H. (1996) Mutator tRNAs are encoded by the *Escherichia coli* mutator genes *mutA* and *mutC*: a novel pathway for mutagenesis. *Proc. Natl Acad. Sci. USA*, **93**, 4380.

81. Michaels, M. L., Cruz, C., and Miller, J. H. (1990) *mutA* and *mutC*: two mutator loci in *Escherichia coli* that stimulate transversions. *Proc. Natl Acad. Sci. USA*, **87**, 9211.

82. Lemontt, J. F. (1971) Mutants of yeast defective in mutation induced by ultraviolet light. *Genetics*, **68**, 21.

83. Lawrence, C. W., Das, G., and Christensen, R. B. (1985) *REV7*, a new gene concerned with UV mutagenesis in yeast. *Mol. Gen. Genet.*, **200**, 80.

84. Lawrence, C. W. and Hinkle, D. C. (1996) DNA polymerase ζ and the control of DNA damage induced mutagenesis in eukaryotes. In *Cancer surveys: genetic instability in cancer*, Vol. 28 (ed. T. Lindahl), p. 21. Cold Spring Harbor Laboratory Press, Cold Spring Harbor, NY.

85. Quah, S. -K., von Borstel, R. C., and Hastings, P. J. (1980) The origin of spontaneous mutations in *Saccharomyces cerevisiae*. *Genetics*, **96**, 819.

86. Roche, H., Gietz, R. D., and Kunz, B. A. (1994) The specificity of the yeast *rev3Δ* antimutator and *REV3* dependency of the mutator resulting from a defect (*rad1Δ*) in nucleotide excision repair. *Genetics*, **137**, 637.

87. Datta, A. and Jinks-Robertson, S. (1995) Association of increased spontaneous mutation rates with high levels of transcription in yeast. *Science*, **268**, 1616.

88. Holbeck, S. L. and Strathern, J. N. (1995) A role for *REV3* in mutagenesis during double-strand break repair in *Saccharomyces cerevisiae*. *Genetics*, **147**, 1017.

89. Lawrence, C. W., Krauss, B. R., and Christensen, R. B. (1985) New mutations affecting induced mutagenesis in yeast. *Mutat. Res.*, **150**, 211.

90. Nisson, P. E. and Lawrence, C. W. (1986) The isolation and characterization of *ngm2*, a mutation that affects nitrosoguanidine mutagenesis in yeast. *Mol. Gen. Genet.*, **204**, 90.

91. Lawrence, C. W. and Christensen, R. B. (1979) Ultraviolet-induced reversion of *cyc1* alleles in radiation-sensitive strains of yeast III. *rev3* mutant strains. *Genetics*, **92**, 397.

92. Lawrence, C. W., O'Brien, T., and Bond, J. (1984) UV-induced reversion of *his4* frameshift mutations in *rad6*, *rev1*, and *rev3* mutants of yeast. *Mol. Gen. Genet.*, **195**, 487.

93. McKee, R. H. and Lawrence, C. W. (1979) Genetic analysis of gamma-ray mutagenesis in yeast II. Allele specific control of mutagenesis. *Genetics*, **93**, 375.

94. Prakash, L. (1976) Effect of genes controlling radiation sensitivity on chemically induced mutations in *Saccharomyces cerevisiae*. *Genetics*, **83**, 285.

95. Cassier, C., Chanet, R., Henriques, J. A. P., and Moustacchi, E. (1980) The effects of three *PSO* genes on induced mutagenesis: a novel class of mutationally defective yeast. *Genetics*, **96**, 841.

96. Baynton, K., Bresson-Roy, A., and Fuchs, R. P. (1998) Analysis of damage tolerance pathways in *Saccharomyces cerevisiae*: a requirement for *REV3* DNA polymerase in translesion synthesis. *Mol. Cell. Biol.*, **18**, 960.

97. Lawrence, C. W., Nisson, P. E., and Christensen, R. B. (1985) UV and chemical mutagenesis in *rev7* mutants of yeast. *Mol. Gen. Genet.*, **200**, 86.

98. Lawrence, C. W. and Christensen, R. B. (1979) Ultraviolet-induced reversion of *cyc1* alleles in radiation-sensitive strains of yeast I. *rev1* mutant strains. *J. Mol. Biol.*, **122**, 1.

99. Cassier-Chauvat, C. and Fabre, F. (1991) A similar defect in UV mutagenesis conferred by the *rad6* and *rad18* mutations in *Saccharomyces cerevisiae*. *Mutat. Res.*, **254**, 247.

100. Torres-Ramos, C. A., Yoder, B. L., Burgers, P. M., Prakash, S., and Prakash, L. (1996) Requirement of proliferating cell nuclear antigen in *RAD6*-dependent post replication DNA repair. *Proc. Natl Acad. Sci. USA*, **93**, 9679.

101. Johnson, R. E., Henderson, S. T., Petes, T. D., Prakash, S., Bankmann, M., and Prakash, L. (1992) *Saccharomyces cerevisiae RAD5*-encoded DNA repair protein contains DNA helicase and zinc-binding sequence motifs and affects the stability of simple repetitive sequences in the genome. *Mol. Cell. Biol.*, **12**, 3807.

102. McDonald, J. P., Levine, A. S., and Woodgate, R. (1997) The *Saccharomyces cerevisiae RAD30* gene, a homologue of *Escherichia coli dinB* and *umuC*, is DNA damage inducible and functions in a novel error-free postreplication repair mechanism. *Genetics*, **147**, 1557.

103. Bailly, V., Lamb., J., Sung, P., Prakash, S., and Prakash, L. (1994) Specific complex formation between yeast *RAD6* and *RAD18* proteins: a potential mechanism for targeting *RAD6* ubiquitin-conjugating activity to DNA damage sites. *Genes Dev.*, **8**, 811.

104. Jentsch, S., McGrath, J. P., and Varshavsky, A. (1987) The yeast DNA repair gene RAD6 encodes a ubiquitin-conjugating enzyme. *Nature*, **329**, 131.

105. Jones, J. S., Weber, S., and Prakash, L. (1988) The *Saccharomyces cerevisiae RAD18* gene encodes a protein that contains potential zinc finger domains for nucleic acid binding and a putative nucleotide binding sequence. *Nucl. Acids Res.*, **16**, 7119.

106. Reynolds, P., Weber, S., and Prakash, L. (1985) *RAD6* gene of *Saccharomyces cerevisiae* encodes a protein containing a tract of 13 consecutive aspartates. *Proc. Natl Acad. Sci. USA*, **82**, 168.

107. Sung, P., Prakash, S., and Prakash, L. (1990) Mutation of cysteine-88 in the *Saccharomyces cerevisiae* RAD6 protein abolishes its ubiquitin-conjugating activity and its various biological functions. *Proc. Natl Acad. Sci. USA*, **87**, 2695.

108. Sung, P., Prakash, S., and Prakash, L. (1991) Stable ester conjugate between the *Saccharomyces cerevisiae* RAD6 protein and ubiquitin has no biological activity. *J. Mol. Biol.*, **221**, 745.

109. Dohmen, R. J., Madura, K., Bartel, B., and Varshavsky, A. (1991) The N-end rule is mediated by the UBC2 (RAD6) ubiquitin-conjugating enzyme. *Proc. Natl Acad. Sci. USA*, **88**, 7351.

110. Watkins, J. F., Sung, P., Prakash, S., and Prakash, L. (1993) The extremely conserved amino terminus of *RAD6* ubiquitin-conjugating enzyme is essential for amino-end rule-dependent protein degradation. *Genes Dev.*, **7**, 250.

111. Madura, K., Prakash, S., and Prakash, L. (1990) Expression of the *Saccharomyces cerevisiae* DNA repair gene *RAD6* that encodes a ubiquitin conjugating enzyme increases in response to DNA damage and in meiosis. *Nucl. Acids Res.*, **18**, 771.

112. Jones, J. S. and Prakash, L. (1991) Transcript levels of the *Saccharomyces cerevisiae* DNA repair gene *RAD18* increase in UV irradiated cells and during meiosis but not during the mitotic cell cycle. *Nucl. Acids Res.*, **19**, 893.

113. Singhal, R. K., Hinkle, D. C., and Lawrence, C. W. (1992) The *REV3* gene of *Saccharomyces cerevisiae* is transcriptionally regulated more like a repair gene than one encoding a DNA polymerase. *Mol. Gen. Genet.*, **236**, 17.

114. Larimer, F. W., Perry, J. R., and Hardigree, A. A. (1989) The *REV1* gene of *Saccharomyces cerevisiae*: isolation, sequence, and functional analysis. *J. Bacteriol.*, **171**, 230.

115. Morrison, A., Christensen, R. B., Alley, J. A., Beck, A. K., Bernstine, E. G., Lemontt, J. F., *et al.* (1989) *REV3*, a *Saccharomyces cerevisiae* gene whose function is required for induced mutagenesis, is predicted to encode a non-essential DNA polymerase. *J. Bacteriol.*, **171**, 5659.

116. Torpey, L. E., Gibbs, P. E. M., Nelson, J., and Lawrence, C. W. (1994) Cloning and sequence of *REV7*, a gene whose function is required for DNA damage induced mutagenesis in *Saccharomyces cerevisiae*. *Yeast*, **10**, 1503.

117. Mendelman, L. V., Petruska, J., and Goodman, M. F. (1990) Base mispair extension kinetics. Comparison of DNA polymerase alpha and reverse transcriptase. *J. Biol. Chem.*, **265**, 2338.

118. Gibbs, P. E. M. and Lawrence, C. W. (1995) Novel mutagenic properties of abasic sites in *Saccharomyces cerevisiae*. *J. Mol. Biol.*, **251**, 229.

119. Hinnebusch, A. G. and Liebman, S. W. (1991) *Protein synthesis and translational control*. In *The molecular biology of the yeast Saccharomyces: genome dynamics, protein synthesis, and energetics* (ed. J. R. Broach, J. R. Pringle, and E. W. Jones), p. 627. Cold Spring Harbor Laboratory Press, Cold Spring Harbor, NY.

120. Strauss, B. S. (1991) The 'A-rule' of mutagen specificity: a consequence of DNA polymerase bypass of non-instructional lesions? *BioEssays*, **13**, 79.

121. Banerjee, S. K., Christensen, R. B., Lawrence, C. W., and LeClerc, J. E. (1988) Frequency and spectrum of mutations produced by a single cis-syn thymine-thymine cyclobutane dimer in a single-stranded vector. *Proc. Natl Acad. Sci. USA*, **85**, 8141.

122. Banerjee, S. K., Borden, A., Christensen, R. B., LeClerc, J. E., and Lawrence, C. W. (1990) SOS-dependent replication past a single *trans-syn* T-T cyclobutane dimer gives a different mutation spectrum and increased error rate compared with replication past this lesion in uninduced cells. *J. Bacteriol.*, **172**, 2105.

123. Lawrence, C. W., Borden, A., Banerjee, S. K., and LeClerc, J. E. (1990) Mutation frequency and spectrum resulting from a single abasic site in a single-stranded vector. *Nucl. Acids Res.*, **18**, 2153.

124. LeClerc, J. E., Borden, A., and Lawrence, C. W. (1991) The thymine–thymine pyrimidine–pyrimidone (6–4) ultraviolet light photoproduct is highly mutagenic and specifically induces 3′ thymine-to-cytosine transitions. *Proc. Natl Acad. Sci. USA*, **88**, 9685.

125. Burnouf, D., Koehl, P., and Fuchs, R. P. P. (1989) Single adduct mutagenesis: strong effect of the position of a single acetylaminofluorene adduct within a mutation hot spot. *Proc. Natl Acad. Sci. USA*, **86**, 4147.

126. Gibbs, P. E. M., Kilbey, B. J., Banerjee, S. K., and Lawrence, C. W. (1993) The frequency and accuracy of replication past a thymine–thymine cyclobutane dimer are very different in *Saccharomyces cerevisiae* and *Escherichia coli*. *J. Bacteriol.*, **175**, 2607.

127. Gibbs, P. E. M., Borden, A., and Lawrence, C. W. (1995) The T–T pyrimidine (6–4) pyrimidinone UV photoproduct is much less mutagenic in yeast than in *Escherichia coli*. *Nucl. Acids Res.*, **23**, 1919.

128. Gibbs, P. E. M. and Lawrence, C. W. (1995) Novel mutagenic properties of abasic sites in *Saccharomyces cerevisiae*. *J. Mol. Biol.*, **251**, 229.

129. Basile, G., Aker, M., and Mortimer, R. K. (1992) Nucleotide sequence and transcriptional regulation of the yeast recombination repair gene *RAD51*. *Mol. Cell. Biol.*, **12**, 3235.

130. Lawrence, C. W. and Christensen, R. B. (1976) UV mutagenesis in radiation sensitive strains of yeast. *Genetics*, **82**, 207.

131. Gibbs, P. E. M., McGregor, W. G., Maher, V. M., Nisson, P., and Lawrence, C. W. (1998) A human homolog of the *Saccharomyces cerevisiae* REV3 gene, which encodes the catalytic subunit of DNA polymerase ζ. *Proc. Natl Acad. Sci. USA*, **95**, 6876.

# 4 | Mismatch repair and cancer

P. KARRAN and M. BIGNAMI

## 1. Introduction

DNA mismatch repair is one of the three known DNA excision repair pathways. Unlike nucleotide and base excision repair, in which the substrate is damaged or altered DNA, mismatch repair acts on undamaged DNA bases which are in inappropriate contexts. Mismatch repair corrects unmodified DNA bases in non-Watson–Crick pairings or small distortions generated by the incorrect alignment of two otherwise normal DNA strands. Structural abnormalities such as these are generated during recombination between DNA molecules that are not perfectly homologous, and the phrase 'mismatch repair' was originally invoked to explain apparently anomalous recombination frequencies. DNA replication errors are also a major source of mismatches. Mismatch repair acts after replication to rectify mismatches that have evaded proofreading by the replication apparatus. Organisms with defective mismatch repair invariably display increased spontaneous mutation rates. Although some of the later steps of the three excision repair pathways share common components, the initial steps, in which the substrate for repair is identified and marked for correction, can be considered unique for each pathway. There is an additional constraint on substrate recognition during mismatch repair. To achieve biologically significant correction, the repair proteins must not only recognize the structurally anomalous mismatches, but must also obtain information as to which of the mispaired DNA strands is incorrect. The type of mismatch repair that utilizes such extrinsic strand discrimination signals is known as long-patch mismatch repair, because removal of the mismatch can involve the excision and resynthesis of a relatively long (hundreds of nucleotides) stretch of DNA. This is in contrast to the relatively short excision tracts (less than 50 bases) produced by the nucleotide or base excision repair pathways. In addition to long-patch repair, there are several examples of more specialized mismatch correction pathways with a restricted specificity. For example, the T residues of G:T mispairs generated by the deamination of 5-methylcytosine are selectively repaired. There may also be a preferential excision of A from the A:G mispairs which are one of the more frequent DNA polymerase errors. Both these repair functions are initiated by DNA glycosylases, and their intrinsic base specificity circumvents the requirement for strand identification. They are really representatives of the base excision repair pathway and we will not consider them further.

Defective mismatch repair is associated with an elevated incidence of cancer. In humans, germline mutations in mismatch repair genes predispose to cancer in the Hereditary nonpolyposis colorectal cancer (HNPCC) syndrome. The loss of a single allele appears to have little effect on mismatch repair proficiency and normal tissues of HNPCC individuals are repair-proficient. Their tumours, which are pre-dominantly colorectal, are defective in mismatch repair, usually because the second repair allele has been inactivated by a somatic mutation. Importantly, similar mis-match repair defects are associated with a significant fraction (5–15%) of apparently sporadic colorectal tumours with no overt genetic involvement. It is generally considered that the loss of both repair alleles and the consequent spontaneous mutator effect is necessary for the accelerated development of both HNPCC and sporadic malignancies. In homozygously repair-defective mice, increased tumour incidence may be accompanied by effects on meiosis which can result in sterility.

In addition to their role in maintaining low levels of spontaneous mutations, mismatch repair also modulates the cytotoxicity of some DNA damaging drugs. The altered sensitivity of mismatch repair-defective cells to various types of DNA damaging agents has direct implications for the use of therapeutic drugs in the treatment of a significant fraction of human tumours.

# 2. Bacterial mismatch repair

## 2.1 The identification of bacterial mismatch repair genes

The bacterium *Streptococcus pneumoniae* naturally takes up exogenous DNA and integrates it into its genome by homologous recombination. The formation of recombination intermediates between donor and recipient DNAs with different sequences (for example one wild-type sequence and one with a point mutation) generates mismatches. There is a wide variation in the integration efficiency of donor DNA segments, which indicates that sometimes the incoming information can be erased. A selective removal of mismatched bases from the donor strand was pro-posed to explain this information loss (reviewed in ref. 1). The possibility of selective excision of incorrectly paired bases had initially been raised to explain similar recombinational anomalies, gene conversion or non-reciprocal exchanges, in fungi (2). Because integration efficiency in *S. pneumoniae* varied from marker to marker, it was further proposed that the efficiency of removal would, to some extent, be dependent on the particular mismatch. Mutants (*hex*) in which the integration efficiency was uniform for all markers were identified. These mutants, *hexA* and *hexB*, were found to have high rates of spontaneous mutation, implicating the postulated heteroduplex repair pathway in avoiding mutations during normal cell growth. The HexA and HexB proteins both participate in a long-patch mismatch correction pathway. Their counterparts in *Escherichia coli* are MutS and MutL. The *mutS* and *mutL* mutant strains were originally identified as mutators with extremely high levels of spontaneous mutation (reviewed in ref.).

## 2.2 The biochemistry of mismatch repair in *Escherichia coli*

The long-patch mismatch repair system of *E. coli* has been characterized in some detail and reconstituted from purified components (4). It is initiated by three dedicated proteins, MutS, MutL, and MutH and subsequent steps are carried out by proteins which are also involved in other DNA transactions (see Fig. 4.1). Mismatch recognition is carried out by MutS which binds selectively to mismatched DNA in the immediate vicinity of the mismatch. The monomeric mass of MutS is 95 kDa but it tends to oligomerize in solution and the functional unit is most likely a homodimer (5). ATP binding and hydrolysis are intrinsic to *E. coli* mismatch repair and MutS has a weak ATPase function which is highly conserved among homologous MutS family members. Mutations within the ATP binding domain of the *E. coli* (or the closely related *Salmonella typhimurium*) protein, which substantially reduce ATP hydrolysis, severely compromise mismatch recognition and mismatch repair activity and act in a dominantly negative fashion *in vivo* (6, 7). Following binding of MutS to the mismatch, the MutL protein is recruited in another ATP-dependent reaction to form a tripartite complex. MutL (a homodimer of 68-kDa subunits) has no recognized overt biochemical function, but in combination with MutS it activates a cryptic endonuclease associated with the third dedicated mismatch repair protein, MutH (8–10). MutH incises mismatched DNA at a specific signal sequence located some distance from the mismatch itself.

Identification of the 'incorrect' partner of the mismatch utilizes the pattern of DNA adenine methylation. Adenine within GATC sequences of *E. coli* DNA is converted to 6-methyladenine by the methylase encoded by the *dam*⁺ gene, and in resting DNA these symmetrical sequences are fully methylated. Adenine modification by the Dam methylase is a postreplicative process, and the daughter DNA strand immediately behind the replication fork is transiently (for about 0.5–3 min) unmethylated (reviewed in ref. 11). Because the parental strand is fully methylated, mismatched bases are in a hemimethylated DNA duplex for this short period and the 'incorrect', misincorporated, partner is always in the non-methylated strand. Hemimethylated DNA is the substrate for the MutH endonuclease which cleaves immediately 5' to the G in an unmethylated GATC sequence to initiate the excision of the incorrect region (10). As expected, DNA in which all the GATC sequences are symmetrically methylated, is not cleaved by MutH *in vitro*. The distribution of GATC signal sequences within *E. coli* DNA is not random, but they are never more than 2000 bases apart (11). Thus, effective surveillance of the whole bacterial genome by the mismatch repair system requires that the excision tract is long (it can be 1000 bases or more). The effectiveness is increased by the bidirectional capacity of mismatch repair. MutH can incise unmethylated GATC sites on either the 5' or the 3' side of the mismatch (see Fig. 4.1). *In vitro* experiments suggest that incision generally occurs at the GATC sequence closest to the mismatch. Excision of the mismatched DNA tract involves the participation of a DNA helicase and, depending on the site of incision, an exo-nuclease with either 5'–3' or 3'–5' polarity. The UvrD protein (DNA helicase II) provides the helicase activity. RecJ exonuclease or exonuclease VII can degrade the displaced strand in a 5'–3' direction. Exonuclease I can perform digestion with the reverse polarity. Resynthesis by DNA polymerase III holoenzyme aided by single-strand binding protein is then followed by DNA ligation to restore the uninterrupted and correct DNA sequence (4).

The use of adenine methylation as a strand signal requires that the mismatch-bound MutS/MutL complex interacts with both MutH and with distant GATC sequences. A plausible model for how this might be achieved has been proposed (12) and is illustrated in Fig. 4.1. The model is based largely on electron microscopic analysis of mismatch repair intermediates formed from purified proteins *in vitro*. MutS binds to linear heteroduplex DNA molecules which contain a single mismatch. In the presence of ATP, the linear molecules adopt a looped configuration in which the MutS dimer remains at the base of the loop but the mismatch has been displaced into the loop. MutL protein, although not absolutely required, accelerates the formation of loops and remains associated with MutS at their base. Loop formation depends on ATP hydrolysis and occurs rapidly enough to represent an intermediate in correction. It is proposed that, in the presence of ATP, the affinity of the MutS/MutL complex for the mismatch is reduced and that ATP hydrolysis then facilitates the movement of flanking DNA on both sides towards the bound complex. This has the effect of feeding flanking DNA through the immobile complex while displacing the original binding site (the mismatch) into a loop. Thus, a hemimethylated GATC sequence, either 3' and 5' to the mismatch, will eventually be drawn through the anchored MutS/MutL complex and will be suitably positioned for cleavage by MutH. Excision of the mismatched section of DNA can then begin at the loop base and will terminate at a point within the loop.

## 3. Human mismatch repair

### 3.1 Identification of human mismatch repair genes

Human mismatch repair genes were identified through a direct connection with human cancer. Genetic linkage analysis of HNPCC pedigrees identified loci on chromosomes 2 (13) and 3 (14), each of which predisposed to early-onset colorectal cancer. Tumours which arose in families with either chromosome 2 or chromosome 3 defects were found to have multiple mutations in simple repeat sequences, known as microsatellites, composed of dinucleotide repeats (for example $(CA)_n$) (13, 15, 16). At around the same time, a survey of randomly selected colorectal carcinomas un-covered a similar pattern of multiple alterations in reiterated sequences in around 12% of cases (17). Thus, most HNPCC tumours, and a significant fraction of apparently sporadic colorectal malignancies, accumulate somatic mutations in microsatellites. This phenotype is now known as RER[+] (for *R*eplication *ER*ror) or MI (for *M*icrosatellite *I*nstability). In parallel studies, screening a human cDNA library

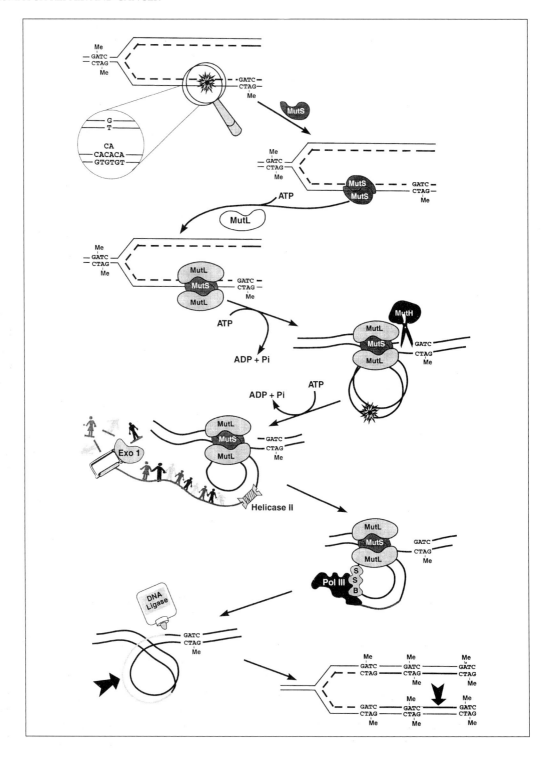

**Fig. 4.1** Postreplicative mismatch repair in *Escherichia coli*. Newly synthesized DNA is scrutinized and mispairs or short unpaired regions are bound by MutS dimers. MutL dimers are recruited in an ATP-dependent reaction. ATP hydrolysis then facilitates extrusion of the mismatched segment into a loop, allowing MutH to induce cleavage of the daughter strand at a hemimethylated GATC sequence. Displacement and destruction of the unmethylated DNA strand bounded by the MutH-induced nick and a site a few nucleotides beyond the mismatch is then carried out by co-ordinated efforts of DNA helicase II and an appropriate exonuclease. DNA polymerase III holoenzyme together with single-strand binding protein (SSB) perform resynthesis of the excised sequences, and repair is completed by DNA ligase-mediated sealing of the single-strand nick. Postrepair methylation by the Dam methylase converts the corrected site to a fully methylated configuration. The position of the previously mismatched sector is indicated ( ↓ ).

with degenerate polymerase chain reaction (PCR) probes, based on regions conserved among the known members of the bacterial and yeast MutS family of proteins, identified a human homologue which was designated *hMSH2* (18). Mutations in *hMSH2* were found in two HNPCC families and the region on chromosome 2p16 to which the HNPCC locus had been mapped was shown to contain *hMSH2* (19). The HNPCC locus on chromosome 3p21–23 was later shown to encode a homologue of the *E. coli* MutL protein which was designated hMLH1 (20, 21). The majority of known familial HNPCC mutations are in either *hMSH2* or *hMLH1* (22). Similar approaches identified two additional MutL homologues, *hPMS2* on chromosome 7p22 and *hPMS1* on chromosome 2q31–33, both of which appear to be mutated much less frequently in HNPCC (23). No biochemical role for hPMS1 has so far been uncovered. Using biochemical approaches, two additional MutS homologues, *hMSH6* (on chromosome 2p16) and *hMSH3* (on chromosome 5q11–13), have been implicated in mismatch repair (24, 25). Although mutations in these genes are found in cell lines established from apparently sporadic colorectal tumours (26, 27), familial HNPCC mutations are apparently rare. No mutations in *hMSH3* have been discovered and there is only a single report of a germline mutation in *hMSH6*, which was found in a Japanese HNPCC family (28).

Colorectal carcinoma is a common disease and mismatch repair-defective colorectal tumours are likely to be numerous. More than 90% of tumours which develop in HNPCC individuals and an estimated 5–15% of apparently non-familial colorectal carcinomas exhibit MI.

## 3.2 Microsatellite instability and mutator effects in repair-defective human cells

Microsatellite sequences are inherently unstable and this accounts for their high degree of polymorphism in the human population. Within somatic cells, however, they are normally stable and this makes them invaluable markers for genetic linkage analysis. Tumours of the MI phenotype have lost the normal somatic stability. An important contribution to understanding the reason for this loss was the observation that repeated sequences exhibited a similar instability in mismatch repair-defective yeast strains (29). The effect was independent of the proofreading function of DNA polymerase and of recombination. Thus strand slippage during replication, rather than persistent misincorporation or unequal crossing over during recombination, was the likely source of variations in repeat number. It was recognized many years ago that DNA sequences composed entirely of simple repeats, such as microsatellites, might be particularly susceptible to misalignment during replication. Because of the sequence reiteration, elongation of the mispaired intermediates is not hampered and loss or gain of single or multiple repeat units is favoured. Within coding sequences, an increase or decrease by one or two bases within a repeated region generates a frameshift mutation. An outline of the process, originally proposed by Streisinger *et al.* (30) to explain the distribution of frameshifts in bacteriophage T4, is shown in Fig. 4.2(a). Over many generations of tumour growth, these simple mutations accumulate (Fig. 4.2(b)). The requirement for the *E. coli* MutS protein to repair loops of one or two unpaired bases was fully consistent with a role for postreplicative mismatch repair in preventing the fixation of frameshifts (31). Thus, in the MI phenotype, alterations in the numbers of the mono-and dinucleotide repeats of microsatellites simply reflects an accumulation of frameshift mutations. The MI phenotype is also observed in cell lines established from tumours (e.g. ref. 32). In this case, the mutational process can be observed as discrete alterations in the numbers of repeat units in subclones derived from the original population (Fig. 4.2(c)). The rate of microsatellite mutation can be estimated from the frequency of the changes and the number of cell divisions.

The accumulation of frameshifts, together with persistent misincorporated bases, contributes to the dramatic increase in spontaneous mutation rates observed in mismatch repair-defective tumour cells. Repair defects can increase the rate of mutation at the hypoxanthine guanine phosphoribosyltransferase (*HPRT*) locus by

**Fig. 4.2** Frameshifts and microsatellite instability. (a)The generation of frameshifts in repeated sequences (30). During replication of reiterated sequences, transient dissociation of template and daughter DNA strands is followed by realignment out of phase. This generates looped regions containing one or more repeat units. If the loop occurs in the template strand (left) the mutation will ultimately be fixed as a – frameshift (–2 in the example shown). Loops in the daughter strand will lead to a +2 frameshift. (b) Microsatellite instability in tumour material. Microsatellite sequences in DNA from the tumour and from normal tissue of the same individual are amplified by a polymerase chain reaction (PCR) using specific primers that flank the microsatellite. The PCR is often carried out with either a radiolabelled primer or deoxynucleoside triphosphate. The PCR products are separated by gel electrophoresis and analysed by autoradiography. In the schematic patterns presented, the predominant alleles are represented by the thicker bands. In the pattern shown in the left-hand panel, the normal tissue is heterozygous for this marker. The longer allele of the same microsatellite is mutated in the tumour material. The individual depicted in the right-hand panel is homozygous for this particular allele. Reduction in the length of one or both copies has occurred in the tumour tissue. Multiple bands seen in both tumour and normal sequences partly reflect slippage by DNA polymerase during the amplification reaction *in vitro*. In the case of the tumour material, extra bands may also be present because the tumours may be imperfectly clonal. (c) Microsatellite instability in cultured tumour cell lines. Single cell clones derived from the mutator population are expanded to about $10^6$ cells (approximately 20 generations) during which time microsatellite mutations occur. Subclones of this population are then analysed for differences in microsatellite length by PCR and gel electrophoresis, as outlined above. In the schematic example shown, parental alleles are shown in grey, mutations in black. Two subclones have retained the parental (WT) pattern, two have mutations in one allele, and the fifth has a mutation in both alleles. This analysis can be adapted to automatic sequencing technology by the use of fluorescently tagged primers.

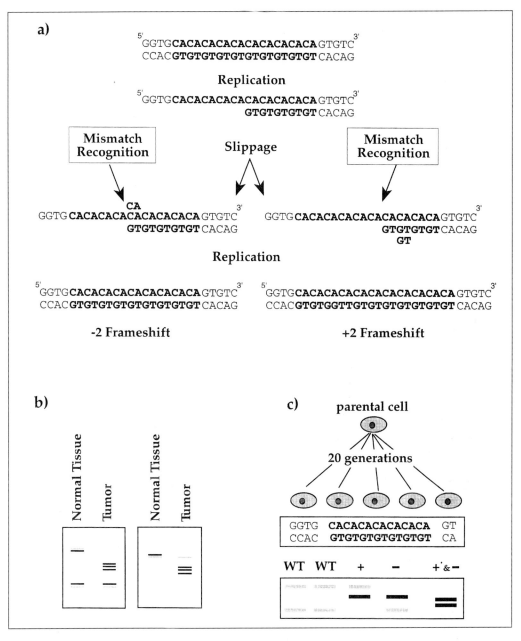

two, or even three, orders of magnitude. In Table 4.1, we summarize the mutator effects in a number of established human tumour cell lines. Although there is a clear consensus for a mutator phenotype, considerable variation exists in estimates of its extent even for a single cell line. Part of the variation in estimating the fold-increase in mutation rate can be accounted for by the uncertainties in measurements of the low rate of spontaneous mutations in repair-proficient cells. Minor technical

**Table 4.1** Mutation rate at the *HPRT* locus in MMR-proficient and -deficient cell lines

| Cell line | Rate | References |
|---|---|---|
| *MMR-proficient cell lines* | $\times 10^{-8}$ | |
| MRC-5 | 4.1 | 33 |
| SW480 | 6.0, 7.5 | 142, 38 |
| HT29 | 5.0 | 105 |
| HL-60 | 4.7 | 143 |
| NHF-1 | <4.7 | 38 |
| SW837 | <1.3 | 142 |
| VACO489 | <0.9 | 142 |
| VACO576 | <0.6 | 142 |
| | | |
| *MMR-deficient cell lines* | Rate $\times 10^{-5}$ | |
| DLD1 (*hMSH6⁻/polẟ⁺/⁻*) | 1.9, 1.5, 0.3, 0.03 | 33, 38, 105, 35 |
| HCT116 (*hMLH1⁻*) | 1.5, 0.9, 0.7, 0.7, 0.1 | 33, 142, 105, 38, 35 |
| AN3CA (*hMSH6/hMLH1⁻*) | 0.4 | 38 |
| SW48 (*hMSH6/hMLH1⁻*) | 0.03 | 105 |
| LoVo (*hMSH2⁻*) | 0.3, 0.1, 0.1 | 105, 35, 143 |
| HEC1A (*hPMS2⁻*) | 1.8, 3.1 | 133, 38 |
| DU145 (*hMLH1/hMSH3⁻*) | 5.6 | 142 |
| RKO | 0.72 | 142 |
| VACO457 | 1.0 | 142 |
| VACO432 | 0.16 | 142 |
| LS180 | 0.06 | 35 |

interlaboratory variations may also play a part, although it should be emphasized that, despite their common source, two separate cultures of a cell line with a mutator phenotype will not be identical. Indeed, populations of any mutator cell line are likely to be highly heterogeneous.

Analysis of the spontaneous *HPRT* mutations in repair-defective cell lines can provide clues to the roles of different proteins in mismatch repair. Two examples are shown in Fig. 4.3. DLD-1 and HCT116 are colorectal carcinoma cell lines which are defective in hMSH6 and hMLH1 respectively. Both have extremely high rates of spontaneous mutation to *HPRT⁻* (increased around 400-fold) (33), indicating that both these proteins participate in the repair of significant premutagenic substrates. The rates of base substitutions (transitions plus transversions) and frameshifts are increased to similar extents in DLD-1 and HCT116, whereas the rates of large deletions are not significantly affected (34, 35). Apparently, hMSH6 and hMLH1 are not involved in the prevention of deletions. Transitions and transversions are affected differently by *hMSH6* and *hMLH1* defects. The increase in the rate of transversions is more marked in DLD-1 than in HCT116, whereas transitions are affected to similar extents. The effect of the *hMLH1* defect on frameshifts is much more dramatic and this class of mutation dominates the spectrum of HCT116. The *hMLH1* defect of HCT116 increases the rate of frameshifts by more than 1000-fold, whereas the increase in DLD-1 is more modest (200-fold) (Fig. 4.4). A similar dominance of frameshifts is observed in the spectrum of mutations in the hMSH2-defective LoVo cell line (35). Supplying a normal copy of the defective repair gene, by transfer of chromosome 2 into DLD-1 (36) or chromosome 3 into HCT116 (37, 38), corrects the repair defect and reduces the mutator effect by more than 90%. This confirms that most of the mutator effect is a direct consequence of their defective mismatch repair functions rather than unrecognized secondary mutations. However, a possible minor contribution of a second defect, for example the known DNA polymerase δ mutation in DLD-1 (39), cannot be excluded and might account for the higher frequency of transversions in the spectrum. In terms of the relative effect on the different types of mutation, the effect of the *hMLH1* mutation in HCT116 is qualitatively similar to that of loss, by a *mutS*, *mutL*, or *mutH* mutation, of the *E. coli* long-patch mismatch repair pathway which principally corrects transition and frameshift mutations (40, 41) (Fig. 4.4). Transversions are corrected mainly by the $dnaQ^+$-encoded ε subunit of the DNA polymerase III holoenzyme. The outstanding feature of the spontaneous mutational spectra in established colorectal and other tumour cell lines with mismatch repair defects and microsatellite instability is the highly significant increase in the rate of frameshift mutation.

Tumours are generally thought to develop by accumulating mutations in genes

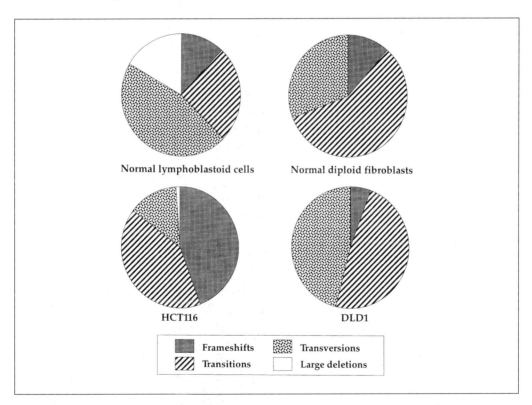

**Fig. 4.3** Mutational spectra in normal and mismatch repair-defective cell lines. Spontaneous mutations in the *HPRT* gene revealed by direct sequencing.

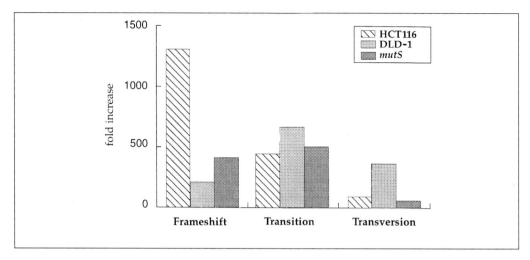

**Fig. 4.4** Increases in the rates of mutation by mutational class. The values for the *MutS* strain of *E. coli* are taken from ref. 40. It should be noted that 70% of frameshifts in the *MutS* strain were found at a single hotspot.

that control cell proliferation or death. Since different tissues respond to different signals for cell proliferation and death, each tissue will have its own set of critical targets relevant to neoplastic transformation. The clear predominance of frameshifts in mismatch repair-defective cells has implications for the development of repair-defective tumours. In particular:

- Frameshifts almost invariably result in a loss of protein function, and are therefore more likely to be associated with inactivation of tumour suppressor genes than the activation of oncogenes.

- Because frameshifts occur in repeated elements, they will selectively accelerate the inactivation of those genes that contain repeats.

- Only those mutations that are relevant to the transformation of a particular tissue will be selected.

Examples of these effects are provided by the insulin-dependent growth factor receptor II gene (IGF-II) (42) and the transforming growth factor β (TGF-β) receptor II gene. TGF-β is an important negative regulator of the growth of many types of epithelial cell, including colon cells, and it induces apoptosis of gastrointestinal tract cells. Colon cells gain a selective growth advantage from the loss of the TGF-β type II receptor, and a loss of responsiveness to TGF-β by colorectal carcinomas is a common clinical observation. The TGF-β receptor type II gene contains an uninterrupted $A_{10}$ sequence which is frequently inactivated by frameshifts in MI colorectal and gastric carcinomas (43). Significantly, normal endometrial tissue does not respond to TGF-β and mutation of the TGF-β type II receptor gene is not found in MI endometrial tumours (44).

Loss of the BAX protein provides another example of a selective growth advantage provided by an inactivating frameshift. BAX acts as a negative regulator of cell growth by providing a dominant signal for apoptosis that can be suppressed by interaction with BCL-2 or Bcl-X$_L$. The *BAX* gene contains a G$_8$ sequence which is frequently mutated by a single G addition or deletion in colorectal (45) and haematopoietic (46) tumours which display a MI phenotype. BAX-mediated apoptosis is important in these cell types and so the absence of this particular pathway of death is likely to confer a significant growth advantage.

Other examples of frameshifts that are likely to confer either direct or indirect proliferative advantages and are particularly prevalent in mismatch repair-deficient tumours are listed in Fig. 4.5. Most repair-defective tumour cell lines have mutations in more than one repair gene. The A$_8$ and C$_8$ tracts in the coding regions of the mismatch repair genes *hMSH3* and *hMSH6* are apparently hot-spots for frameshift mutations in *hMLH1* and *hMSH2* defective tumours (47, 48). This implies that the loss of each repair protein can confer a selective growth advantage—possibly each contributes separately to the overall mutator phenotype. If this is the case, it would indicate additional complexity in the long-patch mismatch repair pathway which is not, so far, apparent from biochemical studies (see below).

A further mechanism that might influence tumour growth is the evasion of immunosurveillance. Since mutated proteins are a source of altered peptides, mismatch repair-defective cells might therefore be particularly vulnerable to attack by cytotoxic T lymphocytes. Loss of the ability to present peptides to cytotoxic T

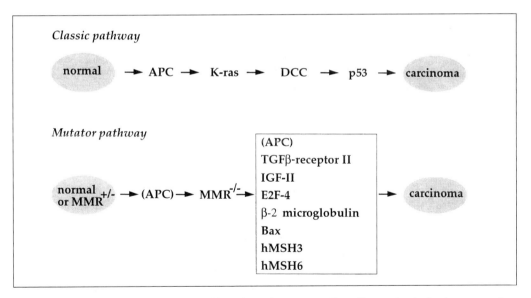

**Fig. 4.5** Pathways of tumour development. The schematic representation of how a developing tumour acquires the necessary five or more mutations. The upper scheme, adapted from ref. 141, is presented for comparison. The postulated mutator pathway takes into account the likelihood that loss of mismatch repair functions (MMR) is an early step in the process, and that MMR defects are likely to have a more pronounced effect on those target genes which contain reiterated regions in expressed (or otherwise essential) sequences. Some known examples are shown boxed.

lymphocytes would confer a selective advantage in the face of an antitumour immune response. Frameshift mutations in the $\beta_2$-microglobulin gene, which is essential for the presentation of peptides to cytotoxic T lymphocytes, are relatively common in MI sporadic colorectal carcinomas (49, 50). This suggests an indirect mechanism by which an increased rate of frameshift mutation can influence tumour cell survival.

The concept that the sequential accumulation of several key mutations is required for the development of neoplasia is well established. Because of the bias towards frameshifts, the type and the order of genomic alterations occurring in the mismatch repair-defective forms of colorectal cancer may differ from those of the repair-proficient cancers. Indeed, there may be two separate pathways by which key mutations can accumulate during tumour development (Fig. 4.5). *APC* mutations, which are so widely prevalent in all human colorectal carcinomas (51, 52), would reflect an early step which is common to both pathways. In contrast, mutations in *p53* or in *Ha-ras*, which are frequent among sporadic colorectal tumours (53), are rare in colorectal carcinomas with MI (17, 54). The two transformation pathways may therefore be alternative. The mutator phenotype in MI tumours may affect a subset of growth control genes, which are normally unaffected by the widespread genome instability that results in the aneuploidy and gross chromosomal rearrangements present in the majority of sporadic cancers. In agreement with this possibility, MI tumours are characteristically close to diploid (15, 55).

## 3.3 The biochemistry of the human mismatch repair pathway

Many features of E. coli long-patch mismatch repair are conserved in human cells. The human pathway is bidirectional (56) and is initiated by a dimeric mismatch recognition complex. The main differences appear to be that the human long-patch repair pathway(s) have some built-in redundancy, employ protein heterodimers rather than homodimers, and do not appear to rely on signal DNA sequences to identify the strand for correction. The current ideas about the initial steps of mismatch repair in human cells are presented in Fig. 4.6.

Human cells encode at least three proteins that participate in mismatch recognition: hMSH2 (105 kDa), hMSH3 (127 kDa), and hMSH6 (160 kDa) are all homologues of the *E. coli* MutS protein. The extent of homology is particularly striking within the ATP binding motif, indicating that ATP utilization is also a key function of human mismatch recognition and repair. The hMSH2 and hMSH6 proteins are the most closely homologous and their genes are syntenic on chromosome 2, which probably reflects an ancient gene duplication event. Although each is homologous to MutS, none of the three human proteins is a significant mismatch recognition factor on its own. Instead, human mismatch repair appears to be initiated by two mismatch binding heterodimers, hMutSα and hMutSβ, which comprise hMSH2:hMSH6 (24, 57) and hMSH2:hMSH3 (25), respectively. Between them, hMutSα and hMutSβ cover the same range of mismatches, single base mispairs, and small loops of 1–4 bases caused by DNA polymerase slippage, as the *E. coli* MutS protein. Single base

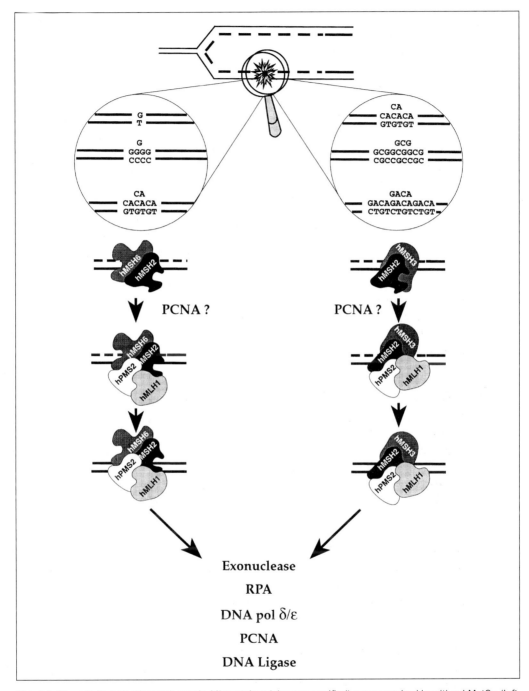

**Fig. 4.6** Steps in human mismatch repair. Mismatches (shown magnified) are recognized by either hMutSα (left-hand fork) or hMutSβ (right-hand fork), which bind to DNA in the region of the mismatch. Subsequent recruitment of hMutLα may require PCNA in an undefined role. After removal of the mismatched DNA section, resynthesis requires DNA polymerase δ (or ε) and PCNA as a processivity factor.

mispairs are apparently recognized by hMutSα which also preferentially binds to DNA containing single base loops. The preferred substrates of hMutSβ appear to be the larger loops of 3–4 unpaired bases. Both complexes are able to initiate the repair of two base loops, indicating a degree of redundancy in correction. The scheme in Fig. 4.6 is based on biochemical data which strongly suggest that hMSH2 is essential for long-patch mismatch repair. The role of ATP in the human pathway is not yet fully elucidated. It has been known for some time that the addition of ATP to the hMutSα:mismatched DNA complex promotes dissociation of hMutSα (57, 58). It was thought likely that, by analogy to *E. coli* MutS protein, this reflected ATP-dependent movement of hMutSα relative to the mismatch until it fell off the end of the short duplex DNA substrates. An alternative mechanism has been proposed in which the dissociation of hMutSα from mismatched DNA is promoted by the exchange of a bound ADP molecule for ATP (59). ATP binding, but not hydrolysis, is required for dissociation. This mismatch recognition switching cycle, differs fundamentally from the proposed action of *E. coli* MutS, in that the free energy of ATP hydrolysis is not used to bring about relative movement of the recognition complex and the mismatch. This is not the only suggested mechanism for hMutSα function, however. An alternative model in which ATP hydrolysis does play a part in complex movement and in which the action of hMutSα is more similar to that proposed for *E. coli* MutS has been advanced by Jiricny and co-workers (60).

The secondary recognition step, analogous to that carried out by the MutL homo-dimer in *E. coli*, is performed by a third heterodimer, hMutLα (61). The components of this complex, hMLH1 and hPMS2, are both homologues of MutL. Just as the precise biochemical function of MutL remains undefined, it is still unclear exactly what hMutLα does. The properties of the analogous complexes in *Saccharomyces cerevisiae* suggest that hMutLα, although itself devoid of mismatch binding activity, might enhance the stability of the hMutSβ or hMutSα:DNA complexes (62). Several genes apparently related to, and syntenic with, *hPMS2* are expressed (23, 63), but their functions remain undefined (64). Experiments using extracts of cells known to be defective in either hMLH1 or hPMS2 suggest that both proteins are essential for long-patch mismatch correction (61, 65). Thus, although there is some apparent redundancy at the mismatch recognition step of long-patch mismatch correction, there seems to be an absolute requirement for the hMutLα complex.

Following excision of the mismatched strand, resynthesis is probably carried out by DNA polymerase δ (66), one of the DNA replication enzymes. DNA polymerase δ uses proliferative cell nuclear antigen (PCNA) as a clamp to increase its processivity. PCNA is also required for nucleotide excision repair (67) and one of the pathways of base excision repair (68), and it is perhaps not surprising that it is also involved in mismatch correction. Interestingly, *in vitro* mismatch correction assays by human cell extracts indicate that PCNA interacts directly with the hMutLα complex, most probably at a step preceding DNA polymerization (69). Since a role in DNA polymerase processivity is not precluded by this interaction, it is possible that PCNA has two distinct functions in human mismatch repair. The human single-strand binding protein, replication protein A (RPA)—which is required for both DNA replication (70) and nucleotide excision repair—(71) is also required for mismatch repair, probably at the stage of DNA repair synthesis (72). At present, it is not known which of the several human DNA ligases catalyses the final ligation of the replaced strand to complete repair.

Human cells do not express a DNA adenine methylase analogous to the *E. coli* Dam protein, and human DNA does not contain 6-methyladenine. The major post-synthetic modification of human DNA is methylation of the 5-position of cytosine in CpG sequences. 5-Methylcytosine has been considered as a possible strand discrimination signal in human mismatch repair (73), but it is presently thought that the signal is most likely to be a free single-strand DNA end. Indeed, single-strand nicks provide full-strand discrimination for correction by human cell extracts. (They can also replace the MutH-mediated incision in the reconstituted *E. coli* repair system and persistent single-strand DNA interruptions can serve as strand signals in intact *E. coli*). The use of free DNA ends as the strand discrimination signal suggests that mismatch repair may be adapted to the discontinuous mode of DNA replication, and that the transient presence of Okazaki fragments denotes the newly replicated DNA strand. The invading strand of the heteroduplex presumably provides the signalling end during recombination.

## 3.4   An alternative mismatch repair pathway

The structure-specific nuclease FEN-1 is involved in processing Okazaki fragments (74, 75). It competes with the p53-inducible, cell-cycle regulatory protein p21[waf-1/cip-1] for binding to PCNA (76, 77). FEN-1 links cell-cycle control in response to DNA damage and the long-patch mismatch repair pathway to a likely alternative repair pathway which processes an important subset of mismatches. The *S. cerevisiae* homologue of FEN-1 is encoded by the *RAD27* gene (78, 79). *RAD27* mutants have a strong spontaneous mutator phenotype dominated by duplication mutations in which the duplicated sequences are flanked by short (3–12 base) direct repeats (80). Both dinucleotide and trinucleotide repeats are also somewhat unstable in *RAD27* mutants in which they undergo mainly expansion (81). This effect is independent of MSH2 and MLH1, implying that the absence of FEN-1, which presumably leads to incorrect processing of Okazaki fragments, can produce a mutator phenotype even in the presence of a functional long-patch mismatch excision pathway. The *RAD27* mutator phenotype can affect the di- and trinucleotide repeats, which are also guarded by long-patch mismatch repair. The substrates for the *RAD27* correction pathway, which is most likely conserved in human cells, are probably recombination intermediates or replication errors of a different type to the DNA polymerase slippage errors that are corrected by the long-patch pathway. Indeed, loss of expression of *hMSH3* is associated with trinucleotide repeat instability (27), but this appears to be a fundamentally different process to trinucleotide expansion which is not seen in MSH3-defective yeast cells (82). The long-patch mismatch repair and *RAD27* pathways appear to protect trinucleotide (and to some extent dinucleotide) repeat sequences in a complementary rather than a redundant fashion. Trinucleotide repeat tract expansion is associated with a number of human genetic disorders, including the fragile X and myotonic dystrophy syndromes, which are not explicable by defects in long-patch mismatch repair (83). Reduced efficiency of the FEN-1 dependent, duplication avoidance pathway is perhaps a more plausible alternative since large expansions of CTG repeats were seen in yeast *RAD27* mutants (81).

## 4. Mismatch repair defects in mouse models

Mice homozygous for targeted inactivations of the *MSH2, MLH1, PMS2, PMS1,* or *MSH6* genes have been constructed (84–90). Their embryonic and immediately post-natal development is apparently normal, indicating that high rates of somatic mutation are not incompatible with development. This also seems to be true of humans. Individuals from rare HNPCC families, where germline *hPMS2* or *hMLH1* mutations appear to have dominant negative effects, do not seem to be particularly disadvantaged when compared to those from families in which the mutations are recessive (91). The most striking feature of the mismatch repair nullizygous mice is infertility. The first division of meiosis is arrested in $MLH1^{-/-}$ males and they produce neither spermatids nor spermatozoa. $MLH1^{-/-}$ females produce oocytes that fail to develop beyond the single cell stage after fertilization (86, 87). Sterility in $PMS2^{-/-}$ mice is confined to males (88). Meiotic abnormalities in $PMS2^{-/-}$ mice affect a later stage than those of $MLH1^{-/-}$ animals. Although the spermatocytes complete meiosis and mature, abnormal spermatozoa are produced. Thus, in both $MLH1^{-/-}$ and $PMS2^{-/-}$ mice, mismatch repair defects affect chromosome synapsis and separation during meiosis—compatible with a recombinational defect in meiotic germ cells. Both the MLH1 and PMS2 proteins participate in meiosis; however, the different meiotic phenotypes suggest that the two MutL homologues do not function in concert but may have alternative partners.

In stark contrast, both male and female $MSH2^{-/-}$ and $MSH6^{-/-}$ mice are fully fertile (84, 90). It is clear that the current outlines of the long-patch mismatch repair in a mitotic cell cycle (Fig. 4.6) do not provide a simple explanation for the effects of mismatch repair defects on meiosis. The most probable explanation is that there are meiosis-specific mismatch recognition factor(s). Two additional MutS homologues, *MSH4* and *MSH5*, which act in the same pathway to facilitate cross-overs of homologous chromosomes at meiosis I, are specifically produced in meiotic yeast cells (92, 93). The human homologue of one of them, *hMSH4*, located on chromosome 1p31, and expressed only in testis and ovary (94), is a good candidate for the meiosis-specific MutS function(s).

Increased mutation rates are also observed in the mismatch repair knock-out animals. $MSH2^{-/-}$ and $MLH1^{-/-}$ embryonic stem cells, normal tissues of $PMS2^{-/-}$ and $MLH1^{-/-}$ animals, tumours arising in $MSH2^{-/-}$, $MLH1^{-/-}$, $PMS2^{-/-}$ mice, as well as $MLH1^{-/-}$ and $PMS2^{-/-}$ spermatozoa all display MI. Instability in $PMS1^{-/-}$ mice is less marked and is confined to $(A)_n$ repeat tracts. *E. coli* transgenes in $MSH2^{-/-}$ and $PMS2^{-/-}$ mouse tissues undergo somatic mutation at increased rates (95, 96). Mutation frequencies are elevated between 5- and 15-fold in the $MSH2^{-/-}$ tissues tested including small intestine, thymus, and brain (Fig. 4.7). Similar increases in mutation rate at the *HPRT* locus or to ouabain resistance are apparent in $MSH2^{-/-}$, embryonic fibroblast cell lines (97). From 30- to 100-fold increases in mutation frequency are observed in a *supF* transgene in $PMS2^{-/-}$ mouse tissues (Fig. 4.7), although the extent is somewhat dependent on the target sequence and, in particular, on the presence of extended G repeats in the target gene (95). The rate of mutation to ouabain resistance is increased about 10-fold in $MLH1^{-/-}$ and $PMS2^{-/-}$ mouse embryonic fibroblast cell lines, but is not significantly altered in $PMS1^{-/-}$ lines (89). Although the data are more limited, the types of mutation that occur in excess in repair-defective somatic mouse cells are generally the same as those in defective human colorectal carcinoma cells. In particular, transitions and frameshifts are the predominant classes in both $MSH2^{-/-}$ and $PMS2^{-/-}$ mouse tissues. In agreement with similar data from human cells,

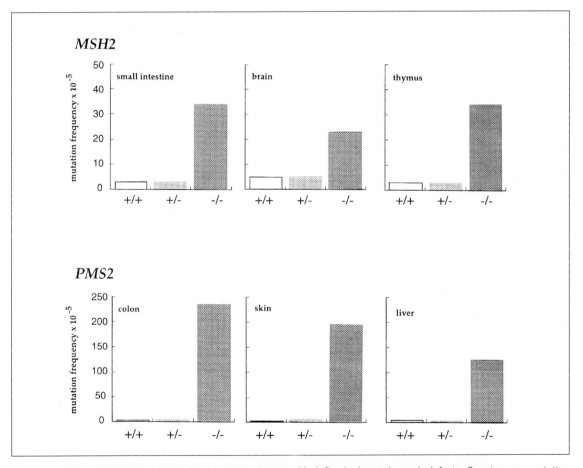

**Fig. 4.7** Mutation frequencies in mouse tissues with defined mismatch repair defects. Spontaneous mutation frequencies were determined in tissues of *lacI* (96) or *supF* (95) transgenic mismatch repair knock-out animals.

although nullizygosity increases the mutation rate between 5- and 100-fold, there is no detectable mutator effect in cells from heterozygous animals.

## 4.1 Mice as models for HNPCC

Lack of mismatch repair is incompatible with a normal lifespan, and homozygous $MSH2^{-/-}$, $MLH1^{-/-}$, $PMS2^{-/-}$, or $MSH6^{-/-}$ mice are all tumour-prone. Most $MSH2^{-/-}$, $MLH1^{-/-}$, and $PMS2^{-/-}$ animals develop thymic lymphoma. Surviving $MSH2^{-/-}$ and $MLH1^{-/-}$ mice—but not $PMS2^{-/-}$ animals—develop intestinal epithelial neoplasms, endometrial or skin tumours between 6 and 12 months of age (89, 98, 99). The intestinal adenomas and carcinomas are clustered in the small intestine, (duodenum, ileum, or jejenum) and not in the colorectum as in HNPCC patients. The skin tumours resemble those associated with the HNPCC variant Muir–Torre syndrome. The median survival of $MSH6^{-/-}$ mice is slightly longer than that of $MSH2^{-/-}$ mice (10 months versus 5–6 months) and the animals develop a wider spectrum of tumours—with not only a predominance of B and T lymphomas and gastrointestinal tumours, but also tumours in other tissues including those of the liver, lung, skin, and soft tissues (90). In agreement with the absence of any defined role for PMS1 in mismatch repair, $PMS1^{-/-}$ animals are not tumour-prone (89). Loss of mismatch repair is therefore generally associated with an elevated risk of tumour development. The differences in the tumour spectra of $MLH1^{-/-}$ and $PMS2^{-/-}$ mice reinforces the impression, from their effects on meiosis, that these two gene products, in addition to acting as a heterodimer, may have additional independent functions.

The vast majority of HNPCC mutations are recessive and there are no known cases with homozygous germline mismatch repair mutations. Dominant negative mutations are probably extremely rare. Mice heterozygous for mismatch repair mutations should therefore represent the most appropriate model for HNPCC. There is some increase in the frequency of malignancies in heterozygous $MSH2^{+/-}$ and $MSH6^{+/-}$ mice, although these are not normally the cause of death (90, 99). The tumours which do develop in $MSH2^{+/-}$ affect multiple sites, but not the colorectum, do not show MI, and retain the functional mismatch repair allele. The heterozygous mismatch repair knock-out mice differ fundamentally from heterozygous HNPCC individuals, in that the loss of mismatch repair is not a significant factor for their tumour development. In this respect, mice provide only a partial model for HNPCC.

## 5. Mismatch repair defects and susceptibility to therapeutic agents

Mismatch repair also contributes to the cytotoxic effects of therapeutic DNA damaging agents (100). The methylating agents temozolomide and Dacarbazine, which can be effective in the treatment of melanoma and brain tumours, are analogues of N-methyl-N′-nitro-N-nitrosoguanidine (MNNG) and N-methyl-N-nitrosourea (MNU). All these agents kill human cells because they methylate the $O^6$

atom of guanine in DNA. Persistent $O^6$-methylguanine ($O^6$-meGua) is a cytotoxic DNA lesion and its toxicity is dependent on a functional mismatch repair pathway. Human cells which have been selected for resistance to MNU or MNNG have mismatch repair defects. Resistance to persistent DNA $O^6$-meGua due to loss of mismatch repair is known as methylation tolerance. Tolerant cell lines with defects in hMutSα (101, 102) and hMutLα (102–104) have been isolated in the laboratory and *hMSH2*, *hMSH6*, *hPMS2* and *hMLH1* tumour cell lines display the methylation-tolerant phenotype (36, 37, 105–107). Restoration of repair by transfer of the appropriate functional human chromosome restores methylating agent sensitivity, confirming the requirement for active mismatch repair in cell killing (36, 37, 108). Methylation-tolerant cell lines are frequently cross-resistant to an unrelated therapeutic agent, 6-thioguanine (6-TG) (for a review see ref. 109) which is incorporated into DNA via the purine salvage pathway. Its cytotoxicity probably derives from its ability to undergo a facile methylation by the intracellular methyl group donor S-adenosylmethionine, and recognition of the resulting DNA 6-methylthioguanine (6-meTG) by the hMutSα mismatch recognition complex (110). As with $O^6$-meGua, recognition by hMutSα initiates a sequence of events that results in cell death. Both 6-meTG and $O^6$-meGua share structural features which provoke miscoding during DNA replication, and this coding ambiguity provides a plausible explanation for their recognition by mismatch repair.

Although not mutually exclusive, two models have been advanced to explain the involvement of mismatch repair in cell death (Fig. 4.8) (111, 112). The first proposes that recognition by hMutSα of replicated $O^6$-meGua or 6-meTG is followed by attempted hMutLα-directed repair. Correction of the mispaired base analogues remains incomplete because 'repair' is always misdirected to the daughter strand. The 'incorrect' base ($O^6$-meGua or 6-meTG) remains in the template strand where it can trigger further futile repair attempts. The incomplete repair events delay replication fork progression (113), which, possibly by inducing double-strand DNA breaks, then triggers cell death. The alternative explanation, that recognition by the hMutSα and hMutLα mismatch repair complexes directly triggers a cell-cycle ($G_2$) checkpoint response which results in cell death, is based on the observation that tolerant cells do not arrest in $G_2/M$ after exposure to 6-TG or MNNG.

Mismatch repair may also modulate the sensitivity to other DNA damaging agents. Cisplatin, which kills cells by introducing cross-links within or between DNA strands, is slightly more cytotoxic towards repair-proficient than closely isogenic-deficient cells (114, 115). Small effects of mismatch repair deficiency on the sensitivity of cells to doxorubicin and etoposide have also been reported (114, 116), as have conflicting effects on their sensitivity to ionizing radiation (97, 117–119) and ultraviolet light (97, 118, 120). It is becoming clear, however, that DNA repair is only one of several cellular responses to DNA damage which impinge on cell survival. It is possible, even likely, that although defective mismatch repair might slightly alter the level of drug resistance, the effect may be obscured by more pronounced changes resulting from defects in unrelated pathways which influence survival. In the absence of any obvious unifying structural similarity between the types of DNA

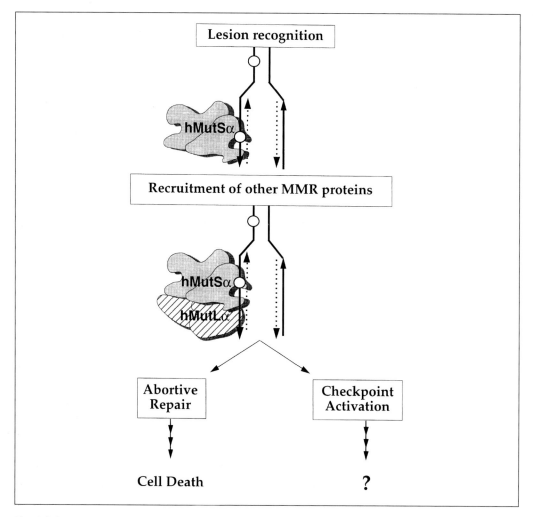

**Fig. 4.8** Possible interactions between mismatch repair and DNA damage. Recognition by hMutSα (or possibly hMutSβ) and the recruitment of hMutLα activates a checkpoint or initiates abortive attempts at excision repair. The abortive repair attempts are postulated to lead to cell death. The outcome of the checkpoint activation depends on the involvement of downstream factors, possibly including DNA repair.

lesion which appear to interact with mismatch repair, it has been suggested that mismatch repair might act as a general sensor of DNA damage and initiate stress responses (101). In view of the wide discrepancies in the extent of protection afforded by loss of mismatch repair (this can be up to 100-fold for methylating agents or 6-TG compared to, at best, 1.5-fold for cisplatin and even less for γ-irradiation) it is perhaps more probable that mismatch repair processes a significant fraction of the potentially lethal DNA lesions produced by methylation or 6TG but a rather minor fraction of those introduced by cisplatin or, possibly, ionizing radiation.

Although it potentiates cell killing by methylating and other agents, mismatch repair protects human tumour cells against the DNA interstrand cross-links (ICLs) formed by therapeutic agents such as *N*-(2-chloroethyl)-*N'*-cyclohexyl-*N*-nitrosourea (CCNU) (107). Sensitivity to CCNU is associated with *hMSH2*, *hMSH6*, and *hMLH1* mutations in a number of cell lines of laboratory or tumour origin. This suggests that mismatch repair might participate directly in the removal of potentially lethal DNA ICLs. Although the mechanism of the increased sensitivity of mismatch repair-defective cells to this important class of therapeutic drugs is not yet understood, the observations have obvious implications for the possible treatment of mismatch repair-defective tumours.

# 6. Important areas for the future

## 6.1 Mismatch repair, transcription-coupled excision repair, and recombinational repair

One intriguing area, in which many questions remain unanswered, is that concerning the interactions among different DNA repair pathways. *E. coli* mismatch repair and nucleotide excision repair (NER) factors participate together in the transcription-coupled repair (TCR) of ultraviolet (UV) light-induced DNA damage. This enhanced excision of damage from the transcribed strand of active genes, which requires the *E. coli* Mfd (mutation frequency decline) protein (121), acts on UV-induced cyclobutane pyrimidine dimers. Mutations in *mutS* and *mutL*, but not in *mutH*, also abolish TCR of UV photoproducts (122). The loss of TCR is reflected in an increased sensitivity to UV, and *E. coli mfd* mutants as well as *mutS* and *mutL* strains are all moderately sensitive to UV. TCR deficiency has been reported in human tumour cell lines bearing *hMLH1*, *hPMS2*, or *hMSH2* mutations (118, 120). Although repair of the overall genome after UV damage is apparently unaffected, TCR is much reduced. Involvement of hMSH2, but not hMLH1, in the TCR of oxidative damage has also been proposed (118). It has been suggested that, by sensing many diverse types of DNA lesion, the recognition system of mismatch correction may facilitate an improved excision repair in human cells. This rather simplistic view is inconsistent with the inability of the purified hMutSα mismatch recognition complex to recognize UV photoproducts (123). Furthermore, both global repair and TCR of photoproducts contribute to the survival of human cells after UV irradiation. Xeroderma pigmentosum group C cells, which are defective in global NER but proficient in TCR, and cells from Cockayne's syndrome A and B patients, which perform global repair but not TCR, are all UV-sensitive. This is in contrast to the absence of significant and predictable changes in UV sensitivity (118, 120) in most mismatch repair-defective human tumour cells. The precise mechanism of TCR is discussed elsewhere in this volume. At present, any involvement of mismatch repair proteins in the process remains obscure.

A better characterized example of overlap between NER and mismatch repair is provided by one of the pathways of recombinational DNA repair in yeast. Genetic evidence indicates that the mismatch repair proteins Msh2 and Msh3 are required for a pathway of double-strand break repair, which also depends on the heterodimeric Rad1/Rad10 endonuclease of NER (124) (the homologue of the human XPF/ERCC1:XPG endonuclease). The removal of non-homologous 3′ ends, formed by the invasion of a partially homologous sequence during repair either by gene conversion or by single-strand annealing, requires the Msh2 and Msh3 proteins (125). It does not appear to involve mismatch correction because Msh6, Mlh1, and Pms1 do not participate (126). The proposed role of the Msh2/Msh3 heterodimer is to recognize, bind, and stabilize the branched structure and help to recruit Rad1/Rad10. The concerted action of Msh2/Msh3 and Rad1/Rad10 is also a requirement for the repair of large loops (26 base pairs long) formed during meiotic recombination in yeast (127).

Whether the interaction between mismatch repair and NER proteins is restricted to the examples we have cited or is part of a more general pattern of collaboration in protecting cells against genetic damage is still an open question.

## 6.2 Adaptive responses and reduced mismatch repair efficiency

In evolutionary terms, spontaneous mutation rates must be finely controlled to minimize the frequency of deleterious mutations yet retain the ability to adapt to changes in the external environment. *E. coli* can adapt to conditions of extreme stress by modulating long-patch mismatch repair. The phenomenon of adaptive mutation, in which life-saving mutations arise in apparently starving or non-growing populations, is an example of such a stress (128). Mutations occur randomly in a very small subpopulation of cells which enter a transient hypermutable state. The signature mutation of these 'adaptive' mutations is the frameshift (129, 130), and it appears that the hypermutating cells modulate their mismatch repair in such a way that MutL becomes limiting (131). The transient suspension of effective mismatch repair is clearly an important adaptive response in *E. coli*. Does it have a counterpart in human cells? The acquisition of the five or more mutations required for the development of a tumour cell is inconsistent with the rates of mutation found in most human tumour cells (132). Only a tiny minority of human tumours has a sustained mutator phenotype through the loss of mismatch repair. The possibility that others might have passed through transient periods of hypermutability, perhaps during stressful periods of growth limitation, is an intriguing one and may be consistent with a report that high rates of mutation might occur selectively in certain non-dividing cells (133). The general question of what factors influence the incidence of spontaneous mutation—including the regulation of mismatch repair in both *E. coli* and human cells—is clearly important.

## 6.3 Epigenetic effects on mismatch repair genes

The cytosine of CpG sequences in human DNA is normally methylated except for the extensive CpG-rich tracts, or islands, which form part of the promoters of many genes. When these CpG islands do become methylated, the event is normally associated with inhibited transcription. Altered levels and patterns of CpG methylation are a common feature of human tumours and contribute to their heritably altered pattern of gene expression. Altered methylation patterns and epigenetic regulation of gene expression seem to be unusually frequent in mismatch repair-defective cells. The overall level of cytosine methylation is higher in HNPCC colorectal tumours than in others (134). Several genes, including the *p16* tumour-suppressor gene, appear to be at risk from hypermethylation in mismatch repair-defective tumours, which also consistently methylate, and thereby inactivate, retrovirally transduced genes at high frequency (135). There is some evidence that the mismatch repair genes themselves might be susceptible to transcriptional silencing, raising the possibility that alterations in the methylation pattern precede the loss of repair capability. The *hMLH1* gene is silenced by hypermethylation of its promoter in some sporadic colorectal tumours (136), suggesting an alternative to mutation as a means of inactivating at least one allele of this mismatch repair gene. More indirectly, for some time it has been recognized that the frequency at which methylation tolerant, mismatch repair-defective variants arise during selection with methylating agents is too high to be the result of two mutational events and is more consistent with the loss or silencing of one allele. Treatments that select for methylation-tolerant, mismatch repair-defective cells may also induce epigenetic changes which reactivate a silent MGMT gene (102). The possibility that epigenetic regulation by altered CpG methylation might offer an alternative to mutation in the silencing of mismatch repair genes, and that this might be a rather frequent event, is an interesting possibility.

## 6.4 Mismatch repair cell-cycle checkpoints and apoptosis

In addition to DNA repair, cell-cycle checkpoints provide an alternative means of protecting the genome. Arrest of the cell cycle immediately before critical phases such as DNA replication (the $G_1$–S checkpoint) or mitosis ($G_2$–M) can help to prevent the fixation of DNA damage. It has been suggested that mismatch repair proteins are part of a cell-cycle checkpoint as a component of a general cellular response to DNA damage (112, 119, 137). Exposure of human cells to 6-TG, methylating agents, or ionizing radiation induces a $G_2$–M arrest. This 'checkpoint response' is impaired in the absence of mismatch correction. One interpretation of this observation is that mismatch repair plays a central role in regulating cell-cycle progression in the event of DNA damage. The presence of a functional $G_2$–M arrest checkpoint does not correlate with the overall sensitivity to DNA damage, however. Mismatch repair-defective cells are extremely *resistant* to methylation and 6-TG damage, but exhibit no major changes in survival after ionizing radiation. A mismatch repair-mediated $G_2$–M checkpoint is clearly not a key factor in determining survival following ionizing radiation. This apparent contradiction might be resolved if mismatch repair interacts with DNA damage in more than one way. For methylation and 6-TG, it acts as a checkpoint (which would tend to enhance survival), but it also promotes cell death by processing DNA $O^6$-meGua and 6-meTG. The observation that mismatch repair is required for the formation of chromosome aberrations (138) or sister chromatid exchanges (108) in cells treated with methylating agents or 6-TG is consistent with this. In the case of ionizing radiation, and probably most other types of DNA damage, it provides the checkpoint function but does not otherwise interact with damage. In these cases, the final survival outcome is probably determined, at least partly, by the efficiency of other DNA repair pathways.

Cell death after methylation damage or 6-TG incorporation is delayed and the initial DNA damage has little immediate impact. The effects of DNA $O^6$-meGua are seen two cell cycles after treatment, so that lethally damaged cells are able to divide just once. Consistent with this, an irreversible commitment to mismatch repair-provoked apoptosis is only evident after 2–3 days (139, 140). Cell death induced by $O^6$-meGua and 6-TG is p53-independent, and mismatch repair stimulated killing can be observed in cells with either wild-type or mutant p53 (107). How DNA lesions generated by mismatch repair-mediated processing of primary DNA damage, such as $O^6$-meGua or 6-meTG, engage the apoptotic pathway is unknown. But it is part of a more general problem of how DNA damage provokes apoptosis, an understanding of which may have important implications for chemotherapy.

# References

1. Claverys, J. P. and Lacks, S. A. (1986) Heteroduplex deoxyribonucleic acid base mismatch repair in bacteria. *Microbiol. Rev.*, **50**, 133.
2. Holliday, R. (1964) A mechanism for gene conversion in fungi. *Genet. Res.*, **5**, 282.
3. Cox, E. C. (1976) Bacterial mutator genes and the control of spontaneous mutation. *Annu. Rev. Genet.*, **10**, 135.
4. Lahue, R. S., Au, K. G., and Modrich, P. (1989) DNA mismatch correction in a defined system. *Science*, **245**, 160.
5. Su, S. S. and Modrich, P. (1986) *Escherichia coli mutS*-encoded protein binds to mismatched DNA base pairs. *Proc. Natl Acad. Sci. USA*, **83**, 5057.
6. Haber, L. T. and Walker, G. C. (1991) Altering the conserved nucleotide binding motif in the *Salmonella typhimurium* MutS mismatch repair protein alters both its ATPase and mismatch binding activities. *EMBO J.*, **10**, 2707.
7. Wu, T.-H. and Marinus, M. G. (1994) Dominant negative mutator mutations in the *mutS* gene of *Escherichia coli*. *J. Bacteriol.*, **176**, 5393.
8. Grilley, M., Welsh, K. M., Su, S. S., and Modrich, P. (1989) Isolation and characterization of the *Escherichia coli* mutL gene product. *J. Biol. Chem.*, **264**, 1000.
9. Welsh, K. M., Liu, A.-L., Clark, S., and Modrich, P. (1987) Isolation and characterization of the *Escherichia coli* mutH gene product. *J. Biol. Chem.*, **262**, 15624.
10. Au, K. G., Welsh, K., and Modrich, P. (1992) Initiation of methyl-directed mismatch repair. *J. Biol. Chem.*, **267**, 12142.
11. Barras, F. and Marinus, M. G. (1997) The great GATC. *Trends Genet.*, **5**, 139.
12. Allen, D. J., Makhov, A., Grilley, M., Taylor, J., Thresher, R., Modrich, P., *et al.* (1997) MutS mediates heteroduplex loop formation by a translocation mechanism. *EMBO J.*, **16**, 4467.
13. Peltomäki, P., Aaltonen, L. A., Sistonen, P., Pylkkänen, L., Mecklin, J.-P., Järvinen, H., *et al.* (1993) Genetic mapping of a locus predisposing to human colorectal cancer. *Science*, **260**, 810.

14. Lindblom, A., Tannergård, P., Werelius, B., and Nordenskjöld, M. (1993) Genetic mapping of a second locus predisposing to hereditary non-polyposis colon cancer. *Nature Genet.*, **5**, 279.

15. Aaltonen, L. A., Peltomäki, P., Leach, F. S., Sistonen, P., Pylkkänen, L., Mecklin, J.-P., *et al.* (1993) Clues to the pathogenesis of familial colorectal cancer. *Science*, **260**, 812.

16. Thibodeau, S. N., Bren, G., and Schaid, D. (1993) Microsatellite instability in cancer of the proximal colon. *Science*, **260**, 816.

17. Ionov, Y., Peinado, M. A., Malkhosyan, S., Shibata, D., and Perucho, M. (1993) Ubiquitous somatic mutations in simple repeated sequences reveal a new mechanism for colonic carcinogenesis. *Nature*, **363**, 558.

18. Fishel, R., Lescoe, M. K., Rao, M. S. R., Copeland, N. G., Jenkins, N. A., Garber, J., *et al.* (1993) The human mutator gene homolog *MSH2* and its association with hereditary nonpolyposis colon cancer. *Cell*, **75**, 1027.

19. Leach, F. S., Nicolaides, N. C., Papadopoulos, N., Liu, B., Jen, J., Parsons, R., *et al.* (1993) Mutations of a mutS homolog in hereditary nonpolyposis colorectal cancer. *Cell*, **75**, 1215.

20. Papadopoulos, N., Nicolaides, N. C., Wei, Y.-F., Ruben, S. M., Carter, K. C., Rosen, C. A., *et al.* (1994) Mutation in a *mutL* homolog in hereditary colon cancer. *Science*, **263**, 1625.

21. Bronner, C. E., Baker, S. M., Morrison, P. T., Warren, G., Smith, L. G., Lescoe, M. K., *et al.* (1994) Mutation in the DNA mismatch repair gene homolog hMLH1 is associated with hereditary non-polyposis colon cancer. *Nature*, **368**, 258.

22. Liu, B., Parsons, R., Papadopoulos, N., Lynch, H. T., Watson, P., Jass, J. R., *et al.* (1996) Analysis of mismatch repair genes in hereditary non-polyposis colorectal cancer patients. *Nature Med.*, **2**, 169.

23. Nicolaides, N. C., Papadopoulos, N., Liu, B., Wei, Y.-F., Carter, K. C., Ruben, S. M., *et al.* (1994) Mutations of two *PMS* homologues in hereditary nonpolyposis colon cancer. *Nature*, **371**, 75.

24. Palombo, F., Gallinari, P., Iaccarino, I., Lettieri, T., Hughes, M., D'Arrigo, A., *et al.* (1995) GTBP, a 160-kilodalton protein essential for mismatch-binding activity in human cells. *Science*, **268**, 1912.

25. Palombo, F., Iaccarino, I., Nakajima, E., Ikejima, M., Shimada, T., and Jiricny, J. (1996) hMutsβ, a heterodimer of hMSH2 and hMSH3, binds to insertion/deletion loops in DNA. *Curr. Biol.*, **6**, 1181.

26. Papadopoulos, N., Nicolaides, N. C., Liu, B., Parsons, R. E., Palombo, F., D'Arrigo, A., *et al.* (1995) Mutations of *GTBP* in genetically unstable cells. *Science*, **268**, 1915.

27. Risinger, J. I., Umar, A., Boyd, J., Berchuck, A., Kunkel, T. A., and Barrett, J. C. (1996) Mutation in *MSH3* in endometrial cancer and evidence for its functional role in heteroduplex repair. *Nature Genet.*, **14**, 102.

28. Miyaki, M., Konishi, M., Tanaka, K., Kikuchi-Yanoshita, R., Muraoka, M., Yasuno, M., *et al.* (1997) Germline mutation of MSH6 as the cause of hereditary nonpolyposis colorectal cancer. *Nature Genet.*, **17**, 271.

29. Strand, M., Prolla, T. A., Liskay, R. M., and Petes, T. D. (1993) Destabilization of tracts of simple repetitive DNA in yeast by mutations affecting mismatch repair. *Nature*, **365**, 274.

30. Streisinger, G., Okada, Y., Emrich, J., Newton, J., Tsugita, A., Terzaghi, E., *et al.* (1966) Frameshift mutations and the genetic code. *Cold Spring Harbor Symp. Quant. Biol.*, **31**, 77.

31. Parker, B. O. and Marinus, M. G. (1992) Repair of heteroduplexes containing small heterologous sequences in *Escherichia coli*. *Proc. Natl Acad. Sci. USA*, **80**, 1730.

32. Parsons, R., Li, G.-M., Longley, M. J., Fang, W.-h., Papadopoulos, N., Jen, J., *et al.* (1993) Hypermutability and mismatch repair deficiency in RER⁺ tumor cells. *Cell*, **75**, 1227.

33. Bhattacharyya, N. P., Skandalis, A., Ganesh, A., Groden, J., and Meuth, M. (1994) Mutator phenotypes in human colorectal carcinoma cell lines. *Proc. Natl Acad. Sci. USA*, **91**, 6319.

34. Bhattacharyya, N. P., Ganesh, A., Phear, G., Richards, B., Skandalis, A., and Meuth, M. (1995) Molecular analysis of mutations in mutator colorectal carcinoma cell lines. *Hum. Mol. Genet.*, **4**, 2057.

35. Malkhosyan, S., McCarty, A., Sawai, H., and Perucho, M. (1996) Differences in the spectrum of spontaneous mutations in the *hprt* gene between tumor cells of the microsatellite mutator phenotype. *Mutat. Res.*, **316**, 249.

36. Umar, A., Koi, M., Risinger, J. I., Glaab, W. E., Tindall, K. R., Kolodner, R. D., *et al.* (1997) Correction of hypermutability, N-methyl-N'-nitro-N-nitrosoguanidine resistance, and defective DNA mismatch repair by introducing chromosome 2 into human tumor cells with mutations in *MSH2* and *MSH6*. *Cancer Res.*, **57**, 3949.

37. Koi, M., Umar, A., Chauhan, D. P., Cherian, S. P., Carethers, J. M., Kunkel, T. A., *et al.* (1994) Human chromosome 3 corrects mismatch repair deficiency and microsatellite instability and reduces N-methyl-N'-nitro-N-nitrosoguanidine tolerance in colon tumor cells with homozygous *hMLH1* mutation. *Cancer Res.*, **54**, 4308.

38. Glaab, W. E. and Tindall, K. R. (1997) Mutation rate at the *hprt* locus in human cancer cell lines with specific mismatch repair-gene defects. *Carcinogenesis*, **18**, 1.

39. da Costa, L. T., Liu, B., El-Deiry, W. S., Hamilton, S. R., Kinzler, K. W., Vogelstein, B., *et al.* (1995) Polymerase δ variants in RER colorectal tumours. *Nature Genet.*, **9**, 10.

40. Schaaper, R. M. and Dunn, R. L. (1987) Spectra of spontaneous mutations in *Escherichia coli* strains defective in mismatch correction: the nature of *in vivo* DNA replication errors. *Proc. Natl Acad. Sci. USA*, **84**, 6220.

41. Rewinski, C. and Marinus, M. G. (1987) Mutation spectrum in *Escherichia coli* DNA mismatch repair deficient (mutH) strain. *Nucl. Acids Res.*, **15**, 8205.

42. Souza, R. F., Appel, R., Yin, J., Wang, S., Smolinski, K. N., Abraham, J. M., *et al.* (1996) Microsatellite instability in the insulin-like growth factor II receptor gene in gastrointestinal tumours. *Nature Genet.*, **14**, 255.

43. Markowitz, S., Wang, J., Myerhoff, L., Parsons, R., Sun, L., Lutterbaugh, J., *et al.* (1995) Inactivation of the type II TGF-β receptor in colon cancer cells with microsatellite instability. *Science*, **268**, 1336.

44. Myeroff, L. L., Parsons, R., Kim, S.-J., Hedrick, L., Cho, K. R., Orth, K., *et al.* (1995) Transforming growth factor β receptor type II gene mutation common in colon and gastric but rare in endometrial cancers with microsatellite instability. *Cancer Res.*, **55**, 5545.

45. Rampino, N., Yamamoto, H., Ionov, Y., Li, Y., Sawai, H., Reed, J. C., *et al.* (1997) Somatic frameshift mutations in the *BAX* gene in colon cancers of the microsatellite mutator phenotype. *Science*, **275**, 967.

46. Brimmel, M., Mendiola, R., Mangion, J., and Packham, G. (1998) Bax frameshift mutations in cell lines derived from human haematopoietic malignancies are associated with resistance to apoptosis and microsatellite instability. *Oncogene*, **16**, 1803.

47. Malkhosyan, S., Rampino, N., Yamamoto, H., and Perucho, M. (1996) Frameshift mutator mutations. *Nature*, **382**, 499.

48. Ikeda, M., Orimo, H., Moriyama, H., Nakajima, E., Matsubara, N., Mibu, R., *et al.* (1998) Close correlation between mutations of *E2F4* and *hMSH3* genes in colorectal cancers with microsatellite instability. *Cancer Res.*, **58**, 594.

49. Branch, P., Bicknell, D., Rowan, A., Bodmer, W., and Karran, P. (1995) Immune surveillance in colorectal carcinoma. *Nature Genet.*, **9**, 231.

50. Bicknell, D. C., Kaklamanis, L., Hampson, R., Bodmer, W. F., and Karran, P. (1996) Selection for β2-microglobulin mutation in mismatch repair-defective colorectal carcinomas. *Curr. Biol.*, **6**, 1695.

51. Powell, S. M., Zilz, N., Beazer-Barclay, Y., Bryan, T. M., Hamilton, S. R., Thibodeau, S. N., *et al.* (1992) *APC* mutations occur early during colorectal tumorigenesis. *Nature*, **359**, 235.

52. Homfray, T. F. R., Cottrell, S. E., Ilyas, M., Rowan, A., Talbot, I. C., Bodmer, W. F., *et al.* (1998) Defects in mismatch repair occur after *APC* mutations in the pathogenesis of sporadic colorectal tumours. *Hum. Mutat.*, **11**, 114.

53. Kinzler, K. W. and Vogelstein, B. (1996) Lessons from hereditary colorectal cancer. *Cell*, **87**, 159.

54. Kim, H., Jen, J., Vogelstein, B., and Hamilton, S. R. (1994) Clinical and pathological characteristics of sporadic colorectal carcinomas with DNA replication errors in microsatellite sequences. *Am. J. Pathol.*, **145**, 148.

55. Lothe, R. A., Peltomäki, P., Meling, G. I., Altonen, L. A., Nyström-Lahti, M., Pylkkänen, L., *et al.* (1993) Genomic instability in colorectal cancer: Relationship to clinicopathological variables and family history. *Cancer Res.*, **53**, 5849.

56. Fang, W. and Modrich, P. (1993) Human strand-specific mismatch repair occurs by a bidirectional mechanism similar to that of the bacterial reaction. *J. Biol. Chem.*, **268**, 11838.

57. Drummond, J. T., Li, G.-M., Longley, M. J., and Modrich, P. (1995) Isolation of an hMSH2-p160 heterodimer that restores DNA mismatch repair to tumor cells. *Science*, **268**, 1909.

58. Hughes, M. J. and Jiricny, J. (1992) The purification of a human mismatch binding protein and identification of its associated ATPase and helicase activities. *J. Biol. Chem.*, **267**, 23876.

59. Gradia, S., Acharya, S., and Fishel, R. (1997) The human mismatch recognition complex hMSH2-hMSH6 functions as a novel molecular switch. *Cell*, **91**, 995.

60. Iaccarino, I., Marra, G., Palombo, F., and Jiricny, J. (1998) hMSH2 and hMSH6 play distinct roles in mismatch binding and contribute differently to the ATPase activity of hMutSα. *EMBO J.*, **17**, 2677.

61. Li, G.-M. and Modrich, P. (1995) Restoration of mismatch repair to nuclear extracts of H6 colorectal tumor cells by a heterodimer of human MutL homologs. *Proc. Natl Acad. Sci. USA*, **92**, 1950.

62. Habraken, Y., Sung, P., Prakash, L., and Prakash, S. (1997) Enhancement of MSH2-MSH3-mediated mismatch recognition by the yeast MLH1-PMS1 complex. *Curr. Biol.*, **7**, 790.

63. Horii, A., Han, H.-J., Sasaki, S., Shimada, M., and Nakamura, Y. (1994) Cloning, characterization and chromosomal assignment of the human genes homologous to yeast PMS1, a member of mismatch repair genes. *Biochem. Biophys. Res. Commun.*, **204**, 1257.

64. Osborne, L. R., Herbrick, J.-A., Greavette, T., Heng, H. H. Q., Tsui, L.-C., and Schere, S. W. (1997) PMS2-related genes flank the rearrangement breakpoints associated with Williams syndrome and other diseases on human chromosome 7. *Genomics*, **45**, 402.

65. Risinger, J. I., Umar, A., Barrett, J. C., and Kunkel, T. A. (1995) A hPMS2 mutant cell line is defective in strand-specific mismatch repair. *J. Biol. Chem.*, **270**, 18183.

66. Longley, M. J., Pierce, A. J., and Modrich, P. (1997) DNA polymerase δ is required for human mismatch repair *in vitro*. *J. Biol. Chem.*, **272**, 10917.

67. Shivji, M. K. K., Kenny, M. K., and Wood, R. D. (1992) Proliferating cell nuclear antigen is required for DNA excision repair. *Cell*, **69**, 367.

68. Frosina, G., Fortina, P., Rossi, O., Carrozzino, F., Raspaglio, G., Cox, L. S., *et al.* (1996) Two pathways for base excision repair in mammalian cells. *J. Biol. Chem.*, **271**, 9573.

69. Umar, A., Buermeyer, A. B., Simon, J. A., Thomas, D. C., Clark, A. B., Liskay, R. M., *et al.*

(1996) Requirement for PCNA in DNA mismatch repair at a step preceding DNA resynthesis. *Cell*, **87**, 65.

70. Kenny, M. K., Lee, S.-H., and Hurwitz, J. (1989) Multiple functions of human single-stranded-DNA binding protein in simian virus 40 DNA replication: single strand stabilization and stimulation of DNA polymerases $\alpha$ and $\delta$. *Proc. Natl Acad. Sci. USA*, **86**, 9757.

71. Coverley, D., Kenny, M. K., Lane, D. P., and Wood, R. D. (1992) A role for the human single-stranded DNA binding protein HSSB/RPA in an early stage of nucleotide excision repair. *Nucl. Acids Res.*, **20**, 3873.

72. Lin, Y.-L., Shivji, M. K. K., Chen, C., Kolodner, R., Wood, R. D., and Dutta, A. (1998) The evolutionary conserved zinc finger motif in the largest subunit of human replication protein A is required for DNA replication and mismatch repair but not for nucleotide excision repair. *J. Biol. Chem.*, **273**, 1453.

73. Hare, J. T. and Taylor, J. H. (1985) One role for DNA methylation in vertebrate cells is strand discrimination in mismatch repair. *Proc. Natl Acad. Sci. USA*, **82**, 7350.

74. Goulian, M., Richards, S. H., Heard, C. J., and Bigsby, B. M. (1990) Discontinuous DNA synthesis by purified mammalian proteins. *J. Biol. Chem.*, **265**, 18461.

75. Waga, S., Bauer, G., and Stillman, B. (1994) Reconstitution of complete SV40 replication with purified replication factors. *J. Biol. Chem.*, **269**, 10923.

76. Chen, J., Chen, S., and Dutta, A. (1996) p21[Cip1/Waf1] disrupts the recruitment of human Fen1 by proliferating-cell nuclear antigen into the DNA replication complex. *Proc. Natl Acad. Sci. USA*, **93**, 11597.

77. Warbrick, E., Lane, D. P., Glover, D. M., and Cox, L. S. (1997) Homologous regions of Fen1 and p21[Cip1] compete for binding to the same site on PCNA: a potential mechanism to co-ordinate DNA replication and repair. *Oncogene*, **14**, 2313.

78. Johnson, R. E., Gopala, K. K., Prakash, L., and Prakash, S. (1995) Requirement of the yeast rth1 5' to 3' exonuclease for the stability of simple repetitive DNA. *Science*, **269**, 238.

79. Reagan, M. S., Pittenge, C., Siede, W., and Friedberg, E. C. (1995) Characterization of a mutant strain of *Saccharomyces cerevisiae* with a deletion of the RAD27 gene, a structural homolog of the RAD2 nucleotide excision repair gene. *J. Bacteriol.*, **177**, 364.

80. Tishkoff, D. X., Filosi, N., Gaida, G. M., and Kolodner, R. D. (1997) A novel mutation avoidance mechanism dependent on *S. cerevisiae* RAD27 is distinct from DNA mismatch repair. *Cell*, **88**, 253.

81. Freudenreich, C. H., Kantrow, S. M., and Zakian, V. A. (1998) Expansion and length-dependent fragility of CTG repeats in yeast. *Science*, **279**, 853.

82. Miret, J. J., Pessoa-Brandao, L., and Lahue, R. S. (1997) Instability of CAG and CTG trinucleotide repeats in *Saccharomyces cerevisiae*. *Mol. Cell. Biol.*, **17**, 3382.

83. Kramer, P. K., Pearson, C. E., and Sinden, R. R. (1996) Stability of triplet repeats of myotonic dystrophy and fragile X loci in human mismatch repair cell lines. *Hum. Genet.*, **98**, 151.

84. de Wind, N., Dekker, M., Berns, A., Radman, M., and te Riele, H. (1995) Inactivation of the mouse *Msh2* gene results in postreplicational mismatch repair deficiency, methylation tolerance, hyperrecombination, and predisposition to tumorigenesis. *Cell*, **82**, 321.

85. Reitmair, A. H., Schmits, R., Ewel, A., Bapat, B., Redson, M., Mitri, A., *et al.* (1995) *MSH2* deficient mice are viable and susceptible to lymphoid tumours. *Nature Genet.*, **11**, 64.

86. Baker, S. M., Plug, A. W., Prolla, T. A., Bronner, C. E., Harris, A. C., Yao, X, *et al.* (1996) Involvement of mouse Mlh1 in DNA mismatch repair and meiotic crossing over. *Nature Genet.*, **13**, 336.

87. Edelmann, W., Cohen, P. E., Kane, M., Lau, K., Morrow, B., Bennett, S., *et al.* (1996) Meiotic pachytene arrest in MLH-1-deficient mice. *Cell*, **85**, 1125.

88. Baker, S. M., Bronner, C. E., Zhang, L., Plug, A. W., Robatzek, M., Warren, G., *et al.* (1995) Male mice defective in the DNA mismatch repair gene *PMS2* exhibit abnormal chromosome synapsis in meiosis. *Cell*, **82**, 309.

89. Prolla, T. A., Baker, S. M., Harris, A. C., Tsao, J.-L., Yao, X., Bronner, C. E., *et al.* (1998) Tumour susceptibility and spontaneous mutation in mice deficient in Mlh1, Pms1 and Pms2 DNA mismatch repair. *Nature Genet.*, **18**, 276.

90. Edelmann, W., Yang, K., Umar, A., Heyer, J., Lau, K., Fan, K., *et al.* (1997) Mutation in the mismatch repair gene *Msh6* causes cancer susceptibility. *Cell*, **91**, 467.

91. Parsons, R., Li, G.-M., Longley, M., Modrich, P., Liu, B., Berk, T., *et al.* (1995) Mismatch repair deficiency in phenotypically normal cells. *Science*, **268**, 738.

92. Ross-Macdonald, P. and Roeder, G. S. (1994) Mutation of a meiosis-specific MutS homolog decreases crossing-over but not mismatch correction. *Cell*, **79**, 1069.

93. Hollingsworth, N. M., Ponte, L., and Halsey, C. (1995) MSH5, a novel mutL homolog, facilitates meiotic reciprocal recombination between homologs in *Saccharomyces cerevisae* but not mismatch repair. *Genes Dev.*, **9**, 1728.

94. Paquis-Flucklinger, V., Santucci-Darmanin, S., Paul, R., Saunieres, A., Turc-Carel, C., and Desnuelle, C. (1997) Cloning and expression analysis of a meiosis-specific MutS homolog: the human MSH4 gene. *Genomics*, **44**, 188.

95. Narayanan, L., Fritzell, J. A., Baker, S. M., Liskay, R. M., and Glazer, P. M. (1997) Elevated levels of mutation in multiple tissues of mice deficient in the DNA mismatch repair gene *Pms2*. *Proc. Natl Acad. Sci. USA*, **94**, 3122.

96. Andrew, S. E., McKinnon, M., Cheng, B. S., Francis, A., Penney, J., Reitmar, A. H., *et al.* (1998) Tissues of MSH2-deficient mice demonstrate hypermutability on exposure to a DNA methylating agent. *Proc. Natl Acad. Sci. USA*, **95**, 1126.

97. Reitmair, A. H., Risley, R., Bristow, R. G., Wilson, T., Ganesh, A., Jang, A., *et al.* (1997) Mutator phenotype in *Msh2*-deficient murine embryonic fibroblasts. *Cancer Res.*, **57**, 3765.

98. Reitmair, A. H., Cai, J.-C., Bjerknes, M., Redson, M., Cheng, H., Pind, M. T. L., *et al.* (1996) MSH2 deficiency contributes to accelerated APC-mediated intestinal tumorigenesis. *Cancer Res.*, **56**, 2922.

99. de Wind, N., Dekker, M., van Rossum, A., van der Valk, M., and te Riele, H. (1998) Mouse models for hereditary nonpolyposis colorectal cancer. *Cancer Res.*, **58**, 248.

100. Karran, P. and Hampson, R. (1996) Genomic instability and tolerance to alkylating agents. *Cancer Surv.*, **28**, 69.

101. Kat, A., Thilly, W. G., Fang, W. H., Longley, M. J., Li, G. M., and Modrich, P. (1993) An alkylation-tolerant, mutator human cell line is deficient in strand-specific mismatch repair. *Proc. Natl Acad. Sci. USA*, **90**, 6424.

102. Hampson, R., Humbert, O., Macpherson, P., Aquilina, G., and Karran, P. (1997) Mismatch repair defects and $O^6$-methylguanine-DNA methyltransferase expression in acquired resistance to methylating agents in human cells. *J. Biol. Chem.*, **272**, 28596.

103. Ceccotti, S., Aquilina, G., Macpherson, P., Yamada, M., Karran, P., and Bignami, M. (1996) Processing of $O^6$-methylguanine by mismatch correction in human cell extracts. *Curr. Biol.*, **6**, 1528.

104. Ciotta, C., Ceccotti, S., Aquilina, G., Humbert, O., Palombo, F., Jiricny, J., *et al.* (1997) Increased somatic recombination in methylation tolerant cells with defective mismatch repair. *J. Mol. Biol.*, **276**, 738.

105. Branch, P., Hampson, R., and Karran, P. (1995) DNA mismatch binding defects, DNA damage tolerance, and mutator phenotypes in human colorectal carcinoma cell lines. *Cancer Res.*, **55**, 2304.

106. Wedge, S., Porteous, J. K., May, B. L., and Newlands, E. S. (1996) Potentiation of temozolomide and BCNU cytotoxicity by $O^6$-benzylguanine: a comparative study *in vitro*. *Br. J. Cancer*, **72**, 482.

107. Aquilina, G., Ceccotti, S., Martinelli, S., Hampson, R., and Bignami, M. (1998) CCNU-sensitivity in mismatch repair defective human cells. *Cancer Res.*, **58**, 135.

108. Aquilina, G., Fiumicino, S., Zijno, A., and Bignami, M. (1997) Reversal of methylation tolerance by transfer of human chromosome 2. *Mutat. Res.*, **385**, 115.

109. Karran, P. and Bignami, M. (1996) Drug-related killings: a case of mistaken identity. *Chem. Biol.*, **3**, 875.

110. Swann, P. F., Waters, T. R., Moulton, D. C., Xu, Y.-Z., Edwards, M., and Mace, R. (1996) Role of postreplicative DNA mismatch repair in the cytotoxic action of thioguanine. *Science*, **273**, 1109.

111. Karran, P. and Bignami, M. (1994) DNA damage tolerance, mismatch repair and genome instability. *BioEssays*, **16**, 833.

112. Hawn, M. T., Umar, A., Carethers, J. M., Marra, G., Kunkel, T. A., Boland, C. R., *et al.* (1995) Evidence for a connection between the mismatch repair system and the $G_2$ cell cycle checkpoint. *Cancer Res.*, **55**, 3721.

113. Ceccotti, S., Dogliotti, E., Gannon, J., Karran, P., and Bignami, M. (1993) $O^6$-methyl-guanine in DNA inhibits replication *in vitro* by human cell extracts. *Biochemistry*, **32**, 13664.

114. Anthoney, D. A., McIlwrath, A. J., Gallagher, W. M., Edlin, A. R. M., and Brown, R. (1996) Microsatellite instability, apoptosis, and loss of p53 function in drug-resistant tumor cells. *Cancer Res.*, **56**, 1374.

115. Fink, D., Nebel, S., Aebi, S., Zheng, H., Cenni, B., Nehmé, A., *et al.* (1996) The role of DNA mismatch repair in platinum drug resistance. *Cancer Res.*, **56**, 4881.

116. Aebi, S., Fink, D., Gordon, R., Kim, H. K., Fink, J. L., and Howell, S. B. (1997) Resistance to cytotoxic drugs in DNA mismatch repair-deficient cells. *Clin. Cancer Res.*, **3**, 1763.

117. Fritzell, J. A., Narayanan, L., Baker, S. M., Bronner, C. E., Andrew, S. E., Prolla, T. A., *et al.* (1997) Role of mismatch repair in the cytotoxicity of ionizing radiation. *Cancer Res.*, **57**, 5145.

118. Leadon, S. A. and Avrutskaya, A. V. (1997) Differential involvement of the human mismatch repair proteins, hMLH1 and hMSH2 in transcription-coupled repair. *Cancer Res.*, **57**, 3784.

119. Davis, T. W., Patten, C. W.-V., Meyers, M., Kunugi, K. A., Cuthill, S., Reznikoff, C., *et al.* (1998) Defective expression of the DNA mismatch repair protein, MLH1, alters G2–M cell cycle checkpoint arrest following ionizing radiation. *Cancer Res.*, **58**, 767.

120. Mellon, I., Rajpal, D. K., Koi, M., Boland, C. R., and Champe, G. N. (1996) Transcription-coupled repair deficiency and mutations in human mismatch repair genes. *Science*, **272**, 557.

121. Selby, C. P., Witkin, E. M., and Sancar, A. (1991) *Escherichia coli mfd* mutant deficient in 'mutation frequency decline' lacks strand-specific repair: *in vitro* complementation with purified coupling factor. *Proc. Natl Acad. Sci. USA*, **88**, 11574.

122. Mellon, I. and Champe, G. N. (1996) Products of DNA mismatch repair genes *mutS* and *mutL* are required for transcription-coupled nucleotide excision repair of the lactose operon of *Escherichia coli*. *Proc. Natl Acad Sci. USA*, **93**, 1292.

123. Mu, D., Tursun, M., Duckett, D. R., Drummond, J. T., Modrich, P., and Sancar, A. (1997) Recognition and repair of compound DNA lesions (base damage and mismatch) by human mismatch repair and excision repair systems. *Mol. Cell. Biol.*, **17**, 760.

124. Saparbaev, M., Prakash, L., and Prakash, S. (1996) Requirement of mismatch repair genes MSH2 and MSH3 in the RAD1-RAD10 pathway of mitotic recombination in *Saccharomyces cerevisiae*. *Genetics*, **142**, 727.

125. Sugawara, N., Paques, F., Colaiacovo, M., and Haber, J. E. (1997) Role of *Saccharomyces cerevisiae* Msh2 and Msh3 repair proteins in double-strand break-induced recombination. *Proc. Natl Acad. Sci. USA*, **94**, 9214.

126. Paques, F. and Haber, J. E. (1997) Two pathways for removal of non-homologous DNA ends during double-strand break repair in *Saccharomyces cerevisiae*. *Mol. Cell. Biol.*, **17**, 6765.

127. Kirkpatrick, D. T. and Petes, T. D. (1997) Repair of DNA loops involves DNA mismatch and nucleotide excision repair proteins. *Nature*, **387**, 929.

128. Cairns, T., Overbaugh, J., and Miller, S. (1988) The origin of mutants. *Nature*, **335**, 142.

129. Foster, P. L. and Trimarchi, J. M. (1994) Adaptive reversion of a frameshift mutation in *Eschericia coli* by simple base deletions in homopolymeric runs. *Science*, **265**, 407.

130. Rosenberg, S. M., Longerich, S., Gee, P., and Harris, R. S. (1994) Adaptive mutation by deletion in small mononucleotide repeats. *Science*, **265**, 405.

131. Harris, R. S., Feng, G., Ross, K. J., Sidhu, R., Thulin, C., Longerich, S., *et al.* (1997) Mismatch repair protein MutL becomes limiting during stationary-phase mutation. *Genes Dev.*, **11**, 2426.

132. Loeb, L. A. (1991) Mutator phenotype may be required for multistage carcinogenesis. *Cancer Res.*, **51**, 3075.

133. Richards, B., Zhang, H., Phear, G., and Meuth, M. (1997) Conditional mutator phenotypes in hMSH2-deficient tumor cell lines. *Science*, **277**, 1523.

134. Ahuja, N., Mohan, A. L., Li, Q., Stolker, J. M., Herman, J. G., Hamilton, S. R., *et al.* (1997) Association between CpG island methylation and microsatellite instability in colorectal cancer. *Cancer Res.*, **57**, 3370.

135. Lengauer, C., Kinzler, K. W., and Vogelstein, B. (1997) DNA methylation and genetic instability in colorectal cancer cells. *Proc. Natl Acad. Sci. USA*, **94**, 2245.

136. Kane, M. F., Loda, M., Gaida, G. M., Lipman, J., Mishra, R., Goldman, H., *et al.* (1997) Methylation of the *hMLH1* promoter correlates with lack of expression of hMLH1 in sporadic colon tumors and mismatch repair-defective human tumor cell lines. *Cancer Res.*, **57**, 808.

137. Carethers, J. M., Hawn, M. T., Chauhan, D. P., Luce, M. C., Marra, G., Koi, M., *et al.* (1996) Competency in mismatch repair prohibits clonal expansion of cancer cells treated with N-methyl-N'-nitro-N-nitrosoguanidine. *J. Clin. Invest.*, **98**, 199.

138. Armstrong, M. J. and Galloway, S. M. (1997) Mismatch repair provokes chromosome aberrations in hamster cells treated with methylating agents or 6-thioguanine, but not with ethylating agents. *Mutat. Res.*, **373**, 167.

139. Tominaga, Y., Tsuzuki, T., Shiraishi, A., Kawate, H., and Sekiguchi, M. (1997) Alkylation-induced apoptosis of embryonic stem cells in which the gene for DNA repair, methyltransferase, has been disrupted by gene targeting. *Carcinogenesis*, **18**, 889.

140. Meikrantz, W., Bergom, M. A., Memisoglu, A., and Samson, L. (1998) O$^6$-alkylguanine DNA lesions trigger apoptosis. *Carcinogenesis*, **19**, 369.

141. Fearon, E. R. and Vogelstein, B. (1990) A genetic model for colorectal tumorigenesis. *Cell*, **61**, 759.

142. Eshleman, J. R., Lang, E. Z., Bowerfind, G. K., Parsons, R., Vogelstein, B., Willson, J. K. V., *et al.* (1995) Increased mutation rates at the hprt locus accompanies microsatellite instability in colon cancer. *Oncogene*, **10**, 33.

143. Drummond, J. T., Genschel, J., Wolf, E., and Modrich, P. (1997) *DHFR/MSH3* amplification in methotrexate-resistant cells alters the hMutSα/hMutSβ ratio and reduces the efficiency of base–base mismatch repair. *Proc. Natl Acad. Sci. USA*, **94**, 10144.

# 5 | Enzymology of human nucleotide excision repair

HANSPETER NAEGELI

## 1. Introduction

Damaged bases are eliminated from human genomes by two completely different excision mechanisms: repair by base excision and repair by nucleotide excision. Base excision involves the hydrolysis of *N*-glycosidic bonds and is restricted to a rather narrow range of defective or inappropriate DNA constituents (1, 2). Nucleotide excision repair (NER), on the other hand, proceeds through dual endonucleolytic cleavage of damaged strands and is capable of removing a nearly infinite variety of base lesions with little regard for the modification itself. The extraordinary substrate versatility of NER is necessary to cope with bulky (helix distorting) injuries to DNA that arise mainly from exposure to UV radiation and chemical carcinogens (3–6). NER is, in fact, the sole mechanism for the error-free removal of bulky DNA adducts in human cells and, as a consequence, provides a key line of defence against such genotoxic adversities. In addition, NER displays a backup role for the repair of apurinic–apyrimidinic (AP) sites, thymine glycol, 8-hydroxyguanine, $O^6$-methyl-guanine, and similar non-bulky base lesions that are normally processed by more specific repair enzymes, primarily AP endonucleases, DNA glycosylases, or $O^6$-methylguanine-DNA methyltransferase (7, 8). To exert these multiple DNA repair functions, human NER lacks selectivity for a particular type of damage, i.e. a limited number of gene products accommodates and processes a large repertoire of unrelated base modifications while ignoring the normal conformational fluctuations of undamaged DNA. This flexibility is achieved by means of a repair strategy that avoids the direct processing of modified deoxyribonucleotide residues, but, instead, operates exclusively on unmodified segments of native DNA around each lesion site.

The nucleotide excision mode of DNA repair consists of a cut-and-patch mechanism that is executed by the hydrolysis of two phosphodiester bonds, one on either side of the targeted lesion, generating oligonucleotide excision products that contain the defective residue (9, 10). Subsequent enzymatic steps include the replacement of released deoxyribonucleotides by DNA repair synthesis and DNA ligation (Fig. 5.1).

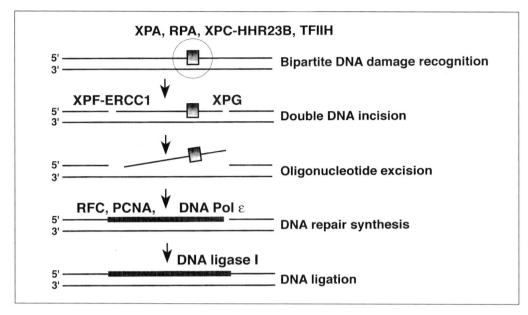

**Fig. 5.1** General scheme of the human NER pathway.

In more detail, the sequence of molecular events during the human NER reaction may be outlined as follows:

- *Bipartite recognition of DNA damage*: abnormal bases are detected by a molecular sensor that probes the secondary structure (or conformation) of the DNA double helix in co-operation with another molecular sensor that probes the primary structure (or chemical composition) of each DNA strand.

- *Double endonucleolytic cleavage*: recruitment of specialized DNA endonucleases leads to the incision of damaged strands on both sides of the offending lesion and at some distance from it (15–25 nucleotides from the lesion on the 5′ side and 5–10 nucleotides on the 3′ side).

- *Oligonucleotide excision*: defective bases are eliminated as the component of 24–32-residue long nucleotide segments, leaving behind single-stranded excision gaps.

- *DNA repair synthesis*: excision repair patches are generated by DNA polymerization using the complementary undamaged strand as a template.

- *DNA ligation*: the newly synthesized repair patches are ligated to the pre-existing DNA strands.

## 2. Human NER factors and general strategies

Several lines of genetic and biochemical investigation have defined a minimal array of proteins that are absolutely required for human NER activity (3, 4, 6, 11, 12). The nomenclature used to indicate these core NER factors is as follows:

- xeroderma pigmentosum complementation group A through G proteins (e.g. XPA, XPC, XPF, and XPG);
- replication protein A (RPA);
- human homologue of Rad23B (HHR23B);
- transcription factor IIH (TFIIH);
- excision repair cross-complementing group 1 protein (ERCC1);
- replication factor C (RFC);
- proliferating cell nuclear antigen (PCNA);
- DNA polymerase $\varepsilon$ (DNA pol $\varepsilon$); and
- DNA ligase I.

Depending on their particular function, these NER subunits can be divided into three broad categories (Fig. 5.1). A first set of factors triggers the NER reaction by localizing damaged sites and inducing the conformational changes in DNA required for double incision. There is a second group of NER subunits that mediate the enzymatic cleavage of damaged strands and, finally, a third set of proteins synthesizes appropriate repair patches.

Presumably, a sophisticated substrate-discrimination and DNA damage-recognition mechanism has evolved because DNA incision is a critical point of no return during the NER pathway. In addition, double DNA incision is a potentially hazardous step as the accumulation of DNA nicks in undamaged chromosomal regions may enhance genomic instability. To minimize the risk of spurious DNA repair reactions at improper (undamaged) sites, double DNA incision by XPF–ERCC1 (5′ to the lesion) and XPG (3′ to the lesion) is tightly regulated by at least four factors (XPA, RPA, TFIIH, XPC–HHR23B) comprising a minimum of 15 different subunits (11, 12). These factors use a bipartite substrate-discrimination mechanism that detects covalent base damage in response to local derangement of Watson–Crick hydrogen bonding and, hence, relies on changes in both the primary and secondary structure of DNA to assemble a preincision complex (13). Upon effective damage recognition and endonucleolytic cleavage, incised DNA is immediately processed by several replicative factors (RFC, PCNA, RPA, DNA pol $\varepsilon$) and by DNA ligase I to re-establish the correct nucleotide sequence as well as strand continuity (12). Tight coupling of DNA incision to DNA repair synthesis and ligation is thought to avoid the exposure of single-stranded gaps or DNA ends to degradation by non-specific cellular nucleases. Like many factors involved in damage recognition and DNA incision, most postincisional components of the human NER system also consist of multiple subunits, such that another 10–15 polypeptides are engaged with completion of DNA excision repair by the proper and rapid processing of incised intermediates.

# 3. The human genetic framework: xeroderma pigmentosum

The availability of naturally occurring mutations associated with clinical syndromes has greatly stimulated studies on human DNA excision repair. About 30 years ago, Cleaver published his discovery that cells of patients suffering from the cancer-prone syndrome xeroderma pigmentosum (XP) carry a severe repair deficiency (14). Individuals afflicted by this rare, recessively inherited disorder are defective or completely deficient in the NER process that normally removes DNA damage caused by UV radiation or bulky chemicals (1, 6, 15). As a consequence, these patients suffer from extreme hypersensitivity to sunlight and present with dry skin (xeroderma), strong pigmentation abnormalities, and malignancies at the exposed areas of the skin (1, 6, 14). These clinical manifestations of the XP syndrome illustrate the disastrous phenotypic consequences of a failure in DNA excision repair and, in particular, provide direct proof for the contention that human NER activity protects carcinogen-exposed tissues from the risk of developing tumours. The frequency of skin cancer in XP patients is—depending on the tumour type, its location, and the patients' age group—up to five orders of magnitude higher than in the normal population (16). Characteristic 'fingerprint' mutations found in tumours of XP patients are tandem CC→TT transitions which, at least in the skin, are predominantly induced by the UV component of sunlight (17). In addition to these neoplastic changes, 20–30% of XP patients exhibit progressive neurological abnormalities associated with neuronal death in the central and peripheral nervous system. This clinical finding suggests that the high metabolic rate combined with the long lifespan of neurones makes them especially dependent on proficient NER activity. In fact, neurones consume large amounts of molecular oxygen, and the reactive oxygen species that are generated as by-products of cellular respiration cause considerable damage to DNA (8, 18).

Based on somatic cell fusion studies, individuals afflicted by XP have been assigned to seven genetic complementation groups (XP-A through XP-G) each containing a mutation in a different NER gene (1, 6). An eighth form of the disease, called xeroderma pigmentosum variant (XP-V), is proficient in NER activity but is thought to display reduced fidelity during the replication of damaged DNA templates. The recommended nomenclature (19) uses a hyphen to indicate cells of a particular complementation group (for example XP-A), distinguishing each cell line from the corresponding gene (for example *XPA*) and its protein product (XPA). In parallel studies, screening mutagenized cultures of Chinese hamster cell lines led to the isolation of UV-sensitive mutants that carry an NER deficiency, and such rodent mutants have been divided into 11 complementation groups designated with Arabic numbers (20). Because rodent cells are more easily transfected than human cells, complementation of NER-deficient hamster cell lines with human genomic libraries, combined with phenotypic selection, was for many years the most successful strategy used to isolate human NER genes. This approach led to the identification of a number of human *ERCC* genes (for *e*xcision *r*epair *c*ross-complementing). *ERCC1* has no *XP* equivalent (21); *ERCC2* was subsequently found to be the same as *XPD*

(22); *ERCC3* is identical to *XPB* (23); *ERCC4* and *ERCC11* are equivalent to *XPF* (24); *ERCC5* is the same as *XPG* (25). *XPA* and *XPC* remain the only human NER genes that were cloned directly by complementation of the repair defect in XP cell lines. *XPA* was originally isolated by transfection of XP-A fibroblasts with mouse genomic DNA (26), and *XPC* was cloned using a human complementary DNA library engineered into an extrachromosomally replicating, Epstein–Barr virus-based vector (27).

# 4. The biochemical framework: *in vitro* reconstitution of human NER activity

The identification of human NER genes led to the overproduction, purification, and biochemical analysis of the respective gene products (summarized in Table 5.1). Using these purified components, human NER activity has been reconstituted in two different laboratories, thereby establishing a set of essential components required for double DNA incision and all subsequent postincisional events. Sancar and collaborators observed that six repair factors consisting of XPA, RPA, XPC–HHR23B, TFIIH, XPF–ERCC1, and XPG are necessary and sufficient to promote the expected DNA incision reactions (11). According to Wood and co-workers, however, there is a requirement for a seventh incision factor (designated IF7), but a gene encoding this additional protein component has not yet been identified (12). It is expected that the isolation and identification of the IF7 subunit will eventually clarify the apparent discrepancy observed between the two laboratories. Importantly, the absence of XPA, RPA, XPC–HHR23B, or TFIIH leads to the complete loss of DNA incision activity, even though XPF–ERCC1 has been directly associated with cleavage 5′ to the lesion and XPG mediates incision on the 3′ side. Omission of XPF–ERCC1 from the reconstituted reaction mixture generates uncoupled 3′ incisions, but omission of XPG results in no incision at all (28). This strict dependence of double DNA incision on the presence of all core NER components may suggest the formation of a large nucleoprotein complex at damaged sites or, alternatively, it could reflect a sequential mechanism that proceeds by the ordered interactions between several NER factors. In any case, the entire human NER pathway could be reconstituted by supplementing the incision reactions with DNA pol $\varepsilon$ in combination with RFC and PCNA (12). Together, these replicative factors are able to convert the single-stranded gaps resulting from double DNA incision and oligonucleotide excision into appropriate repair patches (see Fig. 5.1). Finally, the addition of DNA ligase I promotes ligation of the newly synthesized repair patches. The next few pages will be devoted to reviewing the relevant biochemical properties of these core constituents of the human NER system.

## 4.1 XPA

XPA is a protein of 273 amino acids (and predicted molecular mass of 31 kDa) containing a zinc finger motif, a nuclear localization signal, and at least four

**Table 5.1** Minimal factors required for the in vitro reconstitution of human NER activity (see text for references)

| Factor | Subunits | Main properties | Role in human NER |
|---|---|---|---|
| XPA | 1 Polypeptide (31 kDa) | Sensor of defective base-pairing; zinc finger; interactions with RPA, TFIIH, and ERCC1 | DNA damage recognition. Anchor for nucleoprotein assembly? |
| RPA | p70, p34, p14 | Single-stranded DNA binding; interaction with XPA, XPG, p53 | Partner of XPA in DNA damage recognition. Stabilization of unwound DNA? Protection of single-strand gap? |
| XPC–HHR23B | XPC (125 kDa) HHR23B (58 kDa) | DNA binding; interaction with TFIIH Ubiquitin-like domain | Formation or stabilization of unwound complex? Molecular chaperone? Interaction with chromatin? |
| TFIIH | p89/XPB/ERCC3 p80/XPD/ERCC2 p62/TFB1 p52/TFB2 p44/SSL1 p38/cdk7/MO15 p37/cyclin H p36/MAT1 p34 | 3′→5′ DNA helicase and ATPase 5′→3′ DNA helicase and ATPase Interaction with p53<br><br>Zinc finger Cyclin-dependent protein kinase Regulatory partner of cdk7 Stimulation of cdk7 activity Zinc finger | DNA damage recognition? DNA unwinding in the preincision complex? Coupling NER to transcription? Coupling DNA damage recognition to cell-cycle regulation? |
| XPF–ERCC1 | XPF (112 kDa) ERCC1 (38 kDa) | 5′ Junction-specific single-stranded DNA endonuclease; interaction with XPA | 5′ Incision |
| XPG | 1 Polypeptide (?? kDa) | 3′ Junction-specific DNA endonuclease; interaction with RPA, TFIIH, and PCNA | 3′ Incision |
| RFC | p140, p40, p38, p37, p36 | Primer recognition; DNA-dependent ATPase | Matchmaker function for loading of PCNA on to DNA; recycling of incision factors? |
| PCNA | (p32)$_3$ | DNA polymerase clamp | Processivity of DNA repair synthesis |
| DNA polymerase ε | p258, p55 | DNA polymerase with proofreading activity | Synthesis of NER patches; proofreading |
| DNA ligase I | 1 Polypeptide (102 kDa) | DNA ligase activity; interaction with PCNA | Ligation of repair patches |

individual domains for specific protein–protein interactions with other human NER factors (Fig. 5.2). The DNA binding domain of XPA has been mapped to a 122-amino acid region spanning from Met-98 to Phe-219 (29). The zinc finger of XPA displays the motif Cys-105–$(X)_2$–Cys-108–$(X)_{17}$–Cys-126–$(X)_2$–Cys-129 (26), and is indispensable for normal XPA protein structure and function. Mutant XPA proteins, in which the zinc finger was disrupted by replacing one of the four critical cysteine residues with serine, have a vastly different protein conformation (30) and result in the

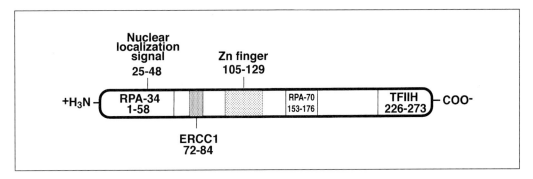

**Fig. 5.2** Domain structure of human XPA protein.

complete loss of NER activity (31). Zinc fingers are well known as DNA binding motifs that can recognize the major groove in a nucleotide sequence-dependent manner. The involvement of a zinc finger motif in the processing of DNA damage, with no requirement for sequence-selectivity, is also a common theme as there are several other zinc finger proteins implicated in contacts with damaged DNA substrates, including UvrA (a subunit of NER in *Escherichia coli*; see ref. 32), Fpg protein (a DNA glycosylase; ref. 33), Rad18 protein from *Saccharomyces cerevisiae* (a DNA-dependent ATPase involved in the replication of damaged DNA in yeast; ref. 34), and PARP (a eukaryotic enzyme that polymerizes ADP-ribose moieties in response to DNA strand breaks; ref. 35). In addition to serving as DNA binding motifs for contacts with the major groove, zinc fingers are also implicated in intermolecular interactions between polypeptides (36), but whether the zinc finger of XPA mediates DNA binding or rather protein binding still has to be elucidated.

Purified or recombinant XPA protein binds DNA with a moderate preference for damaged substrates over undamaged controls (37). Quantitative analysis by electrophoretic mobility shift assays revealed association constants in the range $10^5$ to $10^6 \, \mathrm{M}^{-1}$ (see also Section 5.3), indicating that the nucleic acid binding domain of XPA mediates only weak interactions with DNA substrates (37). Also, XPA shows no overt catalytic activity, at least not as a solitary protein. Nevertheless, its striking ability to undergo multiple protein–protein interactions suggests that XPA may function as a nucleation centre for the assembly of a damage recognition or incision intermediate on DNA. XPA forms a complex with two subunits (p34 and p70) of RPA (38, 39). The portion of XPA mediating binding to p34 was identified within its first 58 residues (Fig. 5.2). A second domain, located between XPA residues 153 and 176, mediates the interaction with p70. Deletion mutants of XPA that fail to bind the large (p70) subunit of RPA confer NER deficiency, whereas binding to the p34 subunit of RPA does not seem to be essential for NER activity (39). The interaction domain with ERCC1 has been mapped to XPA residues 72–84. Again, mutations in XPA that prevent association with ERCC1 confer a defective NER function (40, 41). Finally, using its C-terminal domain (residues 226–273) XPA also interacts with the multi-subunit transcription factor TFIIH (42). In contrast to the various enzymatic steps of the NER pathway—such as DNA incision, DNA repair synthesis, and DNA ligation, which have already been studied in great detail—the mechanism by which XPA and its partner proteins locate damaged substrates is poorly understood. Thus, the question of how XPA may exploit its DNA binding and protein–protein interaction motifs to co-ordinate damage recognition and the assembly of incision complexes will be one of the main subjects of this review.

## 4.2 RPA

Mammalian single-stranded DNA binding protein, referred to as replication protein A (RPA), is an essential factor for several metabolic DNA transactions, including DNA replication (43), NER (11, 12), mismatch repair (44), and recombination (45). RPA is a tight complex made up of three polypeptides of 70, 34, and 14 kDa. The p70 subunit binds to single-stranded DNA with an association constant of approximately $10^9$ $M^{-1}$, while it has a low affinity for native double-stranded DNA. This DNA binding activity has been localized to the central portion of p70 (46), although other domains of RPA may contribute to interactions with DNA as well (47). The largest subunit also contains a putative zinc finger domain near the C terminus, which is essential for its function in DNA replication but is not required for NER activity (44). The other two subunits are thought to provide interfaces for regulatory signals that co-ordinate the different cellular processes involving RPA. For example, p34 is phosphorylated in a cell cycle-dependent manner (48) or upon exposure to genotoxic agents (49), but subsequent studies showed that both the phosphorylated and unphosphorylated forms of RPA are equally active in supporting NER activity *in vitro* (50). RPA also displays interaction domains for the association with other proteins, notably with the NER factors XPA (38, 39) or XPG (38) and with tumour suppressor protein p53 (51).

Based on its single-stranded DNA binding activity, RPA may adopt several potential functions in human NER. As an essential replication accessory factor it may protect the single-stranded gap generated by double DNA incision and stimulate DNA repair synthesis, thereby re-establishing the double helical structure of the substrate (52). Additionally, RPA may act at a biochemical step that precedes dual DNA incision, for example by stabilizing a preincision complex in which DNA is partially denatured (3, 4, 53). RPA may even target the nuclease subunits of human NER (XPF–ERCC1 and XPG) to their specific sites of action (53). Indeed, the footprint generated by RPA on single-stranded DNA reaches an extension of nearly 30 nucleotides (54), which corresponds approximately to the size of single-stranded gaps after double DNA incision in human NER. Long before the involvement of RPA in NER had been unequivocally established, Toulmé and co-workers predicted that factors with affinity for single-stranded DNA may also contribute to DNA damage recognition (55). These authors used fluorescence spectroscopy to investigate the interaction between aromatic amino-acid side chains of gene 32 protein (a single-stranded DNA binding factor of bacteriophage T4) and damaged DNA substrates. Their results indicate that gene 32 protein binds more efficiently to double-stranded

DNA modified with either *cis*-diamminedichloroplatinum(II) or various amino-fluorene derivatives than to native double-stranded DNA, leading to the conclusion that certain single-stranded DNA binding motifs may serve to probe DNA conformation and recognize the single-strand character of sites at which base pairing is disrupted upon formation of highly distorting adducts.

## 4.3   XPC–HHR23B

The XPC–HHR23B heterodimer was purified to homogeneity by *in vitro* complementation of the XP-C excision repair defect in a cell-free repair system (56). The isolated factor consists of two tightly associated proteins of 125 and 58 kDa. The 125-kDa subunit represents the *XPC* gene product, while the 58-kDa partner was designated HHR23B because it is a human homologue of the *S. cerevisiae* NER protein Rad23. A second human homologue of yeast *RAD23* was also identified and designated *HHR23A*, but essentially only the *HHR23B* gene product is found complexed with XPC. HHR23B is not essential for double DNA incision in the reconstituted human NER system (57), although some studies indicate that the addition of this protein (or its homologue HHR23A) stimulates NER activity *in vitro* (58).

The XPC–HHR23B heterodimer binds single-stranded and double-stranded DNA with remarkable affinity (binding constants of about $2 \times 10^8 \, M^{-1}$), but no enzymatic activity could be associated with this factor (57). The only striking domain recognizable in the amino acid sequence of its two components is a ubiquitin-like N terminus in the Rad23 homologue (56). It has been suggested that ubiquitin moieties may act as a molecular chaperone enabling proper folding and assembly in multiprotein complexes (59). This view is supported by the observation that yeast Rad23 protein promotes the association of Rad14 (the *S. cerevisiae* homologue of XPA) with TFIIH *in vitro* (60). The finding that XPC itself interacts with TFIIH lends further support to this multiprotein assembly hypothesis (61). On the other hand, there is also a disassembly hypothesis as the ubiquitin-like domain of either HHR23A or HHR23B has been shown to interact with the human 26S proteasome, a large protein destruction machinery (62). This observation suggests that degradation of HHR23B, mediated by its ubiquitin-like tail, may disrupt the NER complex, thereby accelerating the recycling of those NER factors present only in limiting amounts.

XPC protein is selectively implicated in the repair of non-transcribed bulk DNA. In contrast to other XP mutants that exhibit a general NER defect, XP-C cell lines are deficient in genome overall repair but maintain a normal capacity to process the transcribed strand of active genes (63). Several models have been proposed to explain this unique phenotype. A possible scenario predicts that XPC–HHR23B exploits its affinity for DNA as well as its ubiquitin-like domain to facilitate the assembly of preincision complexes (involving XPA, RPA, and TFIIH), thereby initiating NER on non-transcribed DNA (64). In this model, the repair of transcribed sequences does not require XPC–HHR23B because XPA, RPA, and TFIIH may be recruited directly by the stalled RNA polymerase II or by the products of Cockayne syndrome group A and group B genes (the reader is referred to Chapter 6 where the topic of transcription-coupled repair is discussed in detail). A model substrate consisting of a cyclobutane–pyrimidine dimer flanked on the 3' side by a 10-nucleotide long region of mispaired bases obviates the need for XPC protein in the *in vitro* DNA incision reaction (65). This finding prompted the suggestion that XPC may assist in DNA unwinding and then stabilize the unwound single strands in the preincision complex. Again, the repair of transcribed templates would not require this particular function of XPC–HHR23B because RNA polymerase II enzymes stalled at lesion sites generate a specific nucleoprotein intermediate in which the two strands of DNA are already sufficiently separated by the transcription bubble (65). An alternative hypothesis to explain the phenotype of proficient transcription-coupled repair but deficient global repair, postulates that XPC–HHR23B may function to uncouple TFIIH from the basal transcription initiation machinery and make it available for NER of inactive genomic regions (56). Yet another scenario implicates XPC–HHR23B in the repair of transcriptionally silent DNA segments by altering the more condensed chromatin structure in these regions and providing access of NER factors to the DNA substrate (66).

## 4.4 TFIIH

The involvement of a general transcription factor (TFIIH) in human NER has been independently reported by several laboratories (61, 67, 68). While searching for cellular proteins that promote transcription by RNA polymerase II, these groups stumbled upon a factor that contains, among other polypeptides, the products of the NER genes *XPB* and *XPD*. Further characterization of TFIIH revealed that it is a multimeric and multifunctional complex made up of 7–9 polypeptides, depending on the purification scheme (69). Its subunits include p89/XPB/ERCC3, p80/XPD/ERCC2, p62/TFB1, p52/TFB2, p44/SSL1, p38/cdk7/MO15, p37/cyclin H, p36/MAT1, and p34 (Table 5.1). This complex possesses DNA-dependent ATPase, 5'→3' DNA helicase, 3'→5' DNA helicase (see Fig. 5.3 for the polarity of DNA helicases), and protein kinase activity. TFIIH exploits its different enzymatic activities to regulate transcription initiation, NER, and cell-cycle progression (69). These different functions appear to be associated with the variable formation of several distinct TFIIH complexes that differ qualitatively in their polypeptide composition (70, 71).

In transcription, TFIIH is one of several factors required for the initiation of RNA synthesis by RNA polymerase II. In the absence of TFIIH, RNA polymerase II stalls immediately downstream of the transcription start site and produces aborted transcripts of less than 50 residues in length, indicating that TFIIH relieves a block that prevents the polymerase from entering the elongation phase of RNA synthesis. The term 'promoter clearance' has been introduced to indicate that TFIIH overcomes a functional block to the initial translocation of RNA polymerase II along DNA (72). It is reasonable to assume that TFIIH may catalyse a similar reaction in the context of NER, perhaps by using its 5'→3' and 3'→5' DNA helicase activities to unwind the duplex and promote the translocation of a recognition complex near sites of damage.

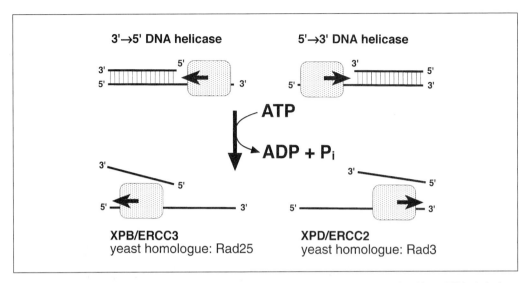

**Fig. 5.3** Reaction mechanism of DNA helicases. These enzymes exploit the energy gained from ATP hydrolysis to separate the two strands of duplex DNA. The helicase reaction is accompanied by protein translocation in either the 3'→5' direction or the 5'→3' direction with respect to the DNA strand to which the enzyme is bound. XPB and its yeast homologue Rad25 have 3'→5' polarity, while XPD and yeast Rad3 protein have 5'→3' polarity. These DNA helicases are components of the multisubunit transcription/repair/cell-cycle factor TFIIH.

Thus, the repair function of TFIIH may constitute a simple extension of its activity during a late step of transcription initiation.

The presence of a protein kinase activity in TFIIH stimulated the attractive idea that this multifunctional complex may provide a link between transcription, excision repair, and cell-cycle regulation (70, 73). The protein kinase component of TFIIH is essential in transcription, where it phosphorylates the C-terminal domain of the large subunit of RNA polymerase II (69). The TFIIH-associated protein kinase seems to be dispensable for NER activity (74), at least when tested in the *in vitro* reconstituted assay system. However, it may play a role in cell-cycle regulation, as the protein kinase complexed with TFIIH is identical to cdk7/MO15, a polypeptide that was previously identified as the catalytic subunit of a cyclin H-dependent kinase (70, 75). Subsequent studies showed that TFIIH also contains cyclin H, the regulatory partner of cdk7/MO15, and MAT1 (for *ménage à trois*), a factor that stimulates the protein kinase activity of cdk7 (70, 76). Since the complex formed by cdk7, cyclin H, and MAT1 is believed to play a role in the regulation of cell-cycle progression, it was postulated that TFIIH may initiate (or interrupt) a signalling cascade in response to DNA damage, thereby arresting downstream cell-cycle events. A follow-up study by Egly and co-workers, showing that the TFIIH-associated cdk7 kinase activity is reduced after exposure to UV light, argues in favour of such a regulatory circuit that depends on phosphorylation signals emanating from TFIIH (76). Tumour suppressor protein p53 is able to interact with the XPB, XPD, and p62 components of TFIIH (77) and therefore represents a likely effector intermediate in this signalling pathway (78).

The cdk7 kinase in conjunction with cyclin H and MAT1 has indeed been shown to phosphorylate p53 (79). Notably, the loss of p53 function in human cells correlates with the reduced repair of UV radiation products seen at the genome overall level (80), but the physical presence of p53 protein in the *in vitro* assay neither stimulates nor inhibits NER activity (81). In combination, these results suggest that p53 may favour DNA excision repair by an indirect mechanism, perhaps by inducing the expression of rate-limiting NER subunits, rather than through direct interactions with individual NER components.

## 4.5  XPF–ERCC1 and XPG

Two separate endonuclease activities are involved in double DNA incision during the human NER process. XPF–ERCC1 incises DNA on the 5′ side of a lesion, while XPG incises DNA on the 3′ side. The products of *XPF* and *ERCC1* genes form a tight heterodimer displaying stoichiometry and single strand-specific endonuclease activity, while either factor alone is enzymatically inert (82). The XPF–ERCC1 complex acts preferentially on single-stranded DNA or the single-stranded region of 'bubble' substrates, i.e. duplex DNA containing a non-complementary sequence of 30 nucleotides in the centre (82). In the presence of RPA, XPF–ERCC1 stops to cut single-stranded DNA indiscriminately and adopts a specific double-stranded/single-stranded DNA junction cutting activity, indicating that RPA confers structure specificity to the XPF–ERCC1 endonuclease (53). Under these conditions, only the strand that undergoes the transition from double-stranded to single-stranded DNA in the 5′ to 3′ direction is cleaved by XPF–ERCC1 (Fig. 5.4). In addition to promoting structure specificity, RPA also exerts quantitative effects by stimulating the DNA incision activity of XPF–ERCC1 (53).

XPG is a member of a family of homologous eukaryotic nucleases that also includes replication-associated enzymes such as flap endonuclease 1 (FEN-1). XPG incises 'Y-shaped' DNA substrates (consisting of a duplex region with two single-stranded arms) by cutting at the boundary between double-stranded and single-stranded DNA (83). The incision is made at the branch point or a few bases into the duplex region (Fig. 5.4). When incubated with a 'bubble' structure of at least 10 mispaired bases, XPG also cleaves at the border between double-stranded and single-stranded DNA, but only at the 3′ side of the non-complementary region. XPG interacts with RPA (38) and, as was shown for XPF–ERCC1 endonuclease, the junction-specific endonuclease activity of XPG is stimulated by RPA (53). Additionally, XPG associates with TFIIH (84).

In summary, studies using 'bubble' or 'Y-shaped' substrates are consistent with a model of double DNA incision where a subset of NER factors co-operate to stabilize a region of unwound DNA around the damaged site and recruit the two endonucleases, XPF–ERCC1 on the 5′ side and XPG on the 3′ side (3, 4). Direct evidence for the formation of a nucleoprotein intermediate in which DNA is at least partially unwound on both sides of a lesion has been provided by permanganate probing, a

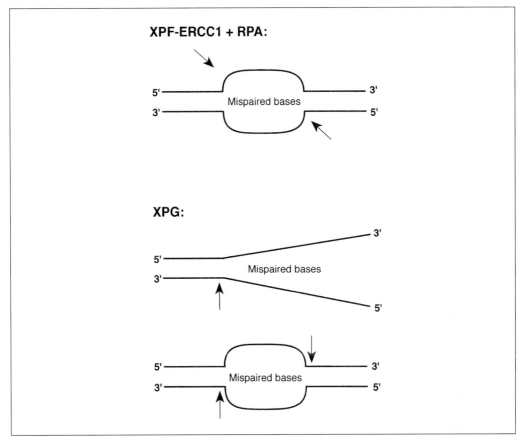

**XPF-ERCC1 + RPA:**

**XPG:**

**Fig. 5.4** Structure-specific endonuclease activity of the XPF–ERCC1 heterodimer and of XPG. The sites of DNA incision are indicated by the arrows.

chemical footprinting technique which oxidizes unpaired thymines and renders them susceptible to cleavage by alkaline treatment (85). Separation of the two DNA strands in this protein–DNA complex requires the addition of XPA, RPA, XPC–HHR23B, TFIIH, and XPG (86). Additionally, strand separation in this intermediate is dependent on the presence of ATP. The polarity of double DNA incision was confirmed by Matsunaga *et al.* (87), who showed that anti-ERCC1 antibodies inhibit the 5′ incision without significantly affecting the 3′ incision. Similarly, anti-XPG antibodies affect the 3′ incision, confirming that XPF–ERCC1 is responsible for the 5′ cut and XPG for the 3′ cut. Subsequent kinetic studies in the reconstituted NER system showed that 3′ incision by XPG precedes 5′ incision by XPF–ERCC1 (74). Using a site-directed cyclobutane–pyrimidine dimer as the substrate, the main sites of DNA incision are at the 22nd phosphodiester bond 5′ and the 6th phosphodiester bond 3′ to the lesion (9) but, with other substrates, some variability in this incision reaction has been observed (10).

## 4.6   RFC, PCNA, DNA pol ε, and DNA ligase I

Double DNA incision produces a single-stranded gap of 24–32 nucleotides in length that is filled in by DNA repair synthesis. The newly formed DNA segments were found to match exactly the gaps generated by oligonucleotide excision without enlargement of the repair patches in either the 3′ or the 5′ direction (88, 89). This DNA resynthesis step of human NER requires RPA, RFC, PCNA, and a PCNA-dependent DNA polymerase (12). PCNA consists of a ring-shaped homotrimeric protein that encircles DNA and functions as a sliding clamp by tethering the DNA polymerase to its template, thereby enhancing the processivity of DNA synthesis (90). In replication, the multisubunit factor RFC recognizes RNA/DNA synthesis primers and then uses the energy gained from ATP hydrolysis to load PCNA on to DNA (91). In NER, RFC and PCNA may have a dual function by first dissociating 5′ incision proteins from the substrate and then promoting the formation of a polymerase clamp. Of the five mammalian DNA polymerases, only DNA pol δ and DNA pol ε require RFC/PCNA, indicating that DNA repair synthesis in NER is carried out by one of these two enzymes. DNA pol δ and DNA pol ε are endowed with proofreading activity to increase template fidelity, suggesting that the resynthesis step should be essentially error-free. In reconstituted *in vitro* systems, both DNA pol δ and DNA pol ε are able to promote repair synthesis, but DNA pol ε is more efficient in generating repair patches that are suitable substrates for subsequent ligation by DNA ligase I (52). An interaction between PCNA and the 3′ nuclease (XPG) suggests that PCNA may also dissociate 3′ incision factors and, hence, promote their turnover (92). There is an additional interaction between PCNA and DNA ligase I that may facilitate the final phosphodiester bond formation (93).

## 5.   The substrate-discrimination problem

During recent years, enormous progress has been made in understanding the principal enzymatic reactions of human NER, primarily those events involving DNA incision, DNA repair synthesis, and DNA ligation. However, the most challenging aspect of this process remains its ability to recognize multiple forms of DNA damage, present in trace amounts, and efficiently reject a very large excess of undamaged DNA nucleotides. Therefore the second part of this review is concerned with the DNA damage-recognition problem associated with human (or mammalian) NER activity.

As suggested by Hanawalt and Haynes more than 30 years ago (5), NER enzymes require extremely high levels of accuracy in the identification of modified residues but, in light of their outstanding substrate versatility, must depend on a damage-recognition mechanism that operates by sensing deviations from the normal conformation or chemistry of DNA without undergoing specific interactions with a particular type of lesion. Which proteins are, in fact, implicated in gauging the 'closeness-of-fit' to the Watson–Crick structure during the preincision steps of human NER? After eliminating those components that catalyse DNA incision, DNA repair synthesis, and DNA ligation, we are left with a minimum of four known factors (and at least 15 polypeptides) that are potentially involved in damage recognition: XPA, RPA, XPC–HHR23B, and TFIIH. However, the mechanism by which three components with DNA binding activity (XPA, RPA, XPC–HHR23B), in combination with a multisubunit complex containing DNA-dependent ATPase and DNA helicase activity (TFIIH), may co-operate to locate a broad range of DNA lesions in the genome is poorly understood. Before summarizing current research devoted to this problem, it is useful to recapitulate some unresolved questions.

## 5.1   Heterogeneity of DNA damage recognition

Although NER essentially eliminates all forms of base adducts, its excision response is highly variable and there is a preference for particular types of lesions. This heterogeneity of human NER is, for example, illustrated by kinetic differences in the removal of UV photoproducts. In the laboratory, substrates for excision repair studies are often generated by exposing cells or DNA to UV sources with peak output in the short-wavelength (190–320 nm) range. The two major types of damage introduced into DNA by this treatment are cyclobutane–pyrimidine dimers and pyrimidine(6–4)pyrimidone photoproducts, typically formed in a ratio of about 3:1 (Fig. 5.5). Human cells are able to process both major UV photoproducts efficiently, but at considerably different rates. Generally, about 50% of (6–4) photoproducts are removed within only 1 h after UV irradiation, but 50% removal of cyclobutane dimers requires a considerably longer postirradiation period of at least 4 h (94, 95).

Detailed analysis of differential repair should eventually lead to the identification of conformational or structural determinants that serve as molecular signals for the recruitment of NER factors. To this end, Gunz et al. (96) developed a repair competition assay that compares the recognition of a series of representative carcinogen–

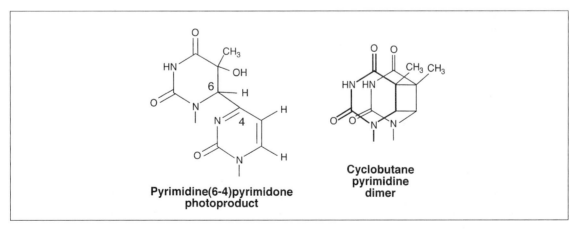

**Pyrimidine(6-4)pyrimidone photoproduct**

**Cyclobutane pyrimidine dimer**

**Fig. 5.5** Predominant base lesions generated by UV irradiation of DNA. The figure shows a pyrimidine(6–4) pyrimidone photoproduct between thymine and cytosine, and a cyclobutane pyrimidine dimer between two adjacent thymines.

DNA adducts in a quantitative manner. Briefly, this assay is based on a site-specifically modified substrate and measures the ability of damaged plasmid DNA to sequester human NER recognition factors and, hence, compete for excision repair activity operating on the site-directed substrate (96, 97). The assay is performed in an NER-proficient human cell extract and involves co-incubating substrate molecules with gradually increasing amounts of covalently closed plasmids carrying a defined number of DNA adducts. This novel approach based on factor sequestration revealed a striking hierarchy of DNA damage recognition with >1000-fold differences between helix-stabilizing and helix-destabilizing lesions. In particular, those adducts that destabilize complementary hydrogen bonding are effective competitors, whereas those adducts that exert opposite effects by stabilizing Watson–Crick geometries display minimal or no competing effects. Thus, analysis of substrate discrimination using repair competition assays indicates that the heterogeneous NER function is determined, at least in part, by damage-induced differences in the local stability of hydrogen bonds. These results suggest the presence of a thermodynamic sensor that recruits the NER machinery by probing Watson–Crick base-pairing configurations.

## 5.2   Damaged DNA binding proteins

There are not only problems in the identification of conformational or structural signals eliciting NER activity, but also in the isolation of NER factors capable of detecting such signals. The techniques used to search for DNA binding proteins, mainly filter retention and electrophoretic mobility shift assays, have been extremely useful in identifying sequence-specific factors. However, these methods have yielded less satisfactory results when applied to the isolation of DNA damage-recognition proteins from crude cell extracts. For example, XPA, RPA, or XPC, which are now believed to constitute damage-recognition subunits of human NER, were completely missed. On the other hand, many proteins that were isolated on the basis of their preferential interaction with damaged DNA, are not or only marginally involved in DNA excision repair processes.

The first human damaged DNA binding (DDB) factor was detected using UV-irradiated DNA probes for filter binding assays (98), and partially purified from human placenta as a 120-kDa polypeptide (99). Presumably it was this same factor that was rediscovered about 12 years later, with the finding that cell lines from 2 out of 13 unrelated XP-complementation group E patients are missing a DDB activity (100). This association with the XP syndrome led to the hypothesis that DDB protein may function as a DNA damage-recognition factor in human NER. The purification of DDB protein was reported almost simultaneously by two laboratories, but with different results. Hwang and Chu (101) obtained a 125-kDa polypeptide that migrated as a monomer on gel-filtration and glycerol-gradient sedimentation. Keeney et al. (102) isolated a heterodimeric protein consisting of two polypeptides of 124 and 41 kDa. Microinjection experiments indicate that this heterodimeric factor corrects the DNA repair defect in those XP-E cells lacking DDB activity (103), but it

**Table 5.2** DNA binding constants ($K_a$) of DDB and XPA proteins (see text for references)

| Factor | Non-damaged DNA | DNA containing cyclobutane pyrimidine dimers | DNA containing (6–4) photoproducts |
|---|---|---|---|
| DDB/XPE | $5.5 \times 10^8 \, M^{-1}$ | $1.7 \times 10^9 \, M^{-1}$ | $1.6 \times 10^{10} \, M^{-1}$ |
| XPA | $\sim 6 \times 10^5 \, M^{-1}$ | $\sim 6 \times 10^5 \, M^{-1}$ | $3 \times 10^6 \, M^{-1}$ |

has not been unequivocally established that one of the two subunits of the isolated DDB factor constitutes the XPE gene product (104).

Using DNA fragments containing one of the major UV photolesions at a unique position, Sancar and collaborators analysed the DNA binding affinity of the DDB heterodimer (105). In mobility shift assays, these authors found that the DDB factor discriminates only modestly between cyclobutane–pyrimidine dimer-containing DNA and undamaged DNA, while the factor binds to pyrimidine(6–4)pyrimidone photoproducts with much higher affinity. Table 5.2 shows that the reaction constants for binding of the DDB factor to DNA are about 3 and 30 times higher in the presence of cyclobutane dimers and (6–4) photoproducts, respectively, than in the absence of DNA damage. These results are consistent with DDB protein being a damage-recognition component of human NER, as cyclobutane dimers are a relatively poor substrate of this process, whereas (6–4) photoproducts are rapidly removed. However, the precise function of this DDB factor remains uncertain. In fact, a complete lack of DDB activity correlates with only an approximately 50% reduction of excision repair activity in the affected XP-E patients (103), indicating that DDB cannot be the only factor responsible for targeting the NER system to DNA damage. This view is supported by the observation that purified DDB fractions are not necessary to reconstitute the human NER reaction (11, 12, 104).

Several hypotheses have been proposed regarding the biochemical function of the DDB factor. It has been noted, for example, that it is a relatively abundant protein and, as a consequence, the DDB factor may constitute a structural element of chromatin that increases the accessibility of NER enzymes to the DNA substrate (102, 106, 107). The DDB factor may also facilitate damage recognition by a mechanism similar to that proposed for the DNA photolyase of *E. coli*, a light-dependent repair enzyme that, in the dark, is able to enhance the rate of DNA incision by the prokaryotic NER system (108). According to this model, DDB interaction with UV photoproducts may provide an antenna that promotes the recruitment of other repair factors by direct protein–protein contacts or, alternatively, DDB protein may induce specific conformational changes of the DNA helix that are more compatible with recognition. Another model proposes that the DDB factor is a molecular chaperone that promotes specific interactions during the assembly of multimeric complexes (104). Of course, the DDB factor may indirectly improve the overall rate of excision repair by facilitating the dissociation of NER complexes after double DNA incision, thereby increasing the turnover of NER subunits. Similar functions have already been tentatively assigned to the XPC–HHR23B heterodimer (see Section 4.3). Finally, a recent study indicates an unexpected activity of DDB protein as the partner of a transcription factor that regulates the expression of DNA replication genes (109).

## 5.3  Shielding from excision repair

Some other DDB proteins that have been isolated from crude extracts by filter-binding or electrophoretic mobility-shift assays may even potentiate genotoxic effects by suppressing rather than promoting DNA repair. For example, proteins containing high-mobility group (HMG) domains bind preferentially to the major DNA adducts formed by the antineoplastic agent cisplatin (cis-diamminedichloro-platinum(II)), i.e. 1,2-intrastrand d(GpG) and d(ApG) cross-links (110). The HMG box is an evolutionary conserved region that displays a characteristic pattern of basic and aromatic amino acids and mediates sequence-specific DNA binding interactions. Proteins containing this motif include HMG-1 and HMG-2, hUBF (human upstream binding factor), the testis determining factor SRY, h-mtTFA (mitochondrial tran-scription factor), LEF-1 (lymphoid enhancer binding factor), and TCF-1α (a T cell-specific transcription factor) (reviewed in ref. 111).

In addition to their physiological binding sites in DNA, HMG box proteins also recognize specific conformational changes generated at intrastrand cisplatin cross-links, typically involving bending and unwinding of the DNA double helix along with exposure of an abnormally wide and hydrophobic minor groove (112). For example, hUBF or SRY bind to DNA molecules containing 1,2-intrastrand d(GpG) cross-links or to their transcriptional target sequences with comparable affinities (association constants in the range of $10^9$ to $10^{10}$ M$^{-1}$), suggesting that cisplatin adducts should be effective competitors of HMG protein-promoter complex formation (110, 113). Support for this hypothesis came from experiments in which hUBF binding to the natural ribosomal RNA promoter was efficiently antagonized by the addition of DNA fragments carrying 1,2-intrastrand d(GpG) cisplatin adducts but not by the addition of undamaged fragments (110). Thus, the term 'transcription factor hijacking' has been introduced to indicate that sequestration at platinated sites may titrate essential transcription factors away from their natural promoter or enhancer sequences (114, 115). Also, the sequestration of regulatory factors at illegiti-mate sites on the chromosomes may lead to the assembly of transcription initiation complexes in the promoter region of genes, including perhaps oncogenes, that are normally tightly regulated or not expressed at all. It has been speculated that this alternative mechanism may result in abnormal oncogene activation by cisplatin cross-links or certain carcinogen–DNA adducts (114).

Another deleterious effect of such DDB proteins is the increased persistence of damaged sites because of protein-induced protection, or shielding, from excision repair processes. In fact, studies on the yeast S. cerevisiae have identified a gene product that confers enhanced sensitivity to cisplatin. IXR1 (for intrastrand cross-link recognition) was isolated from an expression library for its ability to bind to cisplatin-damaged DNA (116). Cloning and sequence analysis of the IXR1 gene showed that it contains two tandemly repeated HMG boxes. When the *IXR1* gene was deleted, the resulting yeast strain grew normally but displayed increased resistance to cisplatin. The possibility that the lower cytotoxicity of cisplatin in this mutant arises from enhanced excision repair was investigated in a follow-up study, where the *IXR1* gene deletion was introduced in NER-deficient rad2, rad4, or rad14 strains (117). These *RAD* genes (for *rad*iation sensitive) encode the *S. cerevisiae* homologues of *XPG* (*RAD2*), *XPC* (*RAD4*), and *XPA* (*RAD14*). In the double mutants carrying deletions of both *IXR1* and a *RAD* gene, the differential sensitivity caused by removing IXR1 protein was nearly abolished, indicating that inhibition of excision repair is indeed the most plausible mechanism by which IXR1 sensitizes yeast cells towards cisplatin.

Protection of platinum adducts from DNA repair processes was confirmed using either NER-proficient human cell extracts (118, 119) or the reconstituted human NER system (120). In these *in vitro* assays, repair of 1,2-intrastrand d(GpG) cisplatin adducts was detected by monitoring oligonucleotide excision from substrates that carry a site-specific lesion (see Figs 5.8 and 5.9 for the excision assay). As expected from the *in vivo* yeast data, excision of d(GpG) platinum adducts was progressively reduced when increasing amounts of HMG-1 or h-mtTFA were added to the repair reactions. It was noted that relatively high concentrations (0.5–5 $\mu$M) of each HMG proteins were required to suppress NER *in vitro*, and these concentrations of individual HMG factors are certainly higher than those found *in vivo*. In mammalian cells, on the other hand, there are many such HMG box proteins that are likely to exert similar inhibitory effects on the repair of cisplatin lesions. Thus, the total cellular concentration of all HMG box proteins may match the concentrations used *in vitro* (118, 120).

The phenomenon of shielding from recognition by excision repair has at least three important clinical implications. First, tissue-specific overexpression of proteins that recognize and protect cisplatin–DNA adducts should enhance the cytotoxic potential of cisplatin and may explain the extraordinary sensitivity of certain malignancies, particularly testicular and ovarian tumours, to this drug (113). Before cisplatin was available, only about 5% of patients with testicular cancer were cured but, with cisplatin treatment, up to 90% of patients can expect long-term survival. Second, cancer cells may avoid this mechanism of cytotoxicity and acquire cisplatin resistance by suppressing the expression of HMG box proteins (118). Third, future pharmacological strategies may exploit the shielding mechanism to increase the therapeutic effectiveness of platinum-based anticancer drugs and generate new derivatives that form bulky base adducts with even higher affinity for cellular HMG domain proteins. Enhanced effectiveness may also be achieved by the overexpression of HMG box proteins in target tumour tissues using appropriate recombinant vectors.

## 5.4 The damaged DNA binding function of XPA protein

These studies conducted on various damaged DNA binding factors (DDB, XPE, HMG box, and, more recently, TATA binding proteins) indicate that, for the initiation of NER, human cells must express additional recognition subunits. One

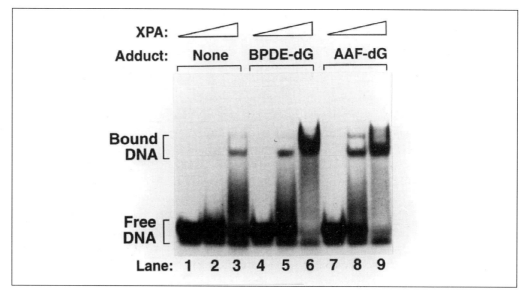

**Fig. 5.6** Electrophoretic mobility shift assay demonstrating the preferential binding of XPA protein to carcinogen-damaged substrates. Increasing amounts of purified recombinant XPA were incubated with short DNA duplexes containing no adduct (lanes 1–3), a site-directed BPDE–dG adduct (lanes 4–6), or a site-directed AAF–dG adduct (lanes 7–9).

such candidate is XPA, a zinc finger protein that is essential for NER activity and was shown by mobility shift assays to bind to DNA fragments with a moderate preference for UV-irradiated over unirradiated DNA (37). XPA protein also binds preferentially to carcinogen-damaged substrates carrying, for example, a single benzo[*a*]pyrene diol-epoxide (BPDE) or a single acetylaminofluorene (AAF) adduct (compare lanes 2, 5, and 8 of Fig. 5.6). XPA shows a binding constant for damaged fragments ($\sim 3 \times 10^6$ M$^{-1}$) which is approximately fivefold higher than its affinity for undamaged controls ($K_a \approx 6 \times 10^5$ M$^{-1}$). Thus, the affinity of XPA is approximately 5000-fold lower than that of DDB/XPE for the same UV-irradiated substrate (see Table 5.2). Furthermore, removal of cyclobutane dimers by treatment with DNA photolyase and visible light does not detectably reduce binding of XPA to the UV-irradiated fragments, indicating that the increased affinity of XPA protein for irradiated DNA is mediated exclusively by (6–4) photoproducts (37).

Although the preference of XPA protein for (6–4) photoproducts seems to correlate again with the higher repair rate of this particular lesion in human cells, it was surprising to find that XPA protein does not bind to cyclobutane pyrimidine dimers at all. Excision of both major UV radiation products is strictly dependent on XPA protein and, in fact, their removal is essentially abolished in XP-A cells (121). However, on the basis of its poor DNA binding capability and its inability to discriminate between normal residues and cyclobutane–pyrimidine dimers, XPA cannot be considered the only damage-recognition subunit of human NER. Also, the mechanism by which cyclobutane dimers (the most abundant lesion generated upon UV radiation) are recognized in human cells remains unknown.

## 5.5 The role of multiprotein assembly in damage recognition

An increasing number of protein–protein interactions between different human NER components are being identified. It remains to be elucidated whether these associations occur in a sequential manner or whether such interactions result in progressively more complex nucleoprotein structures, possibly culminating in the formation of a large 'repairosome' at sites of damage. Importantly, some of these protein–protein interactions may potentiate the selectivity for damaged residues. Bertrand-Burggraf *et al.* (122) introduced the term 'selectivity cascade' to indicate that a series of partly overlapping steps of low selectivity may result in very high specificities for damaged substrates. For example, XPA and RPA interact with each other (see Sections 4.1 and 4.2), and UV–cisplatin- or AAF-damaged DNA fragments display a five- to tenfold higher capacity to bind XPA–RPA complexes than either component (XPA or RPA) alone (38). These results are suggestive of damage recognition by a mechanism that involves the co-operative binding of XPA and RPA to DNA lesions. However, close inspection of the original data in ref. 38 shows that a physical association between XPA and RPA also stimulates the interaction of these proteins with undamaged substrates, such that the co-operative binding of XPA and RPA does not increase their selectivity for damaged deoxyribonucleotides when compared to the undamaged controls. Similarly, XPA associates strongly with ERCC1 (see Section 4.1), and filter-retention assays showed that this interaction stimulates the intrinsic binding of XPA to UV-irradiated DNA up to sevenfold (123). No stimulation of binding activity was observed in the presence of undamaged DNA and, additionally, the ERCC1 protein alone was completely unable to bind DNA, regardless of whether the nucleic acid substrate was damaged or not (123). In combination, these results suggest that selectivity for damaged DNA may be enhanced by XPA-mediated recruitment of XPF–ERCC1 heterodimers to target sites. More recently, however, the significance of this particular interaction has become uncertain because XPF–ERCC1 (the 5' endonuclease) binds only weakly to a nucleoprotein intermediate formed in the presence of all other preincision factors (XPA, RPA, XPC–HHR23B, and TFIIH). Apparently, the XPA subunit residing in such a multiprotein complex is no longer accessible for protein–protein interactions with XPF–ERCC1 (124). Finally, XPA also associates with TFIIH (see Section 4.1), and it has been proposed that this interaction may serve to recruit TFIIH in a DNA damage-dependent manner (125). This hypothesis was prompted by pull-down experiments in which XPA bound to UV-irradiated DNA was able to extract more TFIIH from a human cell lysate than XPA bound to unirradiated DNA. Damage-specific recruitment of TFIIH was then tested directly with purified components (48mer DNA fragments, XPA, and TFIIH), but the results obtained with this reconstituted system rather suggest that XPA attracts similar amounts of TFIIH regardless of whether the DNA substrate is damaged or not (125). In light of these findings, it remains unclear how the sequential assembly of protein–protein complexes may contribute to damage selectivity during the human NER process.

# 6. Bipartite substrate discrimination in human NER

The unresolved problem of DNA damage recognition prompted us to use a completely new approach to dissect the mechanism of substrate discrimination by human NER enzymes. A common feature of essentially all lesions that stimulate NER activity is the covalent modification of DNA bases. Other lesions that are processed by human NER extend to sites at which a base is degraded (for example urea residues), completely lost (apurinic/apyrimidinic sites), or replaced by bulky organic derivatives, such as a cholesterol moiety (7, 8, 57). A major consequence of all these different forms of base modification, base degradation, base loss, or base replacement is the interference with regular Watson–Crick hydrogen bonds between complementary partners (adenine–thymine or guanine–cytosine). For example, AAF–$C^8$-guanine adducts are covalent base modifications that disrupt local base-pairing interactions by forcing the modified guanine into an extrahelical position (126, 127). This conception led us to uncouple covalent DNA lesions from their destabilizing effects on DNA secondary structure by constructing a series of C4'-modified backbone variants.

## 6.1 Analysis of substrate discrimination using C4' backbone modifications

As shown in Fig. 5.7, we manipulated the DNA backbone by adding either a bulky selenophenyl or a bulky pivaloyl group to the C4' position of the deoxyribose moiety. A third non-bulky C4' backbone modification was obtained by inverting the normal geometry of chemical bonds. The C4' position of deoxyribose was chosen for these experiments because of its localization diametrically opposite to the N-glycosidic bond that links the bases to their backbone (Fig. 5.7). Thus, we expected that such C4' modifications would not influence Watson–Crick hydrogen-bonding interactions between complementary bases. In support of this view, crystal and solution structure analysis of C4'-modified nucleosides or the respective nucleotides showed that these residues essentially adopt the same conformation as unmodified nucleotides in double-stranded DNA (128). Also, enzymatic probing with Klenow fragment demonstrated that the coding fidelity of template bases is not disturbed by C4' deoxyribose manipulations, indicating intact base pairing opposite such backbone lesions (129). This conclusion is confirmed by the observation that single C4' modifications induce only marginal changes in the melting temperature of short DNA duplexes (13).

DNA repair activity was measured *in vitro* using the oligonucleotide excision assay originally devised by Huang *et al.* (7, 9), in which the characteristic double DNA incision pattern of human NER is exploited for analytical purposes. As illus-

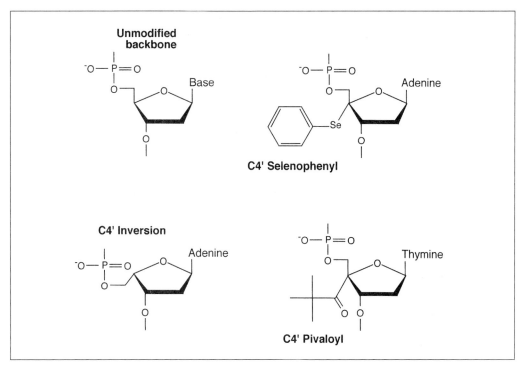

**Fig. 5.7** C4′ backbone variants used as molecular tools to uncouple DNA modifications from their destabilizing effect on DNA secondary structure and, hence, dissect the DNA damage-recognition step of human NER.

trated in Fig. 5.8, site-specifically modified duplex DNA substrates of 147 base pairs were assembled by ligating a 19mer oligonucleotide carrying a single modification with five other oligonucleotides. Prior to ligation, the central 19mer was labelled with [$^{32}$P]ATP at its 5′ end, such that the resulting 147mer substrate contained an internal radiolabel in the vicinity of the site-directed modification (Fig. 5.8). A standard soluble extract from HeLa cells, employed as a source of human NER factors, is able to perform excision repair of damaged DNA molecules when supplemented with ATP and all four deoxyribonucleotides (9, 10, 130). Using internally labelled substrates, damage-specific double DNA incision by human NER generates radioactive products of 24–32 nucleotides in length (Fig. 5.8), that are subsequently resolved by denaturing gel electrophoresis and visualized by autoradiography.

The representative polyacrylamide gel shown in Fig. 5.9 demonstrates the complete lack of oligonucleotide excision during incubation of a HeLa cell extract with DNA substrates containing, for example, a unique C4′ pivaloyl adduct (lane 3). On the other hand, human NER enzymes catalyse very efficient oligonucleotide excision when the same C4′ modification is incorporated into a 3-nucleotide long segment of mispaired bases (lanes 4 and 5), demonstrating that disruption of local hydrogen bonds is indeed an important prerequisite for recognition. The observed excision products are of a similar size range (25–29 residues), but much more abundant than

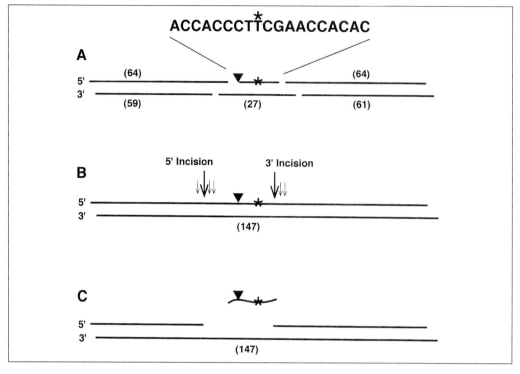

**Fig. 5.8** Oligonucleotide excision assay. (A) Preparation of internally labelled substrates; (B) sites of DNA incision by human NER enzymes; (C) radiolabelled oligonucleotide excision products of 24–32 residues in length carrying the damaged nucleotide. The asterisk denotes a site of covalent DNA modification. The arrowhead indicates the position of radioactive labelling.

those induced by an AAF–C$^8$-guanine lesion in the same sequence (lane 6). A markedly weaker, but nevertheless detectable, excision reaction was also found when C4′-modified deoxyribose residues were combined with a single base mismatch (13). Importantly, sequence heterologies containing either one (13) or three mismatches (lane 2 in Fig. 5.9), in the absence of concurrent C4′ modifications, fail to stimulate oligonucleotide excision.

In summary, we have been able to use C4′ DNA backbone modifications as molecular tools to dissect the recognition problem in human NER into two discrete components. As illustrated in Fig. 5.10, neither disruption of base-pairing complementarity nor C4′ backbone lesions are capable of eliciting NER activity, but the combination of these two substrate alterations constitutes a very potent signal for double DNA incision. We propose the term 'bipartite recognition' to denote this requirement for two separate determinants of recognition, i.e. a base-pair destabilizing defect in the secondary structure (or conformation) of DNA accompanied by a modification of its primary structure (or chemistry). Interestingly, this bipartite mode of substrate discrimination is different from the mechanism of DNA damage recognition in other excision repair pathways. Mismatch repair is active at sites of

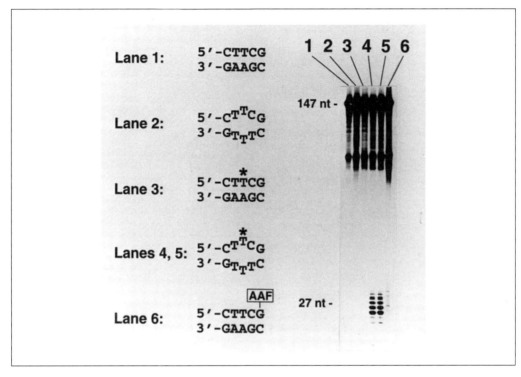

**Fig. 5.9** Bipartite substrate discrimination by human NER. The repair activity was determined in a HeLa cell extract using the excision assay described in Fig. 5.8. The tested lesions and their sequence context in the duplex substrate are indicated, with the asterisks denoting the site of C4' pivaloyl backbone modification. Lane 6 shows a control reaction with a site-specific AAF adduct. Identical results were obtained when we challenged human NER activity with a bulky C4' selenophenyl adduct or a non-bulky C4' inversion.

mispaired or unpaired bases in the absence of covalent changes of DNA constituents (131). In base excision repair, on the other hand, specific DNA glycosylases recognize damaged bases without a strict requirement for duplex destabilization (132). As a consequence of its bipartite damage-recognition strategy, however, human NER is preferentially recruited to carcinogen–DNA adducts that destabilize complementary hydrogen-bonding interactions between partner bases (see Section 6.2). Finally, the identification of a two-component substrate-discrimination strategy indicates that two distinct recognition subunits may coexist in human NER, i.e. a conformational sensor of defective hybridization (see Section 6.3) and a sensor of defective deoxyribonucleotide chemistry (see Section 6.4).

## 6.2 Significance of bipartite recognition in mutagenesis and carcinogenesis

Many different studies converge on the central concept that the mutagenic risk of a particular DNA lesion, and hence its carcinogenic potential, depends on the

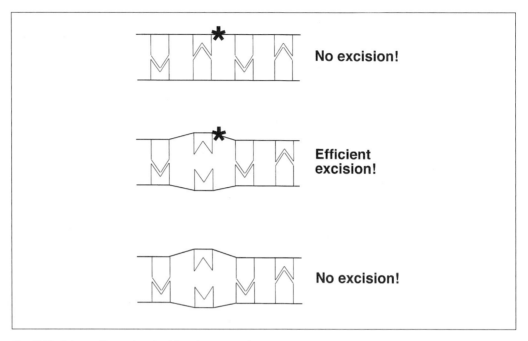

**Fig. 5.10** Scheme illustrating the bipartite mode of substrate discrimination by the human NER system. C4' backbone variants, which do not derange complementary hydrogen bonds on their own, induce NER reactions only in combination with at least one mispaired base. No oligonucleotide excision is detected when DNA contains mispaired bases without concomitant changes in deoxyribose-phosphate composition. Thus, neither C4' backbone lesions (indicated by the asterisk) nor improper base pairing stimulate human NER, but the combination of these two substrate alterations constitute an extremely potent signal for oligonucleotide excision.

efficiency by which this lesion is recognized and eliminated from the genome. DNA excision repair is, in fact, a highly non-uniform process, and lesions that are slowly repaired have been shown to coincide with mutagenesis hotspots in critical target sequences (133). Thus, it is essential to gain an understanding of the conformational factors that determine excision rates in response to a particular carcinogen–DNA adduct. Based on the bipartite mechanism of substrate discrimination, we postulated that disruption of local hydrogen bonds should facilitate recognition by human NER enzymes. DNA adducts generated by the ultimate carcinogen benzo[a]pyrene diol-epoxide (BPDE) offer an excellent experimental system to test this prediction. Benzo[a]pyrene is the prototypical member of a large class of polycyclic aromatic hydrocarbons that are widely distributed in our environment; metabolic activation of these compounds results in the formation of numerous diol-epoxide derivatives that differ in their stereochemistry, but share the common ability to covalently modify DNA (134, 135). For example, reaction of the epoxide moiety of either the (+) or (–)-*anti*-BPDE with the position $N^2$ of guanine generates two pairs of enantiomeric adducts: (+)-*trans*-, (+)-*cis*-, (–)-*trans*-, and (–)-*cis*-*anti*-BPDE–$N^2$-dG.

The solution structures of short DNA duplexes (5'-CCATCGCTACC-3') each containing one of these stereoisomeric guanine adducts have been characterized by

NMR spectroscopy (reviewed in ref. 135). All four BPDE lesions produce unfavourable thermodynamic changes by reducing the stability of duplex DNA but, depending on their stereochemistry, these lesions impose completely different changes on local base pairing. In particular, the (+)-*trans*-and (−)-*trans-anti*-BPDE–$N^2$-dG adducts adopt an external minor groove conformation which weakens, but does not break the hydrogen bonds that mediate Watson–Crick pairing between the modified guanine and its partner cytosine. In contrast to these adducts in *trans*, both *cis-anti*-BPDE–$N^2$-dG adducts adopt a helix-inserted conformation in which the modified guanine and the complementary cytosine are displaced out of the normal intrahelical position, causing severe disruption of their hydrogen bonding interactions. These conformational features of (+)-*cis*- and (−)-*cis-anti*-BPDE–$N^2$-dG are reminiscent of the predominant AAF–$C^8$-dG adduct, as this lesion also involves extrahelical displacement of the modified base (126, 127).

The capacity of human NER to process these stereochemically distinct carcinogen–DNA adducts in the sequence 5′-CCATCGCTACC-3′ has been determined using the *in vitro* oligonucleotide excision assay (134). A direct comparison yielded the following hierarchy of dual DNA incision: AAF–$C^8$-dG$\geq$ (+)-*cis*- = (−)-*cis*- >> (+)-*trans*- = (−)-*trans-anti*-BPDE–$N^2$-dG. Thus, the efficiency of BPDE excision is entirely dictated by damage-induced conformational changes of DNA, and a shift of adduct conformation from *cis* (with local disruption of base pairing) to *trans* (with partially intact base-pairing geometry) produces a 10-fold reduction in repair rates. These results confirm that the recognition and, hence, the repair of carcinogen–DNA adducts is dependent on the degree of residual Watson–Crick hydrogen-bonding interactions at each lesion site.

## 6.3   A potential sensor of defective Watson–Crick hybridization

The finding of a bipartite mode of substrate discrimination invokes the participation of two separate molecular 'callipers' during DNA damage recognition. A molecular 'calliper' for outside measurements senses the secondary structure (or conformation) of DNA. In addition, there is a molecular 'calliper' for inside measurements that senses the primary structure (or chemistry) of DNA. We have used electrophoretic mobility shift assays to examine the potential role of known NER factors in probing the conformation of double helical DNA. Purified XPA protein displays a remarkable ability to discriminate between normal homoduplexes and DNA containing sites of improper base pairing generated by sequence heterologies. In fact, XPA binds to a short duplex with three mispaired bases in the centre even more avidly than to homoduplex substrates of the same length but containing a single BPDE or AAF carcinogen–DNA adduct (N. Buschta-Hedayat *et al.*, in preparation). This result suggests that XPA may indeed act as a conformational sensor that detects defective base-pairing interactions, thereby recruiting the NER system to non-hybridizing helical sites.

It is tempting to postulate that XPA uses its zinc finger, a putative DNA binding domain (see Section 4.1), to discriminate between Watson–Crick conformations in normal duplexes and defective base-pair conformations in damaged duplexes. The zinc finger of XPA is, in principle, comparable to the respective DNA binding motifs of many sequence-specific transcription factors. However, molecular modelling based on nuclear magnetic resonance spectroscopy predicts that, in the vicinity of the zinc finger, there is no clear structural similarity between XPA protein and these various transcription factors (136, 137). To test the possible function of the XPA zinc finger, Ikegami *et al.* (138) extended their biophysical analysis by monitoring specific changes in the nuclear magnetic resonance pattern upon incubation of a minimal DNA binding fragment (XPA$_{98-219}$) with DNA substrate. Surprisingly, these chemical shift perturbation experiments suggest that the zinc finger domain of XPA may not be involved in DNA binding at all (138). On the contrary, XPA appears to detect non-hybridizing sites by a novel mechanism of conformation-dependent protein–DNA interaction. In fact, recent work in our laboratory revealed that XPA displays an extraordinary affinity for DNA fragments containing hydrophobic base analogues, designated 'universal bases' (139), that lack hydrogen bonding capabilities but maintain stacking interactions with appropriate aromatic partners (N. Buschta-Hedayat *et al.*, in preparation). Considering that displacement (or 'flipping out') of natural bases from the normal Watson–Crick conformation is associated with extrahelical exposure of their aromatic ring structures, this finding suggests that XPA may recognize non-hybridizing base pairs by sensing the greater accessibility to hydrophobic residues on the double helical surface. Such an indirect recognition mechanism, mediated by the presence of abnormal hydrophobic attractions on the DNA surface and without any specificity for a particular type of damage, confers very high levels of versatility to the substrate discrimination function of XPA protein, thereby explaining how this subunit may target the human NER system to a wide range of different DNA lesions.

## 6.4   A potential sensor of defective deoxyribonucleotide chemistry

In contrast to the function of XPA as a sensor of base-pair conformation, TFIIH may probe the chemical integrity of DNA and discriminate between normal and defective deoxyribonucleotide constituents. As outlined in Section 4.4, TFIIH is a multimeric complex that includes, as catalytic subunits, two distinct DNA helicases with 3'→5' polarity (XPB) and 5'→3' polarity (XPD). DNA helicases promote strand displacement by disrupting the hydrogen bonds that hold duplex DNA molecules together (see Fig. 5.3). This strand separation reaction is dependent on the hydrolysis of nucleoside 5'-triphosphates and involves unidirectional translocation of helicase enzymes along the DNA substrate. Oh and Grossman (140) were the first to propose that a DNA helicase activity may serve to initiate excision repair mechanisms. In the prokaryotic (A)BC excinuclease system, UvrA and UvrB associate to form a UvrA$_2$B trimer that behaves like a DNA helicase on partial duplex DNA substrates, i.e. it catalyses DNA strand displacement coupled to ATP hydrolysis. The DNA helicase activity of UvrA$_2$B is inhibited by UV radiation damage, implying that translocation of the protein complex is arrested by DNA lesions. On the basis of these observations, Grossman and co-workers proposed that the DNA helicase activity of UvrA$_2$B may serve to scan DNA in search for deoxyribonucleotide damage to be processed by the bacterial NER system (140). The interaction of XPB or XPD with damaged DNA substrates has not yet been investigated, but experiments on Rad3 protein (the highly conserved *S. cerevisiae* homologue of XPD) show that this particular DNA helicase is exceptionally sensitive to covalent modifications affecting the bases (141) or the DNA backbone (142). After arresting its ATPase and DNA helicase activities, Rad3 protein forms stable complexes with damaged DNA strands but not with undamaged DNA (141, 143). Thus, eukaryotic NER systems may have adopted enzymes with DNA helicase activity (Rad3 in yeast and XPD in humans) not only to open (unwind) the double helix but also to promote detection of modified DNA components in a highly specific and accurate manner.

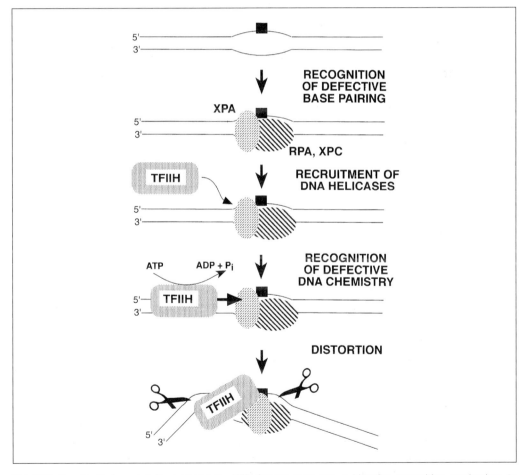

**Fig. 5.11** Hypothetical model illustrating how NER factors may use a bipartite recognition mechanism to discriminate between undamaged and damaged substrates. The rectangle symbolizes a base adduct.

# 7. Conclusions

Our current knowledge of the reaction mechanism of human NER is as follows. Four factors (XPA, RPA, XPC–HHR23B, TFIIH) co-operate to form an unwound pre-incision complex flanked by single-stranded to double-stranded DNA junctions on either side of the offending lesion. Double DNA incision at these junctions by structure-specific endonucleases (XPF–ERCC1 and XPG) generates excision products of 24–32 residues in length containing the damaged site. DNA repair patches are then synthesized and ligated by the action of several replication factors (3, 4, 6, 64, 66, 74). In contrast to these enzymatic events involving either breakage or formation of phosphodiester bonds, the early molecular recognition steps that localize DNA damage in the genome are poorly understood. However, the known biochemical properties of human NER factors, in combination with their bipartite mode of action in discriminating damaged substrates, may be integrated in the following hypothetical model of early NER reactions (Fig. 5.11).

1. *Recognition of defective base-pair conformations.* XPA senses improper hybridization between complementary strands and invades the DNA double helix at sites of defective Watson–Crick base pairing. In particular, XPA uses hydrophobic side chains of its DNA binding domain to monitor the DNA surface in search for abnormal exposure of the aromatic ring component of displaced bases. XPA possesses only low DNA binding affinities, but its function as a conformational sensor of Watson–Crick integrity is potentiated by association with RPA (N. Buschta Hedayat *et al.*, in preparation) and, possibly, XPC–HHR23B.

2. *Recruitment of DNA helicases.* TFIIH is attracted to substrate DNA by the C-terminal domain of XPA. At this stage, the assembly of a multimeric nucleoprotein complex may be facilitated by the presence of HHR23B.

3. *Recognition of defective DNA chemistry.* The DNA helicase activity of XPD and XPB is used to scan short segments of DNA and detect abnormal deoxyribonucleotide residues. Thus, the helicase components of TFIIH act as sensors of defective deoxy-ribonucleotide chemistry. This reaction requires hydrolysis of ATP and involves strand displacement. The opposite polarity of the two DNA helicases ($5' \rightarrow 3'$ and $3' \rightarrow 5'$) indicates that each enzyme translocates along one strand of the antiparallel duplex. Translocation of helicase enzymes is arrested at sites of covalent DNA damage, but only one of the two DNA helicases, for example XPD, is blocked by the lesion.

4. *Distortion.* Studies on Rad3 protein (the highly conserved *S. cerevisiae* homologue of XPD) showed that the damage-induced block of its helicase activity is strictly DNA strand-specific (141). Perhaps, after inhibition of one (either XPD or XPB) of the two DNA helicases, the opposite DNA helicase bound to the undamaged strand may proceed for at least one or a few additional nucleotides. To compensate for this asymmetric movement driven by ATP hydrolysis, DNA is distorted to form a kinked

intermediate. Many other DNA transactions that select highly specific sites in the genome have already been shown to involve distortion of the double helix during an early molecular recognition step (see for example refs 144, 145). DNA bending and unwinding in this kinked recognition intermediate may constitute critical parameters for subsequent protein–DNA and protein–protein interactions. For example, the particular orientation of DNA bending should determine which strand is exposed to double cleavage by endonucleases.

5. *DNA incision*. After these conformational changes, the resulting nucleoprotein structure provides the substrate for DNA incision by structure-specific DNA endo-nucleases, XPG on the one side and XPF–ERCC1 on the other side of the damage.

The bipartite recognition model of Fig. 5.11 accommodates the observation that human NER processes a wide range of structurally and chemically dissimilar DNA lesions without acting on undamaged DNA and, additionally, accounts for the facilitated recognition of those adducts that destabilize complementary base pairing. The model also accommodates the strand specificity associated with NER, i.e. its ability to discriminate between the two strands of duplex DNA and target only the damaged strand for DNA incision. Current research in our and several other laboratories is devoted to elucidating the molecular details of this unique substrate discrimination mechanism.

## Acknowledgements

Research support for the author's laboratory comes from the Swiss National Science Foundation (Grant 31–50518.97).

## References

1. Friedberg, E. C., Walker, G. C., and Siede, W. (1995) *DNA repair and mutagenesis*. American Society for Microbiology, Washington, DC.
2. Dianov, G., Price, A., and Lindahl, T. (1992) Generation of single-nucleotide repair patches following excision of uracil residues from DNA. *Mol. Cell. Biol.*, **12**, 1605.
3. Sancar, A. (1996) DNA excision repair. *Annu. Rev. Biochem.*, **65**, 43.
4. Wood, R. D. (1996) DNA repair in eukaryotes. *Annu. Rev. Biochem*, **65**, 135.
5. Hanawalt, P. C. and Haynes, R. (1965) Repair replication of DNA in bacteria: irrelevance of chemical nature of base defect. *Biochem. Biophys. Res. Commun.*, **19**, 462.
6. Hoeijmakers, J. H. J. (1993) Nucleotide excision repair II: from yeast to mammals. *Trends Genet.*, **9**, 211.
7. Huang, J. C., Hsu, D. S., Kazantsev, A., and Sancar, A. (1994) Substrate spectrum of human excinuclease: repair of abasic sites, methylated bases, mismatches, and bulky adducts. *Proc. Natl Acad. Sci. USA*, **91**, 12213.
8. Reardon, J. T., Bessho, T., Kung, H. C., Bolton, P. H., and Sancar, A. (1997) *In vitro* repair of oxidative DNA damage by human nucleotide excision repair system: possible explanation for neurodegeneration in xeroderma pigmentosum patients. *Proc. Natl Acad. Sci. USA*, **94**, 9463.

9. Huang, J.-C., Svoboda, D. L., Reardon, J. T., and Sancar, A. (1992) Human nucleotide excision nuclease removes thymine dimers from DNA by incising the 22nd phosphodiester bond 5' and the 6th phosphodiester bond 3' to the photodimer. *Proc. Natl Acad. Sci. USA*, **89**, 3664.

10. Moggs, J. G., Yarema, K. J., Essigmann, J. M., and Wood, R. D. (1996) Analysis of incision sites produced by human cell extracts and purified proteins during nucleotide excision repair of a 1,3-intrastrand d(GpTpG)-cisplatin adduct. *J. Biol. Chem.*, **271**, 7177.

11. Mu, D., Park, C.-H., Matsunaga, T., Hsu, D. S., Reardon, J. T., and Sancar, A. (1995) Reconstitution of human DNA repair excision nuclease in a highly defined system. *J. Biol. Chem.*, **270**, 2415.

12. Aboussekhra, A., Biggerstaff, M., Shivji, M. K. K., Vilpo, J. A., Moncollin, V., Podust, V. N., *et al.* (1995) Mammalian DNA nucleotide excision repair reconstituted with purified DNA components. *Cell*, **80**, 859.

13. Hess, M. T., Schwitter, U., Petretta, M., Giese, B., and Naegeli, H. (1997) Bipartite substrate discrimination by human nucleotide excision repair. *Proc. Natl Acad. Sci. USA*, **94**, 6664.

14. Cleaver, J. E. (1968) Defective repair replication of DNA in xeroderma pigmentosum. *Nature*, **218**, 652.

15. Wood, R. D., Robins, P., and Lindahl, T. (1988) Complementation of the xeroderma pigmentosum DNA repair defect in cell-free extracts. *Cell*, **53**, 97.

16. Kraemer, K. H., Lee, M.-M., Andrews, A. D., and Lambert, W. C. (1994) The role of sunlight and DNA repair in melanoma and nonmelanoma skin cancer. *Arch. Dermatol.*, **130**, 1018.

17. Brash, D. E., Rudolph, J. A., Simon, J. A., Lin, A., McKenna, G. J., Baden, H. P., *et al.* (1991) A role for sunlight in skin cancer: UV-induced *p53* mutations in squamous cell carcinoma. *Proc. Natl Acad. Sci. USA*, **88**, 10124.

18. Satoh, M. S., Jones, C. J., Wood, R. D., and Lindahl, T. (1993) DNA excision-repair defect of xeroderma pigmentosum prevents removal of a class of oxygen free radical-induced base lesions. *Proc. Natl Acad. Sci. USA*, **90**, 6335.

19. Lehmann, A. R., Bootsma, D., Clarkson, S. G., Cleaver, J. E., McAlpine, P. J., Tanaka, K., *et al.* (1994) Nomenclature of human DNA repair genes. *Mutat. Res.*, **315**, 41.

20. Collins, A. R. (1993) Mutant rodent cell lines sensitive to ultraviolet light, ionizing radiation and cross-linking agents: a comprehensive survey of genetic and biochemical characteristics. *Mutat. Res.*, **293**, 99.

21. Westerveld, A., Hoeijmakers, J. H. J, van Duin, M., de Wit, J., Odijk, H., Pastink, A., *et al.* (1984) Molecular cloning of a human DNA repair gene. *Nature*, **310**, 425.

22. Flejter, W. L., McDaniel, L. D., Johns, D., Friedberg, E. C., and Schultz, R. A. (1992) Correction of xeroderma pigmentosum complementation group D mutant cell phenotypes by chromosome and gene transfer: involvement of the human *ERCC2* DNA repair gene. *Proc. Natl Acad. Sci. USA*, **89**, 261.

23. Weeda, G., van Ham, R. C. A., Vermeulen, W., Bootsma, D., van der Eb, A. J., and Hoeijmakers, J. H. J. (1990) A presumed DNA helicase encoded by *ERCC-3* is involved in the human repair disorders xeroderma pigmentosum and Cockayne's syndrome. *Cell*, **62**, 777.

24. Sijbers, A. M., de Laat, W. L., Ariza, R. R., Biggerstaff, M., Wei, Y.-F., Moggs, J. G., *et al.* (1996) Xeroderma pigmentosum group F caused by a defect in a structure-specific DNA repair endonuclease. *Cell*, **86**, 811.

25. Scherly, D., Nouspikel, T., Corlet, J., Ucla, C., Bairoch, A., and Clarkson, S. G. (1993) Complementation of the DNA repair defect in xeroderma pigmentosum group G cells by a human cDNA related to yeast *RAD2*. *Nature*, **363**, 182.

26. Tanaka, K., Miura, N., Satokata, I., Miyamoto, I., Yoshida, M. C., Satoh, Y., *et al.* (1990) Analysis of a human DNA excision repair gene involved in group A xeroderma pigmentosum and containing a zinc finger domain. *Nature*, **348**, 73.

27. Legerski, R. and Peterson, C. (1992) Expression cloning of a human DNA repair gene involved in xeroderma pigmentosum group C. *Nature*, **359**, 70.

28. Wakasugi, M., Reardon, J. T., and Sancar, A. (1997) The non-catalytic function of XPG protein during dual incision in human nucleotide excision repair. *J. Biol. Chem.*, **272**, 16030.

29. Kuraoka, I., Morita, E. H., Saijo, M., Matsuda, T., Morikawa, K., Shirakawa, M., *et al.* (1996) Identification of a damaged-DNA binding domain of the XPA protein. *Mutat. Res.*, **362**, 87.

30. Asahina, H., Kuraoka, I., Shirakawa, M., Morita, E. H., Miura, N., Miyamoto, I., *et al.* (1994) The XPA protein is a zinc metalloprotein with an ability to recognize various kinds of DNA damage. *Mutat. Res.*, **315**, 229.

31. Miyamoto, I., Miura, N., Niwa, H., Miyazaki, J., and Tanaka, K. (1992) Mutational analysis of the structure and function of the xeroderma pigmentosum group A complementing protein: identification of essential domains for nuclear localization and DNA excision repair. *J. Biol. Chem.*, **267**, 12182.

32. Van Houten, B. (1990) Nucleotide excision repair in *Escherichia coli*. *Microbiol. Rev.*, **54**, 18.

33. Tchou, J., Michaels, M. L., Miller, J. H., and Grollman, A. P. (1993) Function of the zinc finger in *Escherichia coli* Fpg protein. *J. Biol. Chem.*, **268**, 26738.

34. Jones, J. S., Weber, S., and Prakash, L. (1988) The *Saccharomyces cerevisiae RAD18* gene encodes a protein that contains potential zinc finger domains for nucleic acid binding and a putative nucleotide binding sequence. *Nucl. Acids Res.*, **16**, 7119.

35. Gradwohl, G., Ménissier de Murcia, J., Molinete, M., Simonin, F., Koken, M., Hoeijmakers, J. H. J., *et al.* (1990) The second zinc-finger domain of poly(ADP-ribose) polymerase determines specificity for single-stranded breaks in DNA. *Proc. Natl Acad. Sci. USA*, **87**, 2990.

36. Mackay, J. P. and Crossley, M. (1998) Zinc fingers are sticking together. *Trends Biochem. Sci.*, **23**, 1.

37. Jones, C. J. and Wood, R. D. (1993) Preferential binding of the xeroderma pigmentosum group A complementing protein to damaged DNA. *Biochemistry*, **32**, 12096.

38. He, Z., Henricksen, L. A., Wold, M. S., and Ingles, C. J. (1995) RPA involvement in the damage-recognition and incision steps of nucleotide excision repair. *Nature*, **374**, 566.

39. Li, L., Lu, X., Peterson, C. A., and Legerski, R. J. (1995) An interaction between the DNA repair factor XPA and replication protein A appears essential for nucleotide excision repair. *Mol. Cell. Biol.*, **15**, 5396.

40. Park, C.-H. and Sancar A. (1994) Formation of a ternary complex by human XPA, ERCC1 and ERCC4(XPF) excision repair proteins. *Proc. Natl Acad. Sci. USA*, **91**, 5017.

41. Li, L., Peterson, C. A., and Legerski, R. F. (1995) Mutations in XPA that prevent association with ERCC1 are defective in nucleotide excision repair. *Mol. Cell. Biol.*, **15**, 1993.

42. Park, C.-H., Mu, D., Reardon, J. T., and Sancar, A. (1995) The general transcription-repair factor TFIIH is recruited to the excision repair complex by the XPA protein independent of the TFIIE transcription factor. *J. Biol. Chem.*, **270**, 4896.

43. Hurwitz, J., Dean, F. B., Kwong, A. D., and Lee, S.-H. (1990) The *in vitro* replication of DNA containing the SV40 origin. *J. Biol. Chem.*, **265**, 18043.

44. Lin, Y.-L., Shivji, M. K. K., Chen, C., Kolodner, R., Wood, R. D., and Dutta, A. (1998) The evolutionary conserved zinc finger motif in the largest subunit of human replication protein A is required for DNA replication and mismatch repair but not for nucleotide excision repair. *J. Biol. Chem.*, **273**, 1453.

45. Longhese, M. P., Plevani, P., and Lucchini, G. (1994) Replication factor A is required *in vivo* for DNA replication, repair, and recombination. *Mol. Cell. Biol.*, **14**, 7884.

46. Bochkarev, A., Pfuetzner, R. A., Edwards, A. M., and Frappier, L. (1997) Structure of the single-stranded-DNA-binding domain of replication protein A bound to DNA. *Nature*, **385**, 176.

47. Pfuetzner, R. A., Bochkarev, A., Frappier, L., and Edwards, A. M. (1997) Replication protein A. Characterization and crystallization of the DNA binding domain. *J. Biol. Chem.*, **272**, 430.

48. Din, S.-U., Brill, S. J., Fairman, M. P., and Stillman, B. (1990) Cell-cycle-regulated phosphorylation of DNA replication factor A from human and yeast cells. *Genes Dev.*, **4**, 968.

49. Carty, M. P., Zernik-Kobak, M., McGrath, S., and Dixon, K. (1994) UV light-induced DNA synthesis arrest in HeLa cells is associated with changes in phosphorylation of human single-stranded DNA-binding protein. *EMBO J.*, **13**, 2114.

50. Pan, Z. Q., Park, C.-H., Amin, A. A., Hurwitz, J., and Sancar, A. (1995) Phosphorylated and unphosphorylated forms of human single-stranded DNA-binding protein are equally active in simian virus 40 DNA replication and in nucleotide excision repair. *Proc. Natl Acad. Sci. USA*, **92**, 4636.

51. Abramova, N. A., Russell, J., Botchan, M., and Li, R. (1997) Interaction between replication protein A and p53 is disrupted after UV damage in a DNA repair-dependent manner. *Proc. Natl Acad. Sci. USA*, **94**, 7186.

52. Shivji, M. K. K., Podust, V. N., Hübscher, U., and Wood, R. D. (1995) Nucleotide excision repair DNA synthesis by DNA polymerase ε in the presence of PCNA, RFC, and RPA. *Biochemistry*, **34**, 5011.

53. Matsunaga, T., Park, C.-H., Bessho, T., Mu, D., and Sancar, A. (1996) Replication protein A confers structure-specific endonuclease activities to the XPF–ERCC1 and XPG subunits of human DNA repair excision nuclease. *J. Biol. Chem.*, **271**, 11047.

54. Seroussi, E. and Lavi, S. (1993) Replication protein A is the major single-stranded DNA binding protein detected in mammalian cell extracts by gel retardation assays and UV cross-linking of long and short single-stranded DNA molecules. *J. Biol. Chem.*, **268**, 7147.

55. Toulmé, J. J., Behmoaras, T., Guigues, M., and Hélène, C. (1983) Recognition of chemically damaged DNA by the gene 32 protein from bacteriophage T4. *EMBO J.*, **2**, 505.

56. Masutani, C., Sugasawa, K., Yanagisawa, J., Sonoyama, T., Ui, M., Enomoto, T., *et al.* (1994) Purification and cloning of a nucleotide excision repair complex involving the xeroderma pigmentosum group C protein and a human homolog of yeast RAD23. *EMBO J.*, **13**, 1831.

57. Reardon, J. T., Mu, D., and Sancar, A. (1996) Overproduction, purification, and characterization of the XPC subunit of the human DNA repair excision nuclease. *J. Biol. Chem.*, **271**, 19451.

58. Sugusawa, K., Ng, J. M. Y., Masutani, C., Maekawa, T., Uchida, A., van der Spek, P. J., *et al.* (1997) Two human homologs of Rad23 are functionally interchangeable in complex formation and stimulation of XPC repair activity. *Mol. Cell. Biol.*, **17**, 6924.

59. Wakasugi, M. and Sancar, A. (1998) Assembly, subunit composition, and footprint of human DNA repair excision nuclease. *Proc. Natl Acad. Sci. USA*, **95**, 6669.

60. Guzder, S. N., Bailly, V., Sung, P., Prakash, L., and Prakash, S. (1995) Yeast DNA repair protein RAD23 promotes complex formation between transcription factor TFIIH and DNA damage recognition factor RAD14. *J. Biol. Chem.*, **270**, 8385.

61. Drapkin, R., Reardon, J. T., Ansari, A., Huang, J. C., Zawel, L., Ahn, K., *et al.* (1994) Dual

role of TFIIH in DNA excision repair and in transcription by RNA polymerase II. *Nature*, **368**, 769.

62. Schauber, C., Chen, L., Tongaonkar, P., Vega, I., Lambertson, D., Potts, W., *et al.* (1998) Rad23 links DNA repair to the ubiquitin/proteasome pathway. *Nature*, **391**, 715.

63. Venema, J., van Hoffen, A., Karcagi, V., Natarajan, A. T., van Zeeland, A. A., and Mullenders, L. H. F. (1991) Xeroderma pigmentosum complementation group C cells remove pyrimidine dimers selectively from the transcribed strand of active genes. *Mol. Cell. Biol.*, **11**, 4128.

64. Naegeli, H. (1995) Mechanisms of DNA damage recognition in mammalian nucleotide excision repair. *FASEB J.*, **9**, 1043.

65. Mu, D. and Sancar, A. (1997) Model for XPC-independent transcription-coupled repair of pyrimidine dimers in humans. *J. Biol. Chem.*, **272**, 7570.

66. Hoeijmakers, J. H. J. (1994) Human nucleotide excision repair syndromes: molecular clues to unexpected intricacies. *Eur. J. Cancer*, 30A, 1912.

67. Schaeffer, L., Roy, R., Humbert, S., Moncollin, V., Vermeulen, W., Hoeijmakers, J. H. J., *et al.* (1993) DNA repair helicase: a component of BTF2 (TFIIH) basic transcription factor. *Science*, **260**, 58.

68. Vermeulen, W., van Vuuren, A. J., Chipoulet, M., Schaeffer, L., Appeldoorn, E., Weeda, G., *et al.* (1994) Three unusual repair deficiencies associated with transcription factor BTF2 (TFIIH). Evidence for the existence of a transcription syndrome. *Cold Spring Harb. Symp. Quant. Biol.*, **59**, 317.

69. Drapkin, R. and Reinberg, D. (1994) The multifunctional TFIIH complex and transcriptional control. *Trends Biochem. Sci.*, **19**, 504.

70. Roy, R., Adamczewski, J. P., Seroz, T., Vermeulen, W., Tassan, J.-P., Schaeffer, L., *et al.* (1994) The MO15 cell cycle kinase is associated with the TFIIH transcription-DNA repair factor. *Cell*, **79**, 1093.

71. Svejstrup, J. Q., Wang, Z., Feaver, W. J., Wu, X., Bushnell, D. A., Donahue, T. F., *et al.* (1995) Different forms of TFIIH for transcription and DNA repair: holo-TFIIH and a nucleotide excision repairosome. *Cell*, **80**, 21.

72. Goodrich, J. A. and Tjian, R. (1994) Transcription factor IIE and IIH and ATP hydrolysis direct promoter clearance by RNA polymerase II. *Cell*, **77**, 145.

73. Shiekhattar, R., Mermelstein, F., Fisher, R., Drapkin, R., Dynlacht, B., Wessling, H. C., *et al.* (1995) Cdk-activating kinase (CAK) complex is a component of human transcription factor IIH. *Nature*, **374**, 283.

74. Mu, D., Hsu, D. S., and Sancar, A. (1996) Reaction mechanism of human DNA repair excision nuclease. *J. Biol. Chem.*, **271**, 8285.

75. Fesquet, D., Labbé, J.-C., Derancourt, J., Capony, J.-P., Galas, S., Girard, F., *et al.* (1993) The MO15 gene encodes the catalytic subunit of a protein kinase that inactivates cdc2 and other cyclin-dependent kinases (CDKs) through phosphorylation of Thr161 and its homologues. *EMBO J.*, **12**, 3111.

76. Adamczewski, J. P., Rossignol, M., Tassan, J.-P., Nigg, E. A., Moncollin, V., and Egly, J.-M. (1996) MAT1, cdk7 and cyclin H form a kinase complex which is UV-light sensitive upon association with TFIIH. *EMBO J.*, **15**, 1877.

77. Wang, X. W., Vermeulen, W., Coursen, J. D., Gibson, M., Lupold, S. E., Forrester, K., *et al.* (1995) p53 modulation of TFIIH-associated nucleotide excision repair activity. *Nature Genet.*, **10**, 188.

78. Jones, C. J. and Wynford-Thomas, D. (1995) Is TFIIH an activator of the p53-mediated $G_1/S$ checkpoint? *Trends Genet.*, **11**, 165.

79. Ko, L. J., Shieh, S. Y., Chen, X. B., Jayaraman, L., Tamai, K., Taya, Y., *et al.* (1997) p53 is phosphorylated by CDK7-cyclin H in a p36(MAT1)-dependent manner. *Mol. Cell. Biol.*, **17**, 7220.

80. Ford, J. M. and Hanawalt, P. C. (1995) Li–Fraumeni syndrome fibroblasts homozygous for p53 mutations are deficient in global DNA repair but exhibit normal transcription-coupled repair and enhanced UV resistance. *Proc. Natl Acad. Sci. USA*, **92**, 8876.

81. Léveillard, T., Andera, L., Bissonnette, N., Schaeffer, L., Bracco, L., Egly, J.-M., *et al.* (1996) Functional interactions between p53 and TFIIH complex are affected by tumour-associated mutations. *EMBO J.*, **15**, 1615.

82. Park C.-H., Bessho, T., Matsunaga, T., and Sancar, A. (1995) Purification and characterization of the XPF–ERCC1 complex of human DNA repair excision nuclease. *J. Biol. Chem.*, **270**, 22657.

83. O'Donovan, A. E., Davies, A. A., Moggs, J. G., West, S. C., and Wood, R. D. (1994) XPG endonuclease makes the 3′ incision in human DNA nucleotide excision repair. *Nature*, **371**, 432.

84. Iyer, N., Reagan, M. S., Wu, K. J., Canagarajah, B., and Friedberg, E. C. (1996) Interactions involving the human RNA polymerase II transcription/nucleotide excision repair complex TFIIH, the nucleotide excision repair protein XPG, and Cockayne syndrome group B (CSB) protein. *Biochemistry*, **35**, 2157.

85. Evans, E., Fellows, J., Coffer, A., and Wood, R. D. (1997) Open complex formation around a lesion during nucleotide excision repair provides a structure for cleavage by human XPG protein. *EMBO J.*, **16**, 625.

86. Evans, E., Moggs, J. G., Hwang, J. R., Egly, J.-M., and Wood, R. D. (1997) Mechanism of open complex and dual incision formation by human nucleotide excision repair factors. *EMBO J.*, **16**, 6559.

87. Matsunaga, T., Mu, D., Park, C.-H., Reardon, J. T., and Sancar, A. (1995) Human DNA repair excision nuclease. Analysis of the roles of the subunits involved in dual incisions by using anti-XPG and anti-ERCC1 antibodies. *J. Biol. Chem.*, **270**, 20862.

88. Hansson, J., Munn, M., Rupp, W. D., Kahn, R., and Wood, R. D. (1989) Localization of DNA repair synthesis by human cell extracts to a short region at the site of a lesion. *J. Biol. Chem.*, **264**, 21788.

89. Reardon, J. T., Thompson, L. H., and Sancar, A. (1997) Rodent UV-sensitive mutant cell lines in complementation groups 6–10 have normal general excision repair activity. *Nucl. Acids Res.*, **25**, 1015.

90. Prelich, G., Kostura, M., Marshak, D. R., Mathews, M. B., and Stillman, B. (1987) The cell-cycle regulated proliferating cell nuclear antigen is required for SV40 replication *in vitro*. *Nature*, **326**, 471.

91. Tsurimoto, T. and Stillman, B. (1989) Purification of a cellular replication factor, RF-C, that is required for coordinated synthesis of leading and lagging strands during simian virus 40 DNA replication in *vitro*. *Mol. Cell. Biol.*, **9**, 609.

92. Gary, R., Ludwig, D. L., Cornelius, H. L., MacInnes, M. A., and Park, M. S. (1997) The DNA repair endonuclease XPG binds to proliferating cell nuclear antigen (PCNA) and shares sequence elements with the PCNA-binding regions of FEN-1 and cyclin-dependent kinase inhibitor p21. *J. Biol. Chem.*, **272**, 24552.

93. Levin, D. S., Bai, W., Yao, N., O'Donnell, M., and Tomkinson, A. E. (1997) An interaction between DNA ligase I and proliferating cell nuclear antigen: implications for Okazaki fragment synthesis and joining. *Proc. Natl Acad. Sci. USA*, **94**, 12863.

94. Mitchell, D. L. (1988) The relative cytotoxicity of (6–4) photoproducts and cyclobutane pyrimidine dimers in mammalian cells. *Photochem. Photobiol.*, **48**, 51.

95. Ford, J. M. and Hanawalt, P. C. (1997) Expression of wild-type p53 is required for efficient global genomic nucleotide excision repair in UV-irradiated human fibroblasts. *J. Biol. Chem.*, **272**, 28073.

96. Gunz, D., Hess, M. T., and Naegeli, H. (1996) Recognition of DNA adducts by human nucleotide excision repair: evidence for a thermodynamic probing mechanism. *J. Biol. Chem.*, **271**, 25089.

97. Hess, M. T., Gunz, D., and Naegeli, H. (1996) A repair competition assay to assess recognition by human nucleotide excision repair. *Nucl. Acids Res.*, **24**, 824.

98. Feldberg, R. S. and Grossman, L. (1976) A DNA binding protein from human placenta specific for ultraviolet-damaged DNA. *Biochemistry*, **15**, 2402.

99. Feldberg, R. S., Lucas, J. L., and Dannenberg, A. (1982) A damage-specific DNA binding protein. *J. Biol. Chem.*, **257**, 6394.

100. Chu, G. and Chang, E. (1988) Xeroderma pigmentosum group E cells lack a nuclear factor that binds to damaged DNA. *Science*, **242**, 564.

101. Hwang, B. J. and Chu, G. (1993) Purification and characterization of a human protein that binds to damaged DNA. *Biochemistry*, **32**, 1657.

102. Keeney, S., Chang, G. J., and Linn, S. (1993) Characterization of a human DNA damage binding protein implicated in xeroderma pigmentosum E. *J. Biol. Chem.*, **268**, 21293.

103. Keeney, S., Eker, A. P. M., Brody, T., Vermeulen, W., Bootsma, D., Hoeijmakers, J. H. J., *et al.* (1994) Correction of the DNA repair defect in xeroderma pigmentosum group E by injection of a DNA damage-binding protein. *Proc. Natl Acad. Sci. USA*, **91**, 4053.

104. Kazantsev, A., Mu, D., Nichols, A. F., Zhao, X., Linn, S., and Sancar, A. (1996) Functional complementation of xeroderma pigmentosum complementation group E by replication protein A in an *in vitro* system. *Proc. Natl Acad. Sci. USA*, **93**, 5014.

105. Reardon, J. T., Nichols, A. F., Keeney, S., Smith, C. A., Taylor, J.-S., Linn, S., *et al.* (1993) Comparative analysis of binding of human damaged DNA-binding protein (XPE) and *Escherichia coli* damage recognition protein (UvrA) to the major ultraviolet photo-products: T[c,s]T, T[t,s]T, T[6–4]T, and T[Dewar]T. *J. Biol. Chem.*, **268**, 21301.

106. Rapic, V., Kuraoka, I., Nardo, T., McLenigan, M., Eker, A. P. M., Stefanini, M., *et al.* (1998) Relationship of the xeroderma pigmentosum group E DNA repair defect to the chromatin and DNA binding proteins UV-DDB and replication protein A. *Mol. Cell. Biol.*, **18**, 3182.

107. Hwang, B. J., Toering, S., Francke, U., and Chu, G. (1998) p48 activates a UV-damaged-DNA binding factor and is defective in xeroderma pigmentosum group E cells that lack binding activity. *Mol. Cell. Biol.*, **18**, 4391.

108. Sancar, A., Franklin, K. A., and Sancar, G. B. (1984) *Escherichia coli* DNA photolyase stimulates UvrABC excision nuclease *in vitro*. *Proc. Natl Acad. Sci. USA*, **81**, 7397.

109. Hayes, S., Shiyanov, P., Chen, X., and Raychaudhuri, P. (1998) DDB, a putative DNA repair protein, can function as a transcriptional partner of E2F1. *Mol. Cell. Biol.*, **18**, 240.

110. Treiber, D. K., Zhai, X., Jantzen, H.-M., and Essigmann, J. M. (1994) Cisplatin–DNA adducts are molecular decoys for the ribosomal RNA transcription factor hUBF (human upstream binding factor). *Proc. Natl Acad. Sci. USA*, **91**, 5672.

111. Jantzen, H. M., Admon, A., Bell, S. P., and Tijan, R. (1990) Nuclear transcription factor hUBF contains a DNA-binding motif with homology to HMG proteins. *Nature*, **344**, 830.

112. Takahara, P. M., Rosenzweig, A. C., Frederick, C. A., and Lippard, S. J. (1995) Crystal structure of double-stranded DNA containing the major adduct of the anticancer drug cisplatin. *Nature*, **377**, 649.

113. Trimmer, E. E., Zamble, D. B., Lippard, S. J., and Essigmann, J. M. (1998) Human testis-determining factor SRY binds to the major DNA adduct of cisplatin and a putative target sequence with comparable affinities. *Biochemistry*, **37**, 352.

114. MacLeod, M. C., Powell, K. L., and Tran, N. (1995) Binding of the transcription factor, Sp1, to non-target sites in DNA modified by benzo[*a*]pyrene diol-epoxide. *Carcinogenesis*, **16**, 975.

115. Coin, F., Frit, P., Viollet, B., Salles, B., and Egly, J.-M. (1998) TATA binding protein discriminates between different lesions on DNA, resulting in a transcription decrease. *Mol. Cell. Biol.*, **18**, 3907.

116. Brown, S. J., Kellett, P. J., and Lippard, S. J. (1993) Ixr1, a yeast protein that binds to platinated DNA and confers sensitivity to cisplatin. *Science*, **261**, 603.

117. McA'Nulty, M. M. and Lippard, S. J. (1996) The HMG-domain protein Ixr1 blocks excision repair of cisplatin–DNA adducts in yeast. *Mutat. Res.*, **362**, 75.

118. Huang, J.-C., Zamble, D. B., Reardon, J. T., Lippard, S. J., and Sancar, A. (1994) HMG-domain proteins specifically inhibit the repair of the major DNA adduct of the anticancer drug cisplatin by human excision nuclease. *Proc. Natl Acad. Sci. USA*, **91**, 10394.

119. Szymkowski, D. E., Yarema, K., Essigmann, J. M., Lippard, S. J., and Wood, R. D. (1992) An intrastrand d(GpG) platinum crosslink in duplex M13 DNA is refractory to repair by human cell extracts. *Proc. Natl Acad. Sci. USA*, **89**, 10772.

120. Zamble, D. B., Mu, D., Reardon, J. T., Sancar, A., and Lippard, S. J. (1996) Repair of cisplatin–DNA adducts by the mammalian excision nuclease. *Biochemistry*, **35**, 10004.

121. Tanaka, K. and Wood, R. D. (1994) Xeroderma pigmentosum and nucleotide excision repair. *Trends Biochem. Sci.*, **19**, 83.

122. Bertrand-Burggraf, E., Selby, C. P., Hearst, J. E., and Sancar, A. (1991) Identification of the different intermediates in the interaction of (A)BC excinuclease with its substrate by DNaseI footprinting on two uniquely modified oligonucleotides. *J. Mol. Biol.*, **219**, 27.

123. Nagai, A., Saijo, M., Kuraoka I., Matsuda, T., Kodo, N., Nakatsu, Y., *et al.* (1995) Enhancement of damage-specific DNA binding of XPA by interaction with the ERCC1 DNA repair protein. *Biochem. Biophys. Res. Comm.*, **211**, 960.

124. Mu, D., Wakasugi, M., Hsu, D. S., and Sancar, A. (1997) Characterization of reaction intermediates of human excision repair nuclease. *J. Biol. Chem.*, **272**, 28971.

125. Nocentini, S., Coin, F., Saijo, M., Tanaka, K., and Egly, J.-M. (1997) DNA damage recognition by XPA protein promotes efficient recruitment of transcription factor IIH. *J. Biol. Chem.*, **272**, 22991.

126. O'Handley, S. F., Sanford, S. G., Xu, R., Lester, C. C., Hingerty, B. E., Broyde, S., *et al.* (1993) Structural characterization of an *N*-acetyl-2-aminofluorene (AAF) modified DNA oligomer by NMR, energy minimization, and molecular dynamics. *Biochemistry*, **32**, 2481.

127. Hoffmann, G. R. and Fuchs, R. P. P. (1997) Mechanisms of frameshift mutations: insight from aromatic amines. *Chem. Res. Toxicol.*, **10**, 347.

128. Marx, A., MacWilliams, M. P., Bickle, T. A., Schwitter, U., and Giese, B (1997) 4'-Acetylated thymidines: a new class of DNA chain terminators and photocleavable DNA building blocks. *J. Am. Chem. Soc.*, **119**, 1131.

129. Hess, M. T., Schwitter, U., Petretta, M., Giese, B., and Naegeli, H. (1997) DNA synthesis arrest at C4'-modified deoxyribose residues. *Biochemistry*, **36**, 2332.

130. Manley, J. L., Fire, A., Cano, A., Sharp, P. A., and Gefter, M. L. (1980) DNA-dependent transcription of adenovirus genes in a soluble whole-cell extract. *Proc. Natl Acad. Sci. USA*, **77**, 3855.

131. Jiricny, J. (1994) Colon cancer and DNA repair: have mismatches met their match? *Trends Genet.*, **10**, 164.

132. Mol, C. D., Arvai, A. S., Slupphaug, G., Kavli, B., Alseth, I., Krokan, H. E., *et al.* (1995) Crystal structure and mutational analysis of human uracil-DNA glycosylase: structural basis for specificity and catalysis. *Cell*, **80**, 869.

133. Tornaletti, S. and Pfeifer, G. P. (1994) Slow repair of pyrimidine dimers at *p53* mutation hotspots in skin cancer. *Science*, **263**, 1436.

134. Hess, M. T., Gunz, D., Luneva, N., Geacintov, N. E., and Naegeli, H. (1997) Base pair conformation-dependent excision of benzoapyrene diol-epoxide-guanine adducts by human nucleotide excision repair enzymes. *Mol. Cell. Biol.*, **17**, 7069.

135. Geacintov, N. E., Cosman, M., Hingerty, B. E., Amin, S., Broyde, S., and Patel, D. J. (1997) NMR solution structures of stereoisomeric covalent polycyclic aromatic carcinogen–DNA adducts: principles, patterns, and diversity. *Chem. Res. Toxicol.*, **10**, 111.

136. Morita, E. H., Ohkubo, T., Kuraoka, I., Shirakawa, M., Tanaka, K., and Morikawa, K. (1996) Implications of the zinc-finger motif found in the DNA-binding domain of the human XPA protein. *Genes Cells*, **1**, 437.

137. Buchko, G. W. and Kennedy, M. A. (1997) Human nucleotide excision repair protein XPA: ¹H NMR and CD solution studies of a synthetic peptide fragment corresponding to the zinc-binding domain (101–141). *J. Biomol. Struct. Dynam.*, **6**, 677.

138. Ikegami, T., Kuraoka, I., Saijo, M., Kodo, N., Kyogoku, Y., Morikawa, K., *et al.* (1998) Solution structure of the DNA-and RPA-binding domain of the human repair factor XPA. *Nature Struct. Biol.*, **5**, 701.

139. Loakes, D., Brown, D. M., Linde, S., and Hill, F. (1995) 3-Nitropyrrole and 5-nitroindole are universal bases in primers for DNA sequencing and PCR. *Nucl. Acids Res.*, **23**, 2361.

140. Oh, E. Y. and Grossman, L. (1987) Helicase properties of the *Escherichia coli* UvrAB protein complex. *Proc. Natl Acad. Sci. USA*, **84**, 3638.

141. Naegeli, H., Bardwell, L., and Friedberg, E. C. (1992) The DNA helicase and adenosine triphosphatase activities of yeast Rad3 protein are inhibited by DNA damage. *J. Biol. Chem.*, **267**, 392.

142. Naegeli, H., Bardwell, L., Harosh, I., and Friedberg, E. C. (1992) Substrate specificity of the Rad3 ATPase/DNA helicase of *Saccharomyces cerevisiae* and binding of Rad3 protein to nucleic acids. *J. Biol. Chem.*, **267**, 7839.

143. Naegeli, H., Bardwell, L., and Friedberg, E. C. (1993) Inhibition of Rad3 DNA helicase activity by DNA adducts and abasic sites: implications for the role of a DNA helicase in damage-specific incision of DNA. *Biochemistry*, **32**, 613.

144. Kim, Y., Geiger, J. H., Hahn, S., and Sigler, P. B. (1993) Crystal structure of a yeast TBP/TATA-box complex. *Nature*, **365**, 512.

145. Shi, Q., Thresher, R., Sancar, A., and Griffith, J. (1992) Electron microscopic study of (A)BC excinuclease. DNA is sharply bent in the UvrB–DNA complex. *J. Mol. Biol.*, **226**, 425.

# 6 | Transcription-coupled and global genome repair in yeast and humans

MARCEL TIJSTERMAN, RICHARD A. VERHAGE, and JAAP BROUWER

## 1. Introduction

Nucleotide excision repair (NER) is able to remove a wide variety of DNA lesions from the genome. Incision of the damaged strand on both sides of the lesion is followed by resynthesis of excised DNA using the non-damaged strand as a template. The role of the proteins involved in this cut-and-paste mechanism is studied biochemically, and these studies have recently culminated in the reconstitution of eukaryotic NER *in vitro* using repair components purified from both yeast and human cells (see Chapter 5 for a detailed discussion on the proteins involved). Whereas these *in vitro* assays allowed the identification of a minimal set of proteins (here referred to as core-NER proteins) required to perform the reaction on naked DNA, analysis of repair *in vivo* indicated that additional proteins are necessary for the efficient removal of damage from genomic DNA. These factors are believed to be involved in the first step of the NER pathway: the recognition of different types of DNA damage in a genome that is continuously undergoing various cellular processes (e.g. transcription, replication, recombination) and which, in addition, is folded into chromatin with different levels of complexity. In principle, NER can operate on lesions throughout the genome, but the rate at which this process removes DNA lesions is not the same for different genomic regions. For UV-induced cyclobutane–pyrimidine dimers (CPDs) repair occurs preferentially in transcriptionally active compared to silent DNA (1), and this enhanced repair mainly originates from the increased rate of repair of the transcribed strand versus non-transcribed DNA. This phenomenon is widespread, being found in organisms ranging from *Escherichia coli* to man (2–4), and is designated transcription-coupled repair because it is dependent upon ongoing transcription (5, 6). Lesions in non-transcribed DNA are obviously no target for transcription-coupled repair, but nevertheless are removed by the NER pathway. This process has been termed global (or overall) genome repair, and specific mutants exist that are defective in this mode of repair, demonstrating that transcription-coupled and global genome repair systems can function independently.

This chapter covers the processing of NER substrates *in vivo* and discusses those genes specifically implicated in either one of both subpathways, i.e. transcription coupled- and global genome repair. Since important insight into the complexity of NER was provided by repair analysis of specific subfractions of the genome we will first discuss the methodology used to measure NER inside living cells.

# 2. Techniques

## 2.1 Gene-specific repair analysis

The first direct method demonstrating variations in repair rates of different parts of the genome combined the Southern hybridization technique with the substrate specificity of the phage enzyme T4 endonuclease V (1). This enzyme binds specifically to UV-induced CPDs and incises the DNA strand 5′ of the dimer. Thus, DNA fragments containing such lesions will be digested upon treatment with the enzyme, and this will lead to a loss of intensity of the hybridization signal during Southern analysis (Fig. 6.1). The initial number of CPD lesions in specific DNA fragments induced by UV irradiation is determined by comparing the signal obtained from T4 endo V-treated and mock-treated DNA samples. By allowing cells time to repair the DNA, T4 endo V sensitive sites are removed, and this results in the reappearance of the hybridization signal. Using strand-specific hybridization probes, gene-specific repair analysis was then extended to individual DNA strands, which led to the identification of strand-specific repair (4). More recently, the repair kinetics of many different types of lesions have been studied using variations on this general principle, e.g. by changing the manner in which damage-specific incision is accomplished.

## 2.2 Nucleotide-specific repair analysis

Different methods have been developed to analyse the repair of DNA damage in even more detail, i.e. at the individual nucleotide level. The ability to determine the repair rates of individual bases contributes to our understanding of the NER mechanism *in vivo* as will be discussed later, and, in addition, is of great importance in judging causality between the induction and repair of DNA lesions and the induction of mutations.

At the crossroads of gene- and nucleotide-specific repair analysis, the first example of intragenic repair variations was observed in yeast using a method that combined T4 endo V incision ability with a technique of indirect end-labelling a plasmid repair target (see ref. 3 for details). Repair analysis of genomic targets at nucleotide resolution was first accomplished in *E. coli* (7). By using a sequence-specific, oligonucleotide-directed, end-labelling procedure instead of Southern hybridization, the length of the incised fragments (which corresponds to the position of the original damaged bases) could be determined on a denaturing polyacrylamide gel. In this

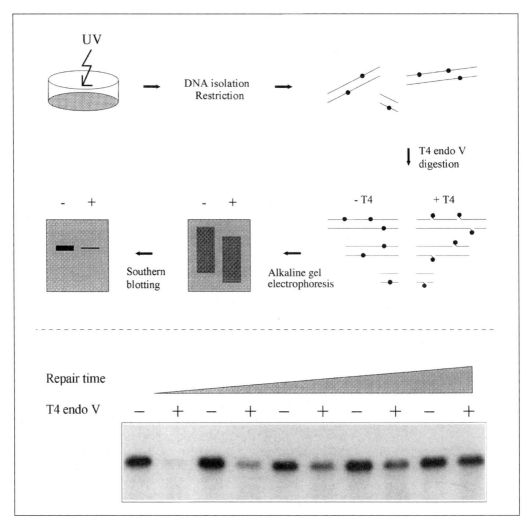

**Fig. 6.1** Schematic representation of the gene-specific repair assay as first described by Bohr *et al.* (1) to analyse the repair of UV-induced CPDs (•). DNA is isolated from cells exposed to UV light and digested with appropriate restriction endonucleases to size the fragment of interest. Samples are treated or mock-treated with the CPD specific enzyme T4 endonuclease V, run in parallel on denaturing agarose gels, and transferred to Southern blots. These blots are then hybridized with a gene- or strand-specific probe. The signal of the full-length fragment in the treated and mock-treated lanes is quantified and the amount of CPDs in the original DNA fragment calculated using the Poisson expression. Repair is measured by comparing the amount of CPDs at each time point after irradiation with the initial induction at time zero.

way, the repair of damaged bases could be individually monitored. Since then, minor alterations in the basic strategy (see Fig. 6.2(A)), have been introduced to either simplify the assay (8) or to increase its specificity to allow repair analysis in yeast (9).

Another method employed in the analysis of yeast repair utilizes the inability of DNA polymerase to bypass DNA lesions (10, 11). In this protocol, the target sequence is determined by the specificity of a selected oligonucleotide, which is extended with DNA polymerase up to the site of base damage. Thus, the length of the obtained fragment determines the position of the DNA damage. In addition, the signal can be amplified in a linear manner by multiple rounds of denaturation, primer annealing, and extension.

To analyse repair at nucleotide resolution in human cells, the ligation-mediated polymerase chain reaction (PCR) strategy was manipulated to visualize strand breaks introduced by the CPD specific T4 endo V (12, 13). However, amplification of the damage specific signal is required when using this technique, as the ratio of the gene of interest versus the bulk DNA is approximately 2–3 $\times$ $10^2$-fold reduced compared to the lower eukaryote or the prokaryote. This powerful tool for quantifying repair rates has been applied to the study of UV-induced CPDs in a number of target genes including the p53 gene (12), although a biased distribution pattern that results from the amplification reaction complicates a quantitative analysis of the damage incidence level (14). As for gene-specific repair analysis, this method has been extended to analyse other (chemically induced) adducts using different enzymes to incise the DNA at sites of base damage (15). The essential elements of this method (reviewed in ref. 16) are illustrated in Fig. 6.2, panel B.

## 2.3 Nucleotide excision repair *in vitro*

Cell-free systems have been developed to dissect the biochemistry of both yeast and human NER and to characterize the proteins required (17–19). One method uses crude-cell lysates incubated with damaged plasmid molecules as substrates in the presence of an energy regenerating system and radiolabelled nucleotides (Fig. 6.3). NER is measured by the incorporation of the labelled nucleotides in the plasmid target during repair synthesis. *In vitro* complementation is accomplished by mixing cell lysates from different NER mutants or by the addition of purified components to cognate mutant-cell extracts. This cell-free system has benefited the purification of various NER proteins, and recently the complete NER reaction has been reconstituted *in vitro* by fractionating human HeLa-cell extracts and subsequently replacing the fractions by highly purified components (20). In other studies, human and yeast repair proteins known to be required for the incision reaction were purified and used to reconstitute incision activity on damaged substrates (21, 22). The high degree of purity of the protein fractions used suggests that the obtained minimal protein combination is sufficient for the incision reaction. Although at present eukaryotic cell-free systems are available that support either NER or transcription, an *in vitro* system exhibiting transcription-coupled repair is still lacking despite extensive research.

## 3. Transcription-coupled nucleotide excision repair

Ever since the first observation of transcription-coupled repair (TCR), i.e. the enhanced repair of transcribed versus non-transcribed DNA, it has been proposed that RNA polymerase molecules arrested at sites of base damage might direct the

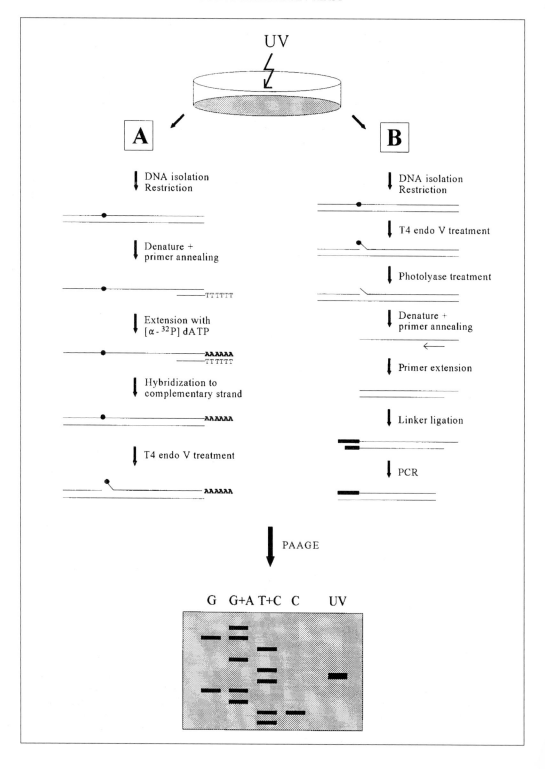

**Fig. 6.2** (A) Nucleotide-specific repair assay using genomic end-labelling (7). DNA is isolated from cells exposed to UV light and digested with appropriate restriction endonucleases to size the fragment of interest. An oligonucleotide is used to 3′ end-label a specific chromosomal fragment by permitting *Taq* DNA polymerase to extend the fragment with radiolabelled dATP using the oligonucleotide's non-complementary 5′ stretch as a template. The ssDNA is converted to dsDNA by hybridization with complementary DNA in order to be incised at the site of a CPD (•) by T4 endonuclease V. Samples are subjected to denaturing PAGE alongside Maxam–Gilbert reactions of the corresponding sequence. The positions of the individual CPDs are indicated by the length of the fragments, and repair is measured by comparing the amount of CPD at each time point with the initial induction at time zero (see also Fig. 6.6). (B) Nucleotide-specific repair assay using ligation-mediated PCR (12, 13). DNA is isolated from cells exposed to UV light and digested with T4 endonuclease V to create ssDNA breaks at CPD sites (•). To create a ligatable 5′ end, samples are treated with photolyase. After denaturation of the DNA, an oligonucleotide complementary to the fragment of interest is annealed and extended to the ssDNA break. Linker DNA is subsequently ligated to the generated blunt ends. The mixture is then amplified by PCR with a linker-specific and a partly overlapping internal oligonucleotide (see ref. 16 for details), and run on denaturing PAGE.

repair machinery to the transcribed strand of active DNA (4). This targeting concept was supported by the observation that CPDs, which are repaired in a strand-specific manner, when situated on the transcribed strand are able to block an active RNA polymerase elongation complex, whereas this lesion had no effect on transcription elongation when located on the non-transcribed strand (23). This hypothesis places the elongating transcription complex in the middle of the NER-transcription

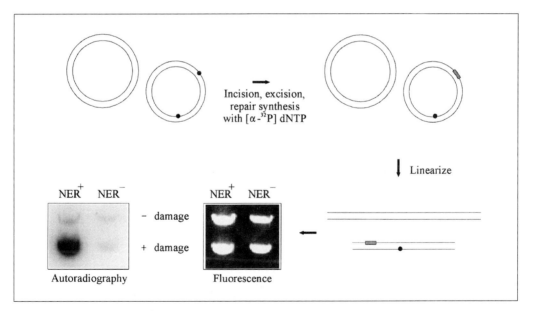

**Fig. 6.3** A schematic representation of the cell-free DNA repair assay as described by Wood *et al.* (17). A mixture of damaged (smaller) and non-damaged (larger) plasmid molecules is incubated with crude cell lysates, radiolabelled nucleotides, and an energy regenerating system. Plasmid DNA is recovered, linearized, and size-fractionated by electrophoresis. Repair is monitored by quantifying the damage-dependent incorporation of radionucleotides in the plasmid target (= DNA damage). By comparing NER-proficient cells (NER⁺) to NER-deficient cells (NER⁻, here *S. cerevisiae rad16Δ*) it is shown that label incorporation is mediated by the NER pathway.

interplay by implying that the initial signal for NER to act upon is a stalled RNA polymerase. Therefore, we will first introduce some of the basic steps involved in transcription by RNA polymerase II before discussing the relationship between this process and that of DNA repair. Our attention will mainly be focused on those features relevant to a more detailed analysis of the close interplay between NER and RNA pol II transcription. For a more extensive coverage of the latter process the reader is directed to recent reviews concerning this subject (see ref. 24 and references therein).

## 3.1 General features of RNA polymerase II transcription

In its most general form, the conventional model for transcription initiation by RNA polymerase II is characterized by the assembly of a multiprotein complex containing several basal transcription factors (TF) including TFIIA, TFIIB, TFIID, TFIIE, TFIIF, and TFIIH. The first step in assembly is the binding of TFIID (the TATA-binding protein (TBP) and associated factors) to the TATA-element, followed by the association of TFIIB, TFIIF, and pol II to the initial protein–DNA complex. Binding of TFIIE, which subsequently recruits TFIIH, completes the preinitiation process (reviewed in ref. 24). The biochemical events that follow the assembly leading to subsequent DNA melting, initiation, and promoter clearance are less clear. TFIIH contains several catalytic activities, including two DNA-dependent ATPase-helicases and a cyclin-dependent kinase (cdk) that is capable of phosphorylating the C-terminal domain of the large subunit of pol II. The DNA helicase activity of TFIIH is thought to be involved in the ATP-dependent formation of the open complex. The kinase activity of TFIIH has been implicated in the transition from transcription initiation to transcription elongation, designated promoter clearance. The carboxy-terminal domain (CTD) of pol II becomes heavily phosphorylated during this transition: whereas only the non-phosphorylated form of RNA pol II can enter the preinitiation complex, the phosphorylated form catalyses elongation (25, 26). Although many kinases are capable of phosphorylating the CTD *in vitro* (27), genetic data suggest that TFIIH is likely to be the prominent factor *in vivo*. Although it has been shown that the polymerase CTD physically interacts with certain general transcription factors in the preinitiation complex, it remains to be established whether phosphorylation of the CTD triggers the release of bound RNA pol II from the promoter.

Another important question concerns the fate of the general transcription factors (GTFs) during the progression of the RNA polymerase complex into the elongation mode. *In vitro* competition experiments have shown that TFIID remains promoter-bound following initiation and promoter clearance (28), whereas other GTFs sequentially dissociate (TFIIB and TFIIE are released before the formation of the tenth phosphodiester bond and TFIIH is released between positions +30 and +68). TFIIF is the only basal transcription factor detected in elongating RNA pol II complexes. The notion that TFIIH is not a component of the transcription elongation complex (28, 29) has important implications for the discussion on the molecular basis of the coupling of RNA pol II transcription to NER (see below).

## 3.2 TFIIH: required for RNA pol II transcription and nucleotide excision repair

The identification of known NER proteins as subunits of the TFIIH complex provided an unexpected direct connection between transcription and NER (30–32). All components of TFIIH have now been identified for both yeast and human, and the composition of the TFIIH complex shows a remarkable degree of evolutionary conservation since each component of one system has a counterpart in the other. Although it has been firmly established that the TFIIH complex is indispensable for NER (for recent reviews see refs 33 and 34), its precise function in the process has yet to be identified (see also Chapter 5).

One of the key questions concerning the function of TFIIH in both transcription and NER is whether this duality can provide a molecular basis for the TCR phenomenon. At this point it should be noted that mutations in TFIIH subunits cripple the NER of both strands (35, 36); demonstrating that TFIIH is not an exclusive TCR factor, but rather is essential for all NER. However, this notion does not exclude TFIIH from being quintessential for TCR: this phenomenon most likely results from enhanced targeting of core NER proteins to base damage in transcribed strands, and that the association of TFIIH with or its affinity for RNA polymerase can provide a rate advantage while still being an essential protein for the repair of non-transcribed DNA. It was initially imagined, subsequent to the obligatory loading of TFIIH on to promoter sites during transcription initiation, that TFIIH remains associated with the elongating transcription complex. This scenario provided a molecular basis for TCR by placing NER proteins in the immediate proximity of RNA polymerase blocking DNA damage. However, as discussed above, TFIIH is released from the transcription machinery *in vitro* soon after or during promoter clearance and no TFIIH has been detected in isolated stalled elongation complexes (28). The possibility that TFIIH itself has an increased affinity specifically for stalled RNA pol II molecules, which would result in the reassociation of this factor upon transcription stalling, lacks experimental support. This suggests that an alternative route is required to target TFIIH and other NER proteins (not necessarily in this order) to stalled elongation complexes. During transcription initiation TFIIH is recruited to the DNA–protein complex by TFIIE, and it has been suggested that this strategy might also apply for stalled RNA polymerases (37). In addition, re-recruitment of TFIIE and TFIIH may be required to resume elongation of the transcript once the damage is repaired, although it is currently unknown whether this resumption of transcription can occur. TFIIE is not a core NER component, since it is not required in reconstituted *in vitro* NER systems (38). However, whether TFIIE is required for TCR is not known. The answer to this question awaits the development of an *in vitro* system capable of performing TCR. A barrier to addressing this problem genetically is that when TFIIE functions identically in NER and transcription, NER-deficient TFIIE mutants will not be viable due to defective transcription (TFIIE alleles specifically disturbed in TCR have not been found). In contrast, factors involved in the transcription process will automatically display a TCR deficiency when transcription is hampered. As an example, in yeast, thermosensitive mutations either in the RNA pol II subunit *RPB1* or TFIIH subunit *KIN28* (responsible for the TFIIH CTD phosphorylation capacity) display a deficiency in TCR *in vivo* at the non-permissive temperature, while general NER is unaffected *in vivo* and *in vitro* (our unpublished observations).

Repair analysis in *E. coli* has provided a molecular model for TCR, which involves additional factors specifically involved in the recruitment of NER proteins towards stalled transcription complexes. Thus far, a detailed analysis of TCR at the bio-chemical level has only been performed using this bacterium, but a rather comprehensive model has emerged (39). This model will be described briefly before considering whether it can serve as a paradigm for eukaryotic systems.

## 3.3 Coupling NER to transcription in *E. coli*: a paradigm for eukaryotes?

In *E. coli*, enhanced repair of the transcribed strand is mediated by the transcription-repair coupling factor (TRCF), which is encoded by the *mfd* gene (39). Cells lacking this gene are deficient in TCR, show a moderate UV sensitivity, and display a shift of UV-induced mutations from the non-transcribed in the direction of the transcribed strand when compared to *mfd*⁺ cells (39–41). The latter is indicative of a reduction in repair of premutagenic lesions from the transcribed strand. *In vitro*, cell-free extracts of an *mfd*⁻ strain are deficient in preferential repair of the transcribed strand (42). In fact, RNA polymerase stalled at a lesion inhibits repair of the transcribed strand, resulting in a net effect of faster repair of non-transcribed DNA (43). This inhibition is relieved by the addition of purified TRCF, which simultaneously confers strand-specificity to the system in favour of the transcribed strand. Using the purified TRCF together with the *E. coli* NER proteins, a defined system capable of strand-specific repair has been reconstituted. The TRCF contains helicase motifs but displays no helicase activity, possesses a weak ATPase activity, and is able to recognize, bind, and displace RNA polymerase stalled at a lesion (39, 44). Together with the ability to bind to the *E. coli* NER protein UvrA, this suggests that the TRCF targets the transcribed strand for repair by recognizing a stalled RNA polymerase and actively recruiting the NER enzymes to the transcription blocking lesion as it dissociates the stalled RNA polymerase (Fig. 6.4). A noteworthy complication concerning this model came from *in vivo* repair analysis, which showed that a TRCF requirement in coupling repair to transcription is not absolute: although the TRCF encoding *mfd* gene is required for TCR of the *E. coli lacI* and *lacZ* genes when transcribed at a basal level, induced *lacZ* transcription leads to strand-specific repair even with a repair-defective *mfd* allele (45).

## 3.4 Transcription-coupled repair in humans

In humans, the hereditary disorder Cockayne syndrome (CS) is associated with defective TCR. CS cells are UV-sensitive and are unable to repair DNA damage in a

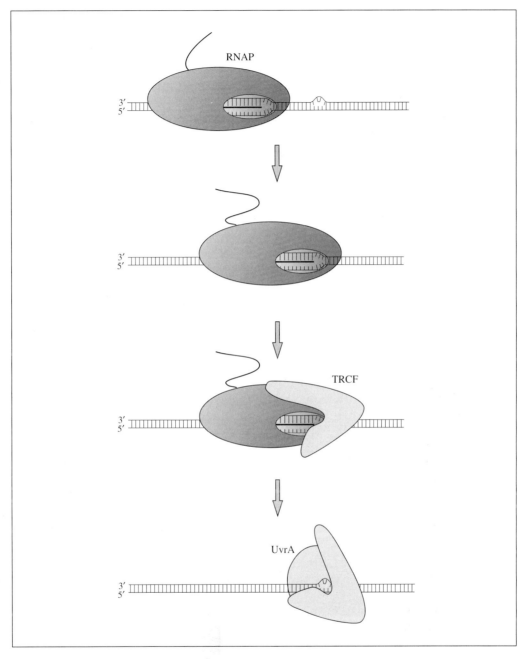

**Fig. 6.4** Model to explain transcription-coupled nucleotide excision repair in the bacterium *Escherichia coli* (39). Elongating RNA polymerase (RNAP) is blocked by a DNA lesion in the transcribed strand. The transcription-repair coupling factor (TRCF) recognizes and displaces the stalled complex, and via interaction with the *E. coli* repair protein UvrA, TRCF directs the NER machinery to the lesion in the transcribed strand.

strand-specific manner (46–48). Uncomplicated CS (CS not associated with xero-derma pigmentosum) comprises two complementation groups: CS-A and CS-B. Both genes have been cloned (49, 50). The *CSA* gene encodes a protein containing WD-motifs, which are found in many proteins with diverse functions in cellular metabolism (51). These proteins, with no apparent catalytic activities, are believed to function by stimulating the formation of multiprotein complexes. Consistent with this notion, CSA appeared to have affinity for CSB and for TFIIH (49). The *CSB* gene encodes a protein that bears a slight resemblance to the *E. coli* TRCF at the amino acid sequence level. The presence of 'helicase motifs' in both CSB and TRCF combined with the phenotypic similarities between *E. coli mfd⁻* and human CS cells led to the suggestion that the CSB protein (or CSB/CSA complex) constitutes a eukaryotic analogue of the *E. coli* TRCF. However, CSB and TRCF are categorized into different subfamilies of the helicase superfamily. Based on the extensive homology in the consensus ATPase/helicase motifs, the CSB protein belongs to the class of proteins known as the Snf2/Swi2 family of DNA-dependent ATPases (52, 53), and, indeed, the purified protein exhibits DNA-dependent ATPase activity *in vitro* (54). This is in contrast to the *E. coli* TRCF which has DNA-independent ATPase activity (44). Another difference between both proteins is that in contrast to the *E. coli* TRCF, recombinant CSB, alone or together with CSA, is unable to displace stalled RNA pol II from the DNA *in vitro* (54). This observation might point to a fundamental difference between the prokaryotic and the eukaryotic mechanism of TCR. Due to the length of certain human genes, abortion of mRNA synthesis each time an RNA polymerase encounters a lesion would seem to be highly inefficient. The notion that some members of the Snf2/Swi2 family are involved in remodelling DNA–protein interactions could hint at a function of CSB in modulating contacts of the elongating transcription complex with the DNA at sites of base damage, allowing NER proteins to reach the damage without displacing the transcription machinery (see refs 55, 56 and Section 3.6 for further discussion).

## 3.5 Transcription-coupled repair in the yeast *Saccharomyces cerevisiae*

### 3.5.1 *RAD26*, the yeast *CSB* homologue

Based on sequence similarity the *S. cerevisiae* homologue of *CSB* has been cloned and designated *RAD26* (57). This gene, like *CSB*, encodes a DNA-dependent ATPase (58) and is at present the only factor known to be specifically involved in TCR in yeast. Mutants lacking *RAD26* display a TCR deficiency, in agreement with the requirement of human CSB for TCR, but are not UV-sensitive. This resistance to UV light, comparable to repair-proficient yeast cells, is thought to originate from efficient overall repair (which will be discussed in Section 4) in this organism (59).

By measuring NER at single nucleotide resolution, a Rad26-independent compo-nent of TCR was revealed for UV-induced CPDs positioned in a region approxi-mately 50 bp downstream of the transcription initiation site (60). The region that is repaired efficiently by Rad26-independent TCR coincides with the region where TFIIH is still present in the transcription initiation complex *in vitro*, before it dissociates and the RNA pol II machinery enters the elongation mode (28). A possible explanation for this inverse correlation between Rad26 requirement for TCR and the association of TFIIH with the transcription machinery emerges when the *E. coli* TRCF model for TCR is considered: Rad26 (or CSB in the human system) recruits TFIIH, alone or associated with other repair proteins, in a so-called repairosome towards RNAPII complexes stalled at the site of the damage. The association of TFIIH with the transcription machinery during the first steps of nascent mRNA synthesis obviates a transcription-repair coupling factor in this region, and a deficiency in TCR due to a defective coupling factor will only be observed downstream of the position where TFIIH is released.

However, this explanation does not account for the observation that the transcriptionally induced *GAL7* gene is repaired entirely in a strand-specific manner in the absence of Rad26 (59). As discussed above, a similar observation was made for the *E. coli lacZ* gene where TCR became independent of the TRCF upon induction (45). These data suggest that high levels of transcription can compensate for the loss of a specific TCR factor. In that respect, it is interesting that TCR in the lowly transcribed *URA3* gene is completely dependent on Rad26 (apart from the region immediately downstream of start), whereas the yeast *RPB2* gene shows an intermediate dependence on Rad26: TCR of the *RPB2* gene is diminished in a *rad26* mutant but is not completely zero (59). Fast repair of only the first 50 nucleotides does not account for the residual TCR level, since, upon closer examination in this mutant, sequences further downstream in the *RPB2* transcribed strand were still repaired in a transcription-coupled manner, albeit more slowly than in wild-type yeast cells (60, and our unpublished observations). Together, these data are important for understanding the role of putative coupling factors, as they imply that neither Rad26 nor TRCF is strictly required for TCR.

Can these observations made in *S. cerevisiae* be extrapolated to the human situation? Although at present no genes have been identified that are repaired in a transcription-coupled manner completely independent of CSB, a recent report demonstrates that a subset of lesions positioned near the transcription initiation site in the transcribed strand of the *cJUN* gene is repaired in *CSB* mutants (61), indicating the generality of this phenomenon.

### 3.5.2  *RAD28*, the yeast *CSA* homologue

The *S. cerevisiae* sequence homologue of CSA has been identified and designated *RAD28*, but, in contrast to human *CS-A* cells, a *rad28* disruption mutant is not TCR-deficient (62). In addition, combining this mutation with a *rad26* gene disruption revealed no functional redundancy between both gene products. Since Rad28, based upon the amino acid sequence, is the WD-repeat protein most homologous to CSA and that the complete yeast genome has been sequenced, it is doubtful whether a functional counterpart of this protein exists in this organism.

## 3.6 Molecular mechanisms of TCR in eukaryotes

The question whether the eukaryotic proteins specifically involved in TCR function as eukaryotic analogues of the *E. coli* TRCF remains a matter of debate. The molecular model proposed by Selby and Sancar (39) for TCR in *E. coli* combines the two features that seem essential to explain the phenomenon: (1) the attraction of NER proteins towards lesions that block transcription; and (2) relief of the inhibition of NER imposed by stalled RNA polymerase complexes. On the basis of this *E. coli* precedent, a eukaryotic model can be envisaged using the properties of CSA, CSB/Rad26, and TFIIH together with our knowledge about the eukaryotic transcription process. In this hypothesis, RNA polymerase II stalls at the site of the damage and serves as a signal for the efficient recruitment of NER proteins including TFIIH, either via the CSA/CSB complex or by another yet unknown mechanism. Subsequently, ATP hydrolysis by CSB is used to modify the mRNA/DNA/RNA pol II complex in order to make the DNA damage accessible for the NER proteins. Whether the complex dissociates or tracks back during this modification and subsequent repair reactions is an intriguing question with respect to the resumption of mRNA synthesis.

Alternatively, a number of observations have supported a view that the CS phenotypes go beyond a NER defect, and the idea that CS cells are impaired in the transcription process itself was first raised by Hoeijmakers and co-workers to explain the phenotypic complexity of patients suffering from CS (63, see also Chapter 7). However, the fact that the CS genes can be disrupted in yeast (57), mice (64), and humans indicates that these proteins are not essential for general transcription. Numerous attempts have been made to distinguish between this 'transcription hypothesis' and the foregoing hypothesis based upon the *E. coli*-TRCF model, but the matter is unresolved at present. Figure 6.5 shows that a distinction between both scenarios is not easily deduced from a TCR phenotype because a defect in transcription as well as in one of the subsequent steps might result in a failure to transduce the 'signal' to the NER machinery.

Recently, a putative function for CSB in the transcription process gained direct experimental support when it was found that the addition of CSB rendered RNA polymerase II more processive in an *in vitro* transcription assay (65) and that CSB can interact with RNA polymerase II even in the absence of DNA damage (66, 67). In addition, it has been shown that transcription is reduced by about 50% in CS cells (68). Also *in vitro*, a reduction has been shown in the transcription capacity of CS cells, although this observation was made only when the substrate DNA molecules contained detectable amounts of base damage (69). Taken together, these observations point to a general role for CSB in transcription by RNA polymerase II. Possibly, elongating RNA polymerases require the action of the CS proteins to counteract structural perturbations in template DNA. These transcription impediments, perhaps transient, can comprise general NER substrates like UV-induced damage, but they can also extend to base excision repair (BER) substrates and even to naturally occurring pause sites. The observation that some forms of oxidative damage (which are generally not repaired by NER but by base excision repair) are not repaired strand-

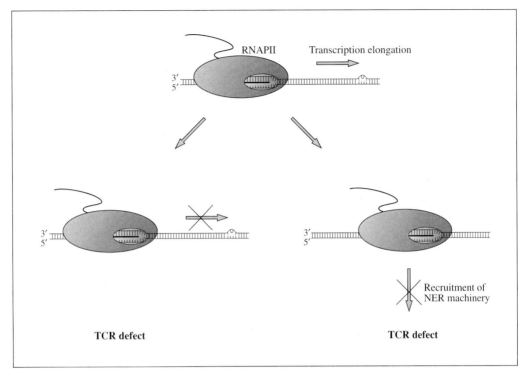

**Fig. 6.5** Schematic representation of the process of transcription elongation by RNA polymerase II (RNAPII) complexes, to illustrate that a decreased efficiency of transcription elongation as well as a defect in the recruitment of NER proteins might lead to defective transcription-coupled repair (TCR).

specifically in CS cells (48) indicates that the function of CS extends to a broader substrate range, which might ultimately be just a stalled RNA polymerase, irrespective the cause of stalling. The reduced capacity of RNA pol II to elongate *in vitro* in the absence of CSB (65) could suggest that the CSB protein facilitates the resumption of transcript elongation, without aborting the associated nascent transcript, perhaps by pushing the stalled RNA polymerase forward or backward to either kick-start the polymerase or allow repair proteins (NER or BER) to enter and repair the damage. This hypothesis implies that CS proteins may function in maintaining transcriptional activity, and that the cause of CS lies with the inability to displace stalled transcription complexes.

A putative role for CSB/Rad26 in the actual recruitment of the NER machinery towards stalled RNA polymerases is hard to reconcile with the observation of TCR in the absence of CSB/Rad26. The latter indicates that NER proteins are targeted preferentially to transcribed DNA also in the absence of CSB/Rad26. We therefore favour an alternative scenario in which NER proteins are recruited to stalled RNA polymerase independent of CSB/Rad26, and where the repair reaction awaits alleviation of the steric hindrance posed by the stalled RNA polymerase II molecule via CSB/Rad26-mediated modification of local mRNA/DNA/RNA pol II contacts.

In this scenario, alternative routes to destabilize (or displace) RNA pol II might compensate for the loss of CSB / Rad26 in TCR. For instance, under high transcription conditions, as for the induced *GAL7* locus, high polymerase density on the template might fulfil these requirements if the forward motion of elongating polymerases is able to destabilize the interaction of a stalled RNA pol II molecule with the DNA, as it can do for nucleosomes.

The ultimate test may require a defined *in vitro* system for TCR that has not yet been established with eukaryotic factors.

## 4. Global genome nucleotide excision repair

DNA repair by the NER pathway operates in both transcribed DNA and in inactive regions of the genome. The term global genome repair, or general repair, is used to describe the removal of DNA damage that is not removed by the TCR pathway. The latter pathway might seem to be a specialized form of repair superimposed upon global genome repair, which leads to the efficient repair of transcribed DNA. However, the following observations indicate that global genome repair and TCR are two distinct NER subpathways. (1) For both yeast and humans, mutant cells have been identified that are impaired specifically in global genome repair. In these mutants, UV-induced CPDs are not removed at all from non-transcribed DNA, while the removal from transcribed DNA is unaffected. (2) The repair kinetics of various types of DNA lesions are different for both pathways. Taken together, this suggests that the recognition process for both repair subpathways may be mechanistically quite different.

One could argue that the initial lesion recognition process (and perhaps even subsequent steps) is influenced by the chromatin environment of the substrate lesion. Indeed, it has been shown that the rate at which individual dimer sites are repaired in non-transcribed DNA differ dramatically. In both human and *S. cerevisiae* cells, CPDs positioned 5' to the transcription start site and on the non-transcribed strand of active genes exhibit a large difference in their repair kinetics (9, 12, 13, 70, see also Fig. 6.6). In general, CPD removal by TCR is more efficient than by global genome repair, but the latter pathway is not intrinsically slow because global genome repair of certain dipyrimidine sequences exhibits repair rates comparable to TCR (9, 70). Significant insight into the cause of this heterogeneity came from two observations: (1) slowly repaired CPDs in certain human promoters correspond to binding sites for transcription factors (reviewed in ref. 71); and (2) repair heterogeneity observed in the non-transcribed strand of the *S. cerevisiae URA3* gene correlates with the positioning of nucleosomes (72), firmly establishing the role of chromatin components in modulating the repair rate of non-transcribed DNA. The heterogeneity observed for non-transcribed DNA is opposite to the uniformity with which CPDs are removed from the transcribed strand. In yeast, independent studies on the active *URA3, RPB2*, and *MFA2* genes have demonstrated that individual CPDs positioned in the transcribed strand display almost no heterogeneity and are removed efficiently irrespective of their position (9, 60, 72, 73). This observation is explained by assuming that

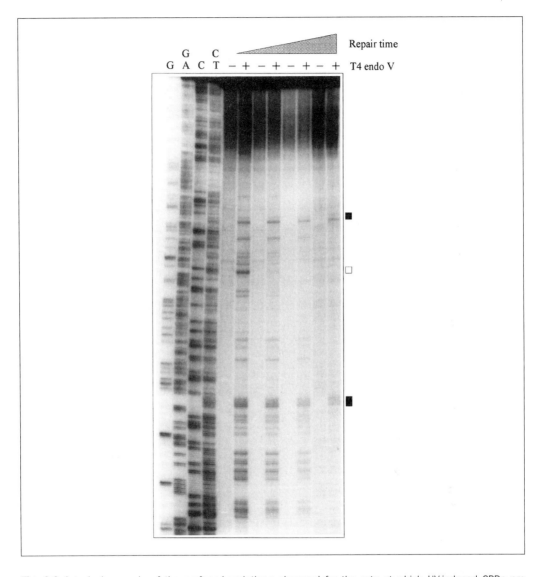

**Fig. 6.6** A typical example of the profound variations observed for the rate at which UV-induced CPDs are removed from non-transcribed DNA sequences in *S. cerevisiae* (the extremes are indicated: ■ for slow repair, and ☐ for fast repair). Repair of individual dinucleotides located upstream of the yeast *RPB2* gene is monitored (0, 20, 40, and 120 minutes) after *S. cerevisiae* cells are exposed to 70 J/m². The CPD incidence level is visualized as described in Section 2.2 and panel A of Fig. 6.2. Samples mock-treated or treated with the dimer-specific enzyme T4 endo V are denoted –and +, respectively. Lanes G, G+A, C, and C+T are Maxam–Gilbert sequencing reactions. CPD bands migrate 1 base slower than the corresponding 5′ nucleotide of the pyrimidine pair in the Maxam–Gilbert lanes.

damage recognition by the RNA polymerase is the rate-determining step in the removal of lesions from transcribed DNA, as discussed in the previous section. Once transcription is initiated, lesions are recognized by the RNA polymerase with equal probability, under conditions where little variations in transcription rate exist throughout the gene. Provided that each CPD blocks transcription to the same extent, this results in identical recognition rates and identical substrates (blocked RNA polymerases) for the subsequent steps in NER. Also, in *E. coli* and human cells homogeneous repair is observed for individual positions in highly transcribed DNA, e.g. the induced *E. coli lacZ* gene (45) and human *cJUN* gene (61, 70).

Another indication that damage recognition by global genome repair is mechanistically distinct from damage recognition by the TCR pathway came from comparing the overall repair rates of both major types of UV-induced photoproducts, i.e. CPDs and pyrimidine (6–4) pyrimidone photoproducts (6–4PPs). As discussed, global genome repair of CPDs is generally slow compared to TCR and, as a consequence of this difference in kinetics, preferential repair of the transcribed strand is observed. On the other hand, UV-induced 6–4PPs are repaired very efficiently throughout the genome. and a contribution of TCR is not detected in NER-proficient human cells when transcribed DNA is compared to non-transcribed DNA (74). However, when repair was analysed in human XP-C cells, which are specifically impaired in global genome repair but maintain TCR (see below), both CPDs and 6–4PPs are removed exclusively from transcribed DNA (75, 76). This indicates that TCR of 6–4PPs is revealed only when the contribution of the global genome repair pathway is eliminated. In addition, the kinetics of 6–4PP removal in XP-C cells are nearly, if not completely, identical to the kinetics of CPD removal (76). Provided that CPDs and 6–4PPs constitute an equal block to transcription, this supports the idea that stalling of RNA polymerase II at the site of base damage determines the rate at which these damages are removed from transcribed DNA. Similar results have been found in human cells for adducts formed by the polycyclic carcinogen N-acetoxy-acetylaminofluorene (NA-AAF) (77, 78). These lesions constitute another example where the overall repair efficiency in normal human fibroblast is dominated by global genome repair. The basis for the more efficient global genome repair of 6–4PPs or AAF compared to CPDs is not known, although it is believed that lesions that profoundly perturb the helical structure of DNA (as is the case for 6–4PPs) are preferentially repaired to lesions that do not, or hardly, affect the helix in this way.

## 4.1 XP-C and the repair of non-transcribed DNA

Cells of human XP-C patients display only a moderate UV-sensitivity. Repair analysis has shown that these cells remove CPDs from the transcribed strand of active genes at a normal rate, while totally lacking in the repair of non-transcribed sequences (75). Furthermore, the residual NER capacity of XP-C cells is abolished in the presence of α-amanitin, a specific inhibitor of RNA pol II transcription elongation (79). These data indicate that the removal of all lesions in these cells is accomplished by TCR and that this process can therefore take place in the absence of the *XPC* gene product.

Despite the fact that a subset of lesions are repaired *in vivo* in the absence of XPC, this gene product is essential for the reconstitution of NER *in vitro*. In an attempt to link the *in vivo* with the *in vitro* data, it has been hypothesized that the function of XPC in NER can be substituted by a stalled RNA polymerase molecule. This apparent suppression might reside either in recognizing the damage, creating a nucleation site for NER complex assembly, and/or in fulfilling structural requirements for NER. Recent *in vitro* analysis, using two different types of DNA structures, shed some light on to this matter. A specific cholesterol-containing substrate could be incised in the absence of XPC, in contrast to other types of lesions (80). In addition, and even more provocatively, in an attempt to mimic the DNA structure when RNA polymerase is blocked by a lesion, several model substrates were used containing unwound DNA sequences 5′ to, 3′ to, or encompassing a pyrimidine dimer in an *in vitro* reconstituted human NER system. These substrate are repaired independently of XPC, while still requiring the other core-NER enzymes and thereby mimicking the *in vivo* TCR repair characteristics (81, 81A).

## 4.2   Rad7 and Rad16 and the repair of non-transcribed DNA

The phenotype of *S. cerevisiae* *rad7* and *rad16* mutants resembles that of human XP-C cells. These mutants display an intermediate UV-sensitivity and are disturbed in the repair of non-transcribed DNA, while the repair of RNA pol II transcribed strands is unaffected (9, 82–84). Recent genetic as well as biochemical data have indicated that *RAD7*- and *RAD16*-encoded products exist as a protein complex (82, 85, 86). In contrast to XPC, this complex can not be considered to be a core-NER component since it is dispensable for NER of naked damaged plasmid DNA in a highly purified and defined reconstituted system (22). On the other hand, cell-free extracts of *rad7* and *rad16* mutants are deficient in performing transcription-independent NER (85, 87, see also Fig. 6.3), demonstrating that the Rad7–Rad16 complex is essential for a less defined *in vitro* system.

Based on the *in vivo* phenotype of the mutants, one might consider the possibility that the Rad7–Rad16 complex functions in damage recognition of non-transcribed DNA and/or making lesions accessible to NER proteins by modifying chromatin at the site of base damage. Recently, direct evidence has been obtained for the hypothesis in which the complex effects the repair of transcriptionally inactive DNA by sensing the DNA damage. By using a DNA mobility shift assay it was demonstrated that the Rad7–Rad16 complex binds UV-irradiated DNA with high specificity in an ATP-dependent manner, and although the complex is not essential to reconstitute NER *in vitro*, it did stimulate the reaction significantly (86). These results, together with the *in vivo* repair characteristics, strongly suggest that the Rad7–Rad16 complex is the DNA damage recognizing entity for non-transcribed DNA in yeast. The substrate used in the *in vitro* reconstitution experiments is naked DNA, therefore these experiments do not address the question of how the complex deals with chromatin inside the cell. Some indications come from the amino acid sequence of Rad16 which predicts two zinc-finger (DNA-binding elements) and DNA-dependent ATPase motifs also found in the Snf2/Swi2 family of proteins. As discussed for CSB, several members of this family are involved in modifying DNA–protein interactions. Swi/Snf2 itself is part of a protein complex that has a role in transcription activation, and is thought to act as a chromatin remodelling machine (52, 88), facilitating the access of transcription factors to nucleosomal DNA in an ATP-dependent manner. Members of the Swi2/Snf2 family of DNA-dependent ATPases might be part of complexes involved in remodelling nucleosome structure to antagonize the repressive effects of chromatin on various processes. Therefore, it is tempting to speculate about a function of Rad7–Rad16 in antagonizing the repressive effects of chromatin for NER. In this light, the observation that the complex is just stimulating NER in a purified system but that it is absolutely essential to NER with cell-free extracts is intriguing, since the latter still contain chromatin components, or at least DNA binding proteins.

How does this relate to damage recognition in humans? Human *RAD7* and *RAD16* homologues have not yet been identified. A *RAD16* homologue exists in fission yeast and this gene is capable of partial complementation of the NER defect in *S. cerevisiae rad16* mutants (89), suggesting that its function is conserved during evolution. As a consequence of the genome sequence projects, mainly that of *Caenorhabditis elegans* and the human project, it is just a matter of time before we will know whether homologues are present in higher eukaryotes. Based on the phenotypic resemblance between *Rad7–Rad16* mutants and XP-C cells one might suspect the XPC protein to be the DNA damage sensor. However, XPC is homologous to Rad4 in yeast, and for both proteins such an activity has not (yet) been identified. Interestingly, complementation studies and two hybrid analysis have identified physical interactions between Rad4 and the Rad7–Rad16 complex (85). The Rad4 protein, like its human counterpart, is essential to reconstitute NER *in vitro* (22), but the phenotypes of human *XPC* and yeast *rad4* mutants differ rather paradoxically. In contrast to human XP-C cells, *rad4* mutants are defective in the repair of both strands of RNA polymerase II transcribed genes (82), indicating that transcription in yeast by RNA polymerase II does not overcome the need for this repair factor, at least not *in vivo*. However, *rad4* mutants are able to repair the transcribed strand but not the nontranscribed strand of rDNA (90), which is transcribed by RNA polymerase I. Since human XP-C cells are deficient for the repair of rDNA (91), it seems that XP-C and *rad4* cells have a reciprocal phenotype. As discussed above, structural requirements fulfilled by RNA polymerase II transcription to obviate XPC in human cells, might in yeast be provided by RNA polymerase I transcription.

Finally, it is important to note that the Rad7–Rad16 complex (and perhaps XPC) is not the only candidate to act as a damage recognition factor in the initiation step of NER. Based upon the higher affinity for damaged DNA than for undamaged DNA, XPA/Rad14 and RPA (discussed in Chapter 5) have been suggested to have this function. Therefore, the two most important questions concerning damage recognition—What entity is responsible for the damage recognizing step in global genome repair? What is the basis for recognition, bearing in mind the substrate specificity of NER?—remain largely unanswered at the present time. Evidently, the matter requires additional biochemical analysis before the mechanism of damage recognition *in vivo* is completely understood.

# 5. Connections between NER and other repair pathways

The discussion has focused primarily on UV-induced photoproducts because the removal of these NER substrates, the CPD in particular, has been studied most extensively. To summarize, two functionally distinct modes of NER have been described, i.e. transcription-coupled and global genome repair, and because different genetic determinants have been found in either one of these modes in both yeast and humans, we have referred to these modes of repair as subpathways. Both subpathways employ the same core-NER proteins, but different accessory proteins are required to operate on lesions in transcribed and non-transcribed DNA. These accessory proteins are most likely to be involved in the very early step of NER: recognition of the damage and subsequent recruitment of the other NER proteins (see Fig. 6.7). *In vivo*, CPDs can not be removed by NER unless they are a target for

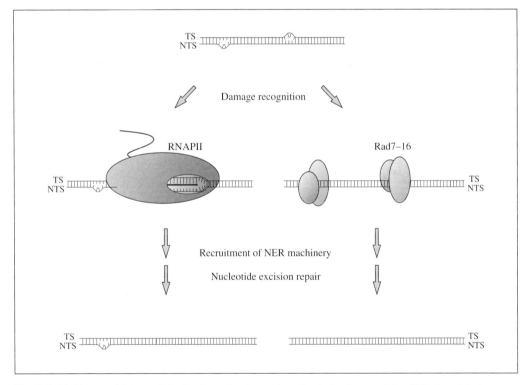

**Fig. 6.7** Working model to explain the two-subpathway hypothesis in *S. cerevisiae*. UV-induced lesions are recognized by the elongating RNA polymerase II complex (RNAPII) leading to transcription-coupled repair, or by the Rad7–Rad16 complex (Rad7–16) leading to global genome repair. Note that in this model the efficiency of recognition determines whether lesions are processed by one or the other repair mode. TS, transcribed strand; NTS, non-transcribed strand.

either one or both subpathways, and the simultaneous inactivation of both subpathways abrogates NER completely, as does mutation of one of the core-NER enzymes.

Although this chapter does not discuss damage to DNA not repaired by NER, some specific cases are of particular relevance here.

## 5.1 NER and direct reversal by DNA photolyases

*S. cerevisiae* cells are capable of removing CPDs from DNA using the enzyme DNA photolyase, which splits the cyclobutane ring using a light-initiated electron transfer reaction (92). In this way, a single enzyme restores the bases to their native form. Using NER-deficient yeast mutants, it has been shown that the reversion rates mediated by photolyase are highly influenced by chromatin for CPDs positioned in non-transcribed DNA (93). An influence of chromatin has also been observed for NER-mediated removal of UV-photoproducts (72). However, in contrast to this similarity, both repair pathways are oppositely affected by transcription: whereas transcription leads to elevated NER rates for CPDs positioned in transcribed DNA, photoreversal is inhibited (94). This observation is consistent with *in vitro* data showing that stalled human RNA polymerase prevents the access of *E. coli* photolyase to DNA damage (23).

## 5.2 NER and mismatch repair

Recently, another interplay between previously considered independent repair pathways became apparent. To remove mismatched bases or small loops that have arisen as a consequence of replication errors, eukaryotic cells use a repertoire of homologues of the *E. coli* mismatch repair proteins MutS and MutL to initiate correction of mismatches (for a mechanistic model for the process see refs 95, 96 and Chapter 4). Surprisingly, *E. coli* and mammalian cells with defects in mismatch repair display a TCR deficiency of UV-induced pyrimidine dimers (97, 98). In contrast, *S. cerevisiae* cells lacking a functional mismatch repair system are fully proficient for TCR (99). The reason for this discrepancy is not known. Since purified mismatch repair proteins do not suppress or enhance NER *in vitro*, and no effect on NER was noticeable when extracts from mismatch repair-deficient cells were compared with mismatch repair-proficient cells (44), further investigation is required to establish a functional connection between both repair processes.

## 5.3 Nucleotide and base excision repair

Another example of the interplay between different repair pathways came from experiments with lesions induced by ionizing radiation. These lesions are primarily repaired by the base excision repair pathway which excises damaged bases by the hydrolysis of the *N*-glycosyl bond linking the modified base to the deoxyribose-phosphate chain (reviewed in refs 100–102). Some lesions induced in DNA by ionizing radiation or by free oxygen radicals, such as thymine glycols, block transcription and are repaired in a transcription-coupled manner (48, 103). These lesions are repaired by means other than NER, since cells mutated in the core-NER proteins XPA or XPF are proficient in removing these lesions and XP-A cells do not display an increased sensitivity to ionizing radiation (103). However, CS cells do display a slightly increased sensitivity towards ionizing radiation, and, moreover, TCR of thymine glycols is absent in these cells, indicating that the CS gene products function aside from NER in other repair pathways as well. Since this type of oxidative damage can also arise from endogenous metabolic processes, it has been suggested that the clinical features observed in CS patients are the result of the failure to overcome the transcription impediment caused by these lesions in transcribed DNA (103).

# Acknowledgements

We thank Dr J. G. Tasseron-de Jong for continuous support and critical reading of the manuscript. Members of the Molecular Genetics laboratory and our colleagues of the DNA repair group of the Medical Genetics Centre are acknowledged for their valuable discussion. R. A. V. and J. B. are supported by the J. A. Cohen Institute for Radiopathology and Radiation Protection.

# References

1. Bohr, V. A., Smith, C. A., Okumoto, D. S., and Hanawalt, P. C. (1985) DNA repair in an active gene: removal of pyrimidine dimers from the DHFR gene of CHO cells is much more efficient than in the genome overall. *Cell*, **40**, 359.
2. Mellon, I. and Hanawalt, P. C. (1989) Induction of the *Escherichia coli* lactose operon selectively increases repair of its transcribed DNA strand. *Nature*, **342**, 95.
3. Smerdon, M. J. and Thoma, F. (1990) Site-specific DNA repair at the nucleosome level in a yeast minichromosome. *Cell*, **61**, 675.
4. Mellon, I. M., Spivak, G. S., and Hanawalt, P. C. (1987) Selective removal of transcription blocking DNA damage from the transcribed strand of the mammalian DHFR gene. *Cell*, **51**, 241.
5. Leadon, S. A. and Lawrence, D. A. (1992) Strand-selective repair of DNA damage in the yeast *GAL7* gene requires RNA polymerase II. *J. Biol. Chem.*, **267**, 23175.
6. Sweder, K. S. and Hanawalt, P. C. (1992) Preferential repair of cyclobutane pyrimidine dimers in the transcribed strand of a gene in yeast chromosomes and plasmids is dependent on transcription. *Proc. Natl Acad. Sci. USA*, **89**, 10696.
7. Kunala, S. and Brash, D. E. (1992) Excision repair at individual bases of the *Escherichia coli lacI* gene: relation to mutation hot spots and transcription coupling activity. *Proc. Natl Acad. Sci. USA*, **89**, 11031.
8. Li, S. and Waters, R. (1996) Nucleotide level detection of cyclobutane pyrimidine dimers using oligonucleotides and magnetic beads to facilitate labelling of DNA fragments incised at the dimers and chemical sequencing reference ladders. *Carcinogenesis*, **17**, 1549.
9. Tijsterman, M., Tasseron-de Jong, J., van de Putte, P., and Brouwer, J. (1996)

Transcription-coupled and global genome repair in the *Saccharomyces cerevisiae RPB2* gene at nucleotide resolution. *Nucl. Acids Res.* **24**, 3499.

10. Axelrod, J. D. and Majors, J. (1989) An improved method for photofootprinting yeast genes *in vivo* using Taq polymerase. *Nucl. Acids Res.*, **17**, 171.

11. Wellinger, R. E. and Thoma, F. (1996) Taq DNA polymerase blockage at pyrimidine dimers. *Nucl. Acids Res.*, **24**, 1578.

12. Tornaletti, S. and Pfeifer, G. P. (1994) Slow repair of pyrimidine dimers at p53 mutation hotspots in skin cancer. *Science*, **263**, 1436.

13. Gao, S., Drouin, R., and Holmquist, G. P. (1994) DNA repair rates mapped along the human PGK1 gene at nucleotide resolution. *Science*, **263**, 1438.

14. Steigerwald, S. D., Pfeifer, G. P., and Riggs, A. D. (1990) Ligation-mediated PCR improves the sensitivity of methylation analysis by restriction enzymes and detection of specific DNA strand breaks. *Nucl. Acids Res.*, **18**, 1435.

15. Wei, D., Maher, V. M., and McCormick, J. J. (1995) Site-specific rates of excision repair of benzo[a]pyrene diol epoxide adducts in the hypoxanthine phosphoribosyltransferase gene of human fibroblasts: correlation with mutation spectra. *Proc. Natl Acad. Sci. USA*, **92**, 2204.

16. Pfeifer, G. P., Drouin, R., and Holmquist, G. P. (1993) Detection of DNA adducts at the DNA sequence level by ligation-mediated PCR. *Mutat. Res.*, **288**, 39.

17. Wood, R. D., Robins, P., and Lindahl, T. (1988) Complementation of the xeroderma pigmentosum DNA repair defect in cell-free extracts. *Cell*, **53**, 97.

18. Sibghat-Ullah, H. I., Carlton, W., and Sancar, A. (1989) Human nucleotide excision repair *in vitro*: repair of pyrimidine dimers, psoralen and cisplatin adducts by HeLa cell-free extract. *Nucl. Acids Res.*, **17**, 4471.

19. Wang, Z., Wu, X., and Friedberg, E. C. (1993) Nucleotide-excision repair of DNA in cell-free extracts of the yeast *Saccharomyces cerevisiae*. *Proc. Natl Acad. Sci. USA*, **90**, 4907.

20. Aboussekhra, A., Biggerstaff, M., Shivji, M. K. K., Vilpo, J. A., Moncollin, V., Produst, V. N., *et al.* (1995) Mammalian DNA nucleotide excision repair reconstituted with purified protein components. *Cell*, **80**, 859.

21. Mu, D., Park, C-H., Matsunaga, T., Hsu, D. S., Reardon, J. T., and Sancar, A. (1995) Reconstitution of human DNA repair excision nuclease in a highly defined system. *J. Biol. Chem.*, **270**, 2415.

22. Guzder, S. N., Habraken, Y., Sung, P., Prakash, L., and Prakash, S. (1995) Reconstitution of yeast nucleotide excision repair with purified Rad proteins, replication protein A, and transcription factor TFIIH. *J. Biol. Chem.*, **270**, 12973.

23. Donahue, B. A., Yin, S., Taylor, J.-S., Reines, D., and Hanawalt, P. C. (1994) Transcript cleavage by RNA polymerase II arrested by a cyclobutane pyrimidine dimer in the DNA template. *Proc. Natl Acad. Sci. USA*, **91**, 8502.

24. Orphanides, G., Lagrange, T., and Reinberg, D. (1996) The general transcription factors of RNA polymerase II. *Genes Dev.*, **10**, 2657.

25. Lu, H., Flores, O., Weinmann, R., and Reinberg, D. (1991) The nonphosphorylated form of RNA polymerase II preferentially associates with the preinitiation complex. *Proc. Natl Acad. Sci. USA*, **88**, 10004.

26. Laybourn, P. J. and Dahmus, M. E. (1989) Transcription-dependent structural changes in the C-terminal domain of mammalian RNA polymerase subunit IIa/o. *J. Biol. Chem.*, **264**, 6693.

27. Dahmus, M. E. (1994) The role of multisite phosphorylation in the regulation of RNA polymerase II activity. *Prog. Nucl. Acid Res. Mol. Biol.*, **48**, 143.

28. Zawel, L., Kumar, K. P., and Reinberg, D. (1995) Recycling of the general transcription factors during RNA polymerase II transcription. *Genes Dev.*, **9**, 1479.

29. Goodrich, J. A. and Tjian, R. (1994) Transcription factors IIE and IIH and ATP hydrolysis direct promoter clearance by RNA polymerase II. *Cell*, **77**, 145.

30. Schaeffer, L., Roy, R., Humbert, S., Moncollin, V., Vermeulen, W., Hoeijmakers, J. H. J., *et al.* (1993) DNA repair helicase: a component of BTF2 (TFIIH) basic transcription factor. *Science*, **260**, 58.

31. Feaver, W. J., Svejstrup, J. Q., Bardwell, L., Bardwell, A. J., Buratowski, S., Gulyas, K. D., *et al.* (1993) Dual roles of a multiprotein complex from *S. cerevisiae* in transcription and DNA repair. *Cell*, **75**, 1379.

32. Drapkin, R., Reardon, J. T., Ansari, A., Huang, J. C., Zawel, L., Ahn, K., *et al.* (1994) Dual role of TFIIH in DNA excision repair and in transcription by RNA polymerase II. *Nature*, **368**, 769.

33. Hoeijmakers, J. H. J., Egly, J-M., and Vermeulen, W. (1996) TFIIH: a key component in multiple DNA transactions. *Curr. Opin. Gen. Dev.*, **6**, 26.

34. Svejstrup, J. Q., Vichi, P., and Egly, J-M. (1996) The multiple roles of transcription/repair factor TFIIH. *Trends Biochem. Sci.*, **21**, 346.

35. Sweder, K. S., Chun, R., Mori, T., and Hanawalt, P. C. (1996) DNA repair deficiencies associated with mutations in genes encoding subunits of transcription initiation factor TFIIH in yeast. *Nucl. Acids Res.*, **24**, 1540.

36. Sweder, K. S. and Hanawalt, P. C. (1994) The COOH terminus of suppressor of stem loop (SSL2/RAD25) in yeast is essential for overall genomic excision repair and transcription-coupled repair. *J. Biol. Chem.*, **269**, 1852.

37. Maxon, M. E., Goodrich, J. A., and Tjian, R. (1994) Transcription factor IIE binds preferentially to RNA polymerase IIa and recruits TFIIH: a model for promoter clearance. *Genes Dev.*, **8**, 515.

38. Park, C. H., Mu, D., Reardon, J. T., and Sancar, A. (1995) The general transcription-repair factor TFIIH is recruited to the excision repair complex by the XPA protein independent of the TFIIE transcription factor. *J. Biol. Chem.*, **270**, 4896.

39. Selby, C. P. and Sancar, A. (1993) Molecular mechanism of transcription-repair coupling. *Science*, **260**, 53.

40. Selby, C. P., Witkin, E. M., and Sancar, A. (1991) *Escherichia coli mfd* mutant deficient in 'mutation frequency decline' lacks strand-specific repair: *in vitro* complementation with purified coupling factor. *Proc. Natl Acad. Sci. USA*, **88**, 11574.

41. Oller, A. R., Fijalkowska, I. J., Dunn, R. L., and Schaaper R. M. (1992) Transcription-repair coupling determines the strandedness of ultraviolet mutagenesis in *Escherichia coli*. *Proc. Natl Acad. Sci. USA*, **89**, 11036.

42. Selby, C. P. and Sancar, A. (1991) Gene- and strand-specific repair *in vitro*: partial purification of a transcription-repair coupling factor. *Proc. Natl Acad. Sci. USA*, **88**, 8232.

43. Selby, C. P. and Sancar, A. (1990) Transcription preferentially inhibits nucleotide excision repair of the template DNA strand *in vitro*. *J. Biol. Chem.*, **265**, 21330.

44. Selby, C. P. and Sancar, A. (1995). Structure and function of transcription-repair coupling factor. II. Catalytic properties. *J. Biol. Chem.*, **270**, 4890.

45. Kunala, S. and Brash, D. E. (1995) Intragenic domains of strand-specific repair in *Escherichia coli*. *J. Mol. Biol.*, **246**, 264.

46. Van Hoffen, A., Natarajan, A. T., Mayne, L. V., van Zeeland, A. A., Mullenders, L. H. F., and J. Venema. (1993) Deficient repair of the transcribed strand of active genes in Cockayne's syndrome cells. *Nucl. Acids Res.*, **21**, 5890.

47. Venema, J., Mullenders, L. H. F., Natarajan, A. T., van Zeeland, A. A. and Mayne, L. V. (1990) The genetic defect in Cockayne syndrome is associated with a defect in repair of UV-induced DNA damage in transcriptionally active DNA. *Proc. Natl Acad. Sci. USA*, **87**, 4707.

48. Leadon, S. A. and Cooper, P. K. (1993) Preferential repair of ionizing radiation-induced damage in the transcribed strand of an active human gene is defective in Cockayne syndrome. *Proc. Natl Acad. Sci. USA*, **90**, 10499.

49. Henning, K. A., Li, L., Iyer, N., McDaniel, L. D., Reagan, M. S., Legerski, R., *et al.* (1995) The Cockayne syndrome group A gene encodes a WD repeat protein that interacts with CSB protein and a subunit of RNA polymerase II TFIIH. *Cell*, **82**, 555.

50. Troelstra, C., van Gool, A., de Wit, J., Vermeulen, W., Bootsma, D., and Hoeijmakers, J. H. J. (1992) *ERCC6*, a member of a subfamily of putative helicases, is involved in Cockayne's syndrome and preferential repair of active genes. *Cell*, **71**, 939.

51. Neer, E. J., Schmidt, C. J., Nambudripad, R., and Smith, T. F. (1994) The ancient regulatory-protein family of WD-repeat proteins *Nature*, **371**, 297.

52. Pazin, M. J. and Kadonaga, J. T. (1997) SWI2/SNF2 and related proteins: ATP-driven motors that disrupt protein–DNA interactions? *Cell*, **88**, 737.

53. Eisen, J. A., Sweder, K. S., and Hanawalt, P. C. (1995) Evolution of the SNF2 family of proteins: subfamilies with distinct sequences and functions. *Nucl. Acids Res.*, **23**, 2715.

54. Selby, C. P. and Sancar, A. (1997) Human transcription-repair coupling factor CSB/ERCC6 is a DNA-stimulated ATPase but is not a helicase and does not disrupt the ternary transcription complex of stalled RNA polymerase II. *J. Biol. Chem.*, **272**,1885.

55. Hanawalt, P. C. (1994) Transcription-coupled repair and human disease. *Science*, **266**, 1957.

56. Hanawalt, P. C., Donahue, B. A., and Sweder, K. S. (1994) Repair and transcription. Collision or collusion? *Curr. Biol.*, **4**, 518.

57. van Gool, A., Verhage, R., Swagemakers, S. M. A., van de Putte, P., Brouwer, J., Troelstra, C., *et al.* (1994) *RAD26*, the functional *S. cerevisiae* homolog of the Cockayne syndrome B gene *ERCC6*. *EMBO J.*, **13**, 5361.

58 Guzder, S. N., Habraken, Y., Sung, P., Prakash, L., and Prakash, S. (1996) *RAD26*, the yeast homolog of human Cockayne's syndrome group B gene, encodes a DNA-dependent ATPase. *J. Biol. Chem.*, **271**, 18314.

59. Verhage, R. A., van Gool, A. J., de Groot, N., Hoeijmakers, J. H. J., van de Putte, P., and Brouwer, J. (1996) Double mutants of *Saccharomyces cerevisiae* with alterations in global genome and transcription-coupled repair. *Mol. Cell. Biol.*, **16**, 496.

60. Tijsterman, M., Verhage, R. A., van de Putte, P., Tasseron-de Jong, J. G., and Brouwer, J. (1997) Transitions in the coupling of transcription and nucleotide excision repair within RNA polymerase II transcribed genes of *Saccharomyces cerevisiae*. *Proc. Natl Acad. Sci. USA*, **94**, 8027.

61. Tu, Y., Bates, S., and Pfeifer, G. P. (1997) Sequence-specific and domain-specific DNA repair in xeroderma pigmentosum and Cockayne syndrome cells. *J. Biol. Chem.*, **272**, 20747.

62. Bhatia, P. K., Verhage, R. A., Brouwer, J., and Friedberg, E. C. (1996) Molecular cloning and characterization of *S. cerevisiae RAD28*: the yeast homolog of the human Cockayne syndrome A (*CSA*) gene. *J. Bacteriol.*, **178**, 5977.

63. Vermeulen, W., van Vuuren, A. J., Chipoulet, M., Schaeffer, L., Appeldoorn, E., Weeda, G., *et al.* (1994) Three unusual repair deficiencies associated with transcription factor BTF2(TFIIH): evidence for the existence of a transcription syndrome. *Cold Spring Harb. Symp. Quant. Biol.*, **59**, 317.

64. van der Horst, G. T., van Steeg, H., Berg, R. J., van Gool, A. J., de Wit, J., Weeda, G., *et al.* (1997) Defective transcription-coupled repair in Cockayne syndrome B mice is associated with skin cancer predisposition. *Cell*, **89**, 425.

65. Selby, C. P. and Sancar, A. (1997) Cockayne syndrome group B protein enhances elongation by RNA polymerase II. *Proc. Natl Acad. Sci. USA*, **94**, 11205.

66. van Gool, A. J., Citterio, E., Rademakers, S., van Os, R., Vermeulen, W., Constantinou, A., *et al.* (1997) The Cockayne syndrome B protein, involved in transcription-coupled DNA repair, resides in an RNA polymerase II-containing complex. *EMBO J.*, **16**, 5955.

67. Tantin, D., Kansal, A., and Carey, M. (1997) Recruitment of the putative transcription-repair coupling factor CSB/ERCC6 to RNA polymerase II elongation complexes. *Mol. Cell. Biol.*, **17**, 6803.

68. Balajee, A. S., May, A., Dianov, G. L., Friedberg, E. C., and Bohr, V. A. (1997) Reduced RNA polymerase II transcription in intact and permeabilized Cockayne syndrome group B cells. *Proc. Natl Acad. Sci. USA*, **94**, 4306.

69. Dianov, G. L., Houle, J. F., Iyer, N., Bohr, V. A., and Friedberg E. C. (1997) Reduced RNA polymerase II transcription in extracts of Cockayne syndrome and xeroderma pigmentosum/Cockayne syndrome cells. *Nucl. Acids Res.*, **25**, 3636.

70. Tu, Y., Tornaletti, S., and Pfeifer, G. P. (1996) DNA repair domains within a human gene: selective repair of sequences near the transcription initiation site. *EMBO J.*, **15**, 675.

71. Pfeifer, G. P. (1997) Formation and processing of UV photoproducts: effects of DNA sequence and chromatin environment. *Photochem. Photobiol.*, **65**, 270.

72. Wellinger, R. E. and Thoma, F. (1997) Nucleosome structure and positioning modulate nucleotide excision repair in the non-transcribed strand of an active gene. *EMBO J.*, **15**, 5046.

73. Teng, Y., Li, S., Waters, R., and Reed, S. H. (1997) Excision repair at the level of the nucleotide in the *Saccharomyces cerevisiae MFA2* gene: mapping of where enhanced repair in the transcribed strand begins or ends and identification of only a partial rad16 requisite for repairing upstream control sequences. *J. Mol. Biol.*, **267**, 324.

74. Vreeswijk, M. P. G., van Hoffen, A., Westland, B. E., Vrieling, H., van Zeeland, A. A., and Mullenders, L. H. F. (1994) Analysis of repair of cyclobutane pyrimidine dimers and pyrimidine 6–4 pyrimidone photoproducts in transcriptionally active and inactive genes in chinese hamster cells. *J. Biol. Chem.*, **269**, 31858.

75. Venema, J., van Hoffen, A., Karcagi, V., Natarajan, A. T., van Zeeland, A. A., and Mullenders, L. H. F. (1991) Xeroderma pigmentosum complementation group C cells remove pyrimidine dimers selectively from the transcribed strand of active genes. *Mol. Cell. Biol.*, **11**, 4128.

76. Van Hoffen, A., Venema, J., Meschini, R., van Zeeland, A. A., and Mullenders, L. H. F. (1995) Transcription-coupled repair removes both cyclobutane pyrimidine dimers and 6–4 photoproducts with equal efficiency and in a sequential way from transcribed DNA in xeroderma pigmentosum group C fibroblasts. *EMBO J.*, **14**, 360.

77. Tang, M.-S., Bohr, V. A., Zhang, X.-S., Pierce, J., and Hanawalt, P. C. (1989) Quantification of aminofluorene adduct formation and repair in defined DNA sequences in mammalian cells using the UVRABC nuclease. *J. Biol. Chem.*, **264**, 14455.

78. Van Oosterwijk, M. F., Filon, R., Kalle, W. H. J., Mullenders, L. H. F., and van Zeeland, A. A. (1996) The sensitivity of human fibroblasts to N-acetoxy-2-acetylaminofluorene is determined by the extent of transcription-coupled repair, and/or their capability to counteract RNA synthesis inhibition. *Nucl. Acids Res.*, **24**, 4653.

79. Carreau, M. and Hunting, D. (1992) Transcription-dependent and independent DNA excision repair pathways in human cells. *Mutat. Res.*, **274**, 57.

80. Mu, D., Hsu, D. S., and Sancar, A. (1996) Reaction mechanism of human DNA repair excision nuclease. *J. Biol. Chem.*, **271**, 8285.

81. Mu, D. and Sancar, A. (1997) Model for XPC-independent transcription-coupled repair of pyrimidine dimers in humans. *J. Biol. Chem.*, **272**, 7570.

81a. Mu, D., Wakasugi, M., Hsu, D. S., and Sancar, A. (1997) Characterization of reaction intermediates of human excision repair nuclease. *J. Biol. Chem.*, **272**, 28971.

82. Verhage, R., Zeeman, A-M., de Groot, N., Gleig, F., Bang, D. D., van de Putte, P., *et al.* (1994) The *RAD7* and *RAD16* genes, which are essential for pyrimidine dimer removal from the silent mating type loci, are also required for repair of the nontranscribed strand of an active gene in *Saccharomyces cerevisiae*. *Mol. Cell. Biol.*, **14**, 6135.

83. Bang, D. D., Verhage, R., Goosen, N., Brouwer, J., and van de Putte, P. (1992) Molecular cloning of *RAD16*, a gene involved in differential repair in *Saccharomyces cerevisiae*. *Nucl. Acids Res.*, **20**, 3925.

84. Terleth, C., Schenk, P., Poot, R., Brouwer, J., and van de Putte, P. (1990) Differential repair of UV damage in *rad* mutants of *Saccharomyces cerevisiae*: a possible function of $G_2$ arrest upon UV irradiation. *Mol. Cell. Biol.*, **10**, 4678.

85. Wang, Z., Wei, S., Reed, S. H., Wu, X., Svejstrup, J. Q., Feaver, W. J., *et al.* (1997) The *RAD7*, *RAD16*, and *RAD23* genes of *Saccharomyces cerevisiae*: requirement for transcription-independent nucleotide excision repair *in vitro* and interactions between the gene products. *Mol. Cell. Biol.*, **17**, 635.

86. Guzder, S. N., Sung, P., Prakash, L., and Prakash, S. (1997) Yeast Rad7–Rad16 complex, specific for the nucleotide excision repair of the nontranscribed DNA strand, is an ATP-dependent DNA damage sensor. *J. Biol. Chem.*, **272**, 21665.

87. He, Z., Wong, J. M. S., Maniar, H. S., Brill, S. J., and Ingles, C. J. (1996) Assessing the requirements for nucleotide excision repair proteins of *Saccharomyces cerevisiae* in an *in vitro* system. *J. Biol. Chem.*, **271**, 28243.

88. Kingston, R. E., Bunker, C. A., and Imbalzano, A. N. (1996) Repression and activation by multiprotein complexes that alter chromatin structure. *Genes Dev.*, **10**, 905.

89. Bang, D. D., Ketting, R., de Ruijter, T., Brandsma, J. A., Verhage, R. A., van de Putte, P., *et al.* (1996) Cloning of *Schizosaccharomyces pombe rhp16⁺*, a gene homologous to the *Saccharomyces cerevisiae RAD16* gene. *Mutat. Res.*, **364**, 57.

90. Verhage, R. A., van de Putte, P., and Brouwer, J. (1996) Repair of rDNA in *Saccharomyces cerevisiae*: *RAD4*-independent strand-specific nucleotide excision repair of RNA polymerase I transcribed genes. *Nucl. Acids Res.*, **24**, 1020.

91. Christians, F. C. and Hanawalt, P. C. (1994) Repair in ribosomal RNA genes is deficient in xeroderma pigmentosum group C and in Cockayne's syndrome cells. *Mutat. Res.*, **323**, 179.

92. Sancar, A. (1994) Structure and function of DNA photolyase. *Biochemistry*, **33**, 2.

93. Suter, B., Livingstone-Zatchej, M., and Thoma, F. (1997) Chromatin structure modulates DNA repair by photolyase *in vivo*. *EMBO J.*, **16**, 2150.

94. Livingstone-Zatchej, M., Meier, A., Suter, B., and Thoma, F. (1997) RNA polymerase II transcription inhibits DNA repair by photolyase in the transcribed strand of active yeast genes. *Nucl. Acids Res.*, **25**, 3795.

95. Modrich, P. and Lahue, R. (1996) Mismatch repair in replication fidelity, genetic recombination, and cancer biology. *Annu. Rev. Biochem.*, **65**, 101.

96. Kolodner, R. (1996) Biochemistry and genetics of eukaryotic mismatch repair. *Genes Dev.*, **10**, 1433.

97. Mellon, I. and Champe, G. N. (1996) Products of DNA mismatch repair genes *mutS* and *mutL* are required for transcription-coupled nucleotide-excision repair of the lactose operon in *Escherichia coli*. *Proc. Natl Acad. Sci. USA*, **93**, 1292.

98. Mellon, I., Rajpal, D. K., Koi, M., Boland, C. R., and Champe, G. N. (1996) Transcription-coupled repair deficiency and mutations in human mismatch repair genes. *Science*, **272**, 557.

99. Sweder, K. S., Verhage, R. A., Crowley, D. J., Crouse, G. F., Brouwer, J., and Hanawalt, P. C. (1996) Mismatch repair mutants in yeast are not defective in transcription-coupled DNA repair of UV-induced DNA damage. *Genetics*, **143**, 1127.

100. Seeberg, E., Eide, L., and Bjoras, M. (1995) The base excision repair pathway. *Trends Biochem. Sci.*, **20**, 391.

101. Wood, R. D. (1996) DNA repair in eukaryotes. *Annu. Rev. Biochem.*, **65**, 135.

102. Sancar, A. (1996) DNA excision repair. *Annu. Rev. Biochem.*, **65**, 43.

103. Cooper, P. K., Nouspikel, T., Clarkson, S. G., and Leadon, S. A. (1997) Defective transcription-coupled repair of oxidative base damage in Cockayne syndrome patients from XP group G. *Science*, **275**, 990.

# 7 | The *ATM* gene and stress response

MARTIN F. LAVIN and KUM KUM KHANNA

## 1. Introduction

Reactive oxygen intermediates (ROI) are generated as by-products of normal oxidative metabolism in the mitochondria and peroxisomes, and are associated with specific cell functions such as cell killing by cytotoxic T cells and natural killer (NK) cells (1, 2). While ROI can, in some cases, be utilized as part of a destructive mechanism (cell killing) there is increasing evidence that they also function as second messengers in signal transduction pathways (3, 4). Under normal conditions the potential damaging effects of the free radicals generated are neutralized by enzymes such as catalase, peroxidases, and superoxide dismutase (1). Perhaps more critical to cell survival are the damaging effects of ROI produced by exogenous agents such as ionizing radiation (5), hypoxia (6), and a variety of other compounds. ROI so produced can interact with cellular macromolecules in abstraction or addition reactions to generate additional radicals. These can either have direct effects, as in damage to DNA, or indirect effects, by initiating a cascade of reactions giving rise to membrane changes or damage (7).

DNA appears to be the major target for radiation-induced cell killing (8). The lesions induced by ionizing radiation include single- and double-strand breaks in the phosphodiester backbone of DNA, base and sugar modifications, as well as cross-links between DNA strands and between DNA and proteins (8–10). The strand breaks in DNA arise either directly by interaction with atoms in the phosphodiester backbone causing the ejection of electrons from these atoms, or indirectly through water radiolysis which generates free radical species capable of abstracting hydrogen atoms from DNA, resulting in strand breakage (7). While a variety of lesions occur in DNA postirradiation, it seems likely that the double-strand break is the most significant lesion for cell killing (11–13). The description of a number of human and rodent cell lines characterized by their hypersensitivity to radiation and reduced ability to repair double-strand breaks in DNA, postirradiation, provides further support for the importance of this lesion (14, 15).

Hypersensitivity to radiation is associated with 11 complementation groups in rodent cell mutants, four of which are complemented by human genes involved in double-strand break repair (16, 17). Mutations in other groups include those in the DNA-dependent protein kinase (DNA-PK) multiprotein complex; the Ku subunits (Ku 70 and Ku 80) that bind to the free ends of DNA and the catalytic subunit of DNA-PK, DNA-PKcs, which is recruited to DNA breaks and sites of damage by the Ku heterodimer (18–20). The *scid* mutation that occurs in DNA-PKcs (19, 21–23) is characterized not only by a defect in double-strand break repair, and, as a consequence, hypersensitivity to ionizing radiation (24), but also by defective V(D)J recombination and immunodeficiency (25–27). Another gene predisposing to radiation hypersensitivity is the gene (*ATM*) mutated in the human genetic disorder ataxia–telangiectasia (A–T) (28, 29). The defect in A–T appears to be largely due to a failure to respond to or recognize DNA damage, and, as a consequence, a fraction of double-strand breaks remains unrepaired (30, 31). It is evident that breaks in DNA are responsible for initiating those pathways responsible for the activation of checkpoints to control the passage of cells through the cell cycle (32). The best described of these is the $G_1/S$ checkpoint, which is primarily activated in response to DNA damage by the p53 pathway (33). Stabilization of p53 leads to the induction of a number of proteins including WAF1/CIPI/p21, which is a cyclin kinase inhibitor, preventing the phosphorylation of key substrates and consequently the ability of cells to pass from $G_1$ to S phase (34).

In addition to the role of DNA damage in initiating signalling, it is also clear that extranuclear pathways of stress-induced signalling exist (35–37). Membrane proteins, including the epidermal growth factor receptor (EGFR), are capable of acting as receptors for transmitting signals arising from both UV and ionizing radiation exposure (38–40). Agents such as UV light and $H_2O_2$ inhibit receptor tyrosine dephosphorylation in a thiol-sensitive reversible manner (41). Inhibition of dephosphorylation is by oxidation of an SH-group of the protein phosphatase regulating receptor activation. The end result of $H_2O_2$ treatment is receptor activation by this mechanism. In order to dissect the signal transduction pathways activated by stress it is important to identify intermediates in these pathways, delineate what mechanisms of activation are involved, and establish the functional consequence, e.g. transcriptional activation leading to a specific phenotype. The intention of this contribution is not to review all the stress-activated pathways *per se* but rather to focus attention on a single gene that appears to play a central role in responding to specific forms of stress. This gene is *ATM*, mutated in the human genetic disorder ataxia–telangiectasia (29). Data from human and from *Atm*-deficient mice also point to a more general signalling role for the ATM protein.

## 2. Ataxia–telangiectasia

The human genetic disorder ataxia–telangiectasia (A–T) has attracted considerable interest since the gene involved was described by Savitsky *et al.* (28). A–T was first described by Syllaba and Henner in 1926 (42), rediscovered by Madame Louis-Bar (43), and elevated to the level of a defined disease entity by Boder and Sedgwick (44). A–T is an autosomal recessive syndrome characterized by progressive neurological

**Table 7.1** Clinical, cellular, and molecular features of ataxia–telangiectasia

|  |  | **Reference** |
| --- | --- | --- |
| *Clinical* | | |
| Cerebellar ataxia | Diminished deep reflexes | Boder 1985 (45) |
| Choreoathetosis | Oculomotor abnormalities | Sedgwick and Boder 1991 (46) |
| Dysarthric speech | Telangiectasia | |
| Growth retardation | Sinopulmonary infection | |
| | | |
| *Cellular* | | |
| Apoptosis susceptibility | Chromosome rearrangements | Shiloh 1995 (58) |
| Chromosomal instability | Radiosensitivity | Taylor *et al.* 1995 (53) |
| Radioresistant DNA | Cell-cycle checkpoint | Houldsworth and Lavin 1980 |
| synthesis | defects | (154) |
| | | Beamish and Lavin 1994 (152) |
| | | |
| *Molecular* | | |
| Mapping gene(s) for A–T | | Gatti *et al.* 1988 (63) |
| Mutation in *ATM* gene | | Savitsky *et al.* 1995 (28) |
| Defective p53/WAF1 response | | Kastan *et al.* 1992 (158) |
| | | Khanna *et al.* 1995 (163) |
| | | |
| Lack of interaction with c-Abl | | Khanna *et al.* 1997 (173) |
| | | Baskaran *et al.* 1997 (174) |

degeneration, immunodeficiency, growth retardation, hypogonadism, and thymic dysplasia (45). Ataxia, which becomes evident when the child begins to walk, is progressive and affects the extremities—involuntary movements are evident by intentional tremor and the child becomes immobile by the end of the first decade of life. Degenerative changes are most evident in the cerebellum as a loss of Purkinje cells and also granular and basket cells, but more widespread pathological changes are evident throughout the CNS (46).

The telangiectases are more of a mystery, but, given what we now know about the identity and likely role of the gene, they may arise as a consequence of oxidative stress. Ischaemic changes in some blood vessels, triggered by light or some other agent, may lead to the dilation of other vessels and thus the characteristic telangiectases. Exposure to high doses of radiation during radiotherapy gives rise to telangiectases (47), which may reflect what is happening in radiosensitive A–T patients in the absence of such exposure.

A–T cells in culture are also characterized by a number of hallmarks that include slow growth rate, extreme radiosensitivity, radioresistant DNA synthesis, chromosomal instability, specific translocations, defective repair of a subtype of DNA strand breaks, and defective cell-cycle checkpoint activation (29; Table 7.1). This suite of abnormalities, that reveal a defect in DNA damage recognition and as a consequence genome instability, can be correlated with the increased incidence of malignancy recorded for A–T patients (48–50).

The characteristic that has been most widely exploited to investigate the defect in

A–T is radiosensitivity. While clinical radiosensitivity was first described for A–T patients in the 1960s (51, 52) it was some time later before it was reported for cells in culture (53, 54). A–T cells are three to four times more sensitive to ionizing radiation and radiomimetic agents than control cells (53–55). Although the exact basis of this radiosensitivity has not yet been established, several reports point to a slower rate of repair of DNA double-strand breaks or the continued existence of residual breaks (30, 56). More recently, it has been shown that approximately 10% of radiation-induced double-strand breaks remain in A–T DNA for at least 96 h postirradiation (31). This extent of residual damage in DNA could well explain the increased sensitivity of these cells to ionizing radiation, since as little as one unrepaired DNA double-strand break is sufficient to kill a cell (11). Now that we know the identity of the A–T gene, *ATM* (28), its possible mode of action in cell-cycle control, and in the detection of DNA damage (28, 57) another explanation is possible. If *ATM* is a 'sensor' of DNA damage then it is likely that the inability to recognize the damage is responsible for the residual DNA breaks and their translation into chromosome breaks rather than a repair defect *per se*. Abnormalities in chromatin structure might be indirect due to the absence of *ATM* or the presence of a mutated form of *ATM*. Either way, the residual chromosome breaks may be at least partly responsible for the radiosensitivity in A–T and may also contribute to the genome instability and cancer predisposition.

## 3. Cloning of *ATM*

Approaches to the identification of the gene responsible for the defect in A–T were based on two strategies: positional cloning and complementation cloning. Complementation involves correction of the cellular phenotype of patient cells (based on radiosensitivity) by the introduction of a library of cDNAs and subsequent identification of cDNA clones that complement this phenotype. Efforts over many years using this approach failed to identify a gene directly responsible for the disease (58). A major obstacle to this approach was posed by the ability of various cDNAs, even in truncated forms, to increase the resistance of A–T cells to ionizing radiation or radiomimetic drugs (59–61). In these experiments cDNAs which expressed only 3' untranslated regions (UTR) of different genes were capable of complementing the radiosensitivity in A–T cells. This was previously observed when the 3'UTR of muscle structural genes partially corrected differentiation-defective mutants (62). In the case of radiosensitivity, the 3'UTR sequences may have indirect effects on signal transduction pathways, for example by sequestering RNA binding proteins.

Linkage analysis is utilized to detect a disease gene based on its genetic locus. This locus is initially mapped to a specific chromosomal region, then narrowed by repeated genetic analysis. Subsequent steps involve long-range cloning and physical mapping, followed by the identification of transcribed sequences, and, finally, identification of mutations in the candidate genes in patients. The localization of an A–T locus by Gatti *et al.* (63) to chromosome 11q22–23 spurred extensive positional cloning efforts that initially lacked pace due to the apparent genetic heterogeneity of

A–T (four separate complementation groups) and the scarcity of genetic markers at the A–T region on chromosome 11. Additional studies on A–T families from several countries (64–66) pointed to a single A–T locus at this region, possibly containing the mutations responsible for all four complementation groups. Exhaustive cloning and physical mapping of the A–T region (67, 68) and the generation of a high-density microsatellite map spanning the A–T locus (69) facilitated narrowing down the search interval for the A–T gene to approximately 850 kb of DNA by a consortium of three laboratories (70). Several candidate genes were recognized within this interval before one, *ATM* (ataxia–telangiectasia mutated), was shown to be mutated in 14 patients with A–T (28).

## 3.1 Mutation analysis

The majority of changes initially described in the *ATM* gene were deletions predicted to give rise to truncated proteins, while in the other cases 3–9 nucleotide inframe deletions were observed. To date, over 300 mutations have been described in *ATM* (see the A–T mutation database on http://www.vmmc.org/vmrc/atm.htm— P. Concannon and R. Gatti). These mutations are broadly distributed along the *ATM* gene (Fig. 7.1), most patients are compound heterozygotes, and approximately 70% of these mutations are predicted to give rise to truncated proteins (71–75). A significant number of the mutations affect mRNA splicing, with up to 50% of the coding sequences undergoing exon skipping (76). As indicated above there is a wide distribution of mutations in *ATM*, most unique to single families, but a single mutation has been reported in 32/33 defective *ATM* alleles in North African Jews

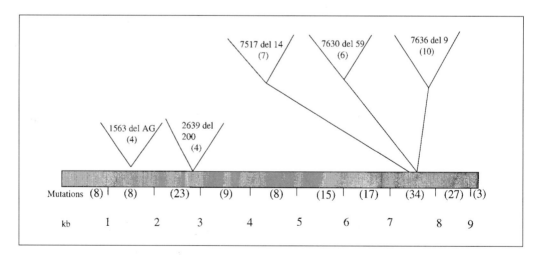

**Fig. 7.1** Some common mutations and overall distribution of mutations in the *ATM* gene. Approximately 70% of these mutations are predicted to give rise to a truncated protein, which is unstable and not detected by immunoblotting. The figures in parentheses below the line refer to the number of mutations detected in 1-kb intervals along the cDNA. Those above the line are numbers of the same mutations at specific sites. (Data taken from P. Concannon and R. Gatti on http://www.vmmc.org/vmrc/atm.htm)

thus providing evidence for a founder effect (77). An inframe deletion of nine nucleotides, 7636 del 9, is the most common *ATM* mutation reported, and is enriched in residents of Britain and Ireland (76). Common mutations have also been recorded in the Japanese population (78) in Puerto Rica, amongst other ethnic populations (79).

## 3.2 ATM cDNA

The initial 7–9 cDNA described by Savitsky *et al.* (28), 5921 bp with an open-reading frame (ORF) of just over 5 kb, was clearly a partial clone since the mRNA was shown to be about 13 kb in size. Successive screening of cDNA libraries identified a number of partly overlapping clones that allowed for the construction of a cDNA contig of around 10 kb (80). Constructing the full-length cDNA in this way was relatively easy, but cloning a full-length cDNA proved to be more elusive. This was largely due to the instability of the clone at the 5′ end of the ORF when transmitted through bacteria. Eventually, two laboratories succeeded in cloning full-length recombinant ATM cDNA that was stably expressed in human cells (81, 82). Figure 7.2 depicts the cloning of full-length ATM cDNA in pMEP4, an Epstein–Barr virus (EBV)-based vector with an inducible metallothionein promoter (81). In both cases, the expression

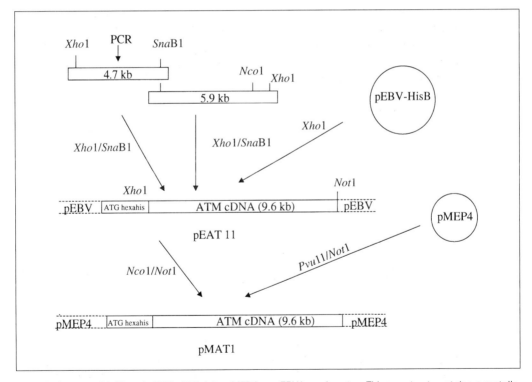

**Fig. 7.2** Cloning of full-length ATM cDNA into pMEP4, an EBV-based vector. This construct contains a metallo-thionein promoter, which allows for the induction of ATM by CdCl2, and also a hexahistidine insert to facilitate protein purification. A full description of this cloning procedure has been published (81).

of this cDNA corrected several aspects of the radiosensitive phenotype in A–T cells. Survival of transfected A–T cells was comparable to that of a normal control postirradiation, radiation-induced chromosome aberrations were reduced to levels of the same order as in controls and the extent of $G_2$-phase delay at longer times after irradiation was reduced. The $G_1/S$ checkpoint was also restored to A–T cells after the introduction of full-length ATM cDNA (81). Recombinant ATM cDNA was also expressed in insect cells using a baculovirus vector, but the amount of ATM protein produced was very low (10–20 ng/100 ml cells 2–3 $\times$ $10^6$ cells/ml) (82, 83). This was a considerable disappointment, since it was expected that this approach would provide sufficient starting material for a relatively straightforward purification procedure.

## 3.3 Genomic organization of *ATM*

The ORF for the *ATM* transcript is 9168 nucleotides, consisting of 3056 amino acids (80). The gene extends over 150 kb of genomic DNA with a total transcript size of about 13 kb, representing 66 exons that vary in size from 43 to 3800 nucleotides, the largest exon corresponding predominantly to the 3′UTR region (84). Introns vary in size from 100 bp to about 11 kb, with the majority in the 1–3 kb range. The compactness of the *ATM* gene, 66 exons contained in 150 kb of DNA, is comparable to that of the Huntington's disease gene which accommodates 67 exons in 180 kb (85). The entire coding region of the mouse homologue, *Atm*, has been reported (86). The total ORF is 9201 nucleotides, with an overall identity of 85% with the human transcript, and encodes a protein of 3066 amino acids.

# 4. The ATM protein

## 4.1 Detection and importance for radiosensitivity

ATM mRNA has an ORF of 9.168 kb and is predicted to code for a protein of 350.6 kDa (3056 amino acids) in size (80). To verify that this was the case antibodies were prepared to different regions of the protein in several laboratories (74, 87–91). In all cases, a protein of around 350 kDa was detected, which corresponded well to the size predicted from the ORF. This protein was absent from cell extracts of A–T patients where the mutations detected would be predicted to give rise to truncated proteins (74, 87, 91). In general, it appears that truncated forms of the ATM protein are highly unstable regardless of where the truncation occurs. A good example to illustrate this is AT9RM, that has a mutation predicted to terminate ATM 9 amino acids from the C-terminus. In these cells ATM protein was detectable at considerably reduced levels (17%) compared to normal (92).

The instability of the ATM protein is also apparent where short inframe deletions occur. A case in point is 7636 del 9 (homozygote genotype), where the amino acids SRI are deleted upstream of the PI3-kinase domain and is predicted to give rise to near full-length protein (28). Two reports demonstrate the presence of ATM in extracts from patients with this mutation, albeit at reduced levels compared to controls (74, 91). While only a few ATM missense mutations were reported in A–T patients in earlier studies, it is evident that there are a number of these mutations particularly towards the 3' end of the molecule (76). Recent data reveal that missense mutations in the PI3-kinase domain of *ATM* also destabilize the protein (Shiloh, unpublished). The picture that emerges is that most mutations in the *ATM* gene destabilize the protein to different extents. This raises the question as to whether the absolute level of protein contributes to the phenotype. In this context A–T heterozygotes would be expected to have approximately 50% of the ATM protein levels of normal controls, and if the amount of protein was important they would be expected to display some aspects of the phenotype. This is certainly the case since there are several reports, utilizing different methodologies, that demonstrate intermediate sensitivity to ionizing radiation in A–T heterozygotes (54, 93–98). More recently, Zhang *et al.* (99) have expressed full-length ATM cDNA in the opposite orientation to sensitize control cells to radiation. Under these conditions the amount of ATM protein in the cell was reduced to very low levels. Finally, AT9RM cells (deletion of nine amino acids at the C-terminus) that show reduced levels of ATM are intermediate in their sensitivity to ionizing radiation, display reduced amounts of radiation-induced chromosomal breakage compared to 'classical' A–T cells, and have a normal DNA synthesis inhibition pattern after exposure to radiation (92, 100). In summary, the amount of cellular ATM protein reflects the degree of radiosensitivity and may also be related to the severity of other characteristics of the disease.

## 4.2   Cellular localization

Since cells from A–T patients are characterized by radiosensitivity, reduced DNA double-strand break repair capacity, and anomalies in cell-cycle control it was expected that ATM would be a nuclear protein. Using immunoblotting and immunolocalization techniques, most studies to date have confirmed that ATM is predominantly located in the nucleus (74, 87–91). However, cell fractionation followed by immunoblotting revealed the presence of 5–20% of ATM in a microsomal fraction (74, 87, 91 and Fig. 7.3). Immunofluorescence confirmed that ATM was predominantly nuclear in fibroblasts, with a relatively uniform distribution in the nucleus except for nucleoli (74). A distinct pattern of punctate labelling was also seen in the cytoplasm. The nature of the cytoplasmic labelling was further investigated by immunogold immunoelectronmicroscopy, revealing that ATM was localized to vesicles ranging in size from 60 to 250 nm (74). In some cases it was possible to detect a single distinct membrane on these vesicles characteristic of peroxisomes. Co-localization of ATM and catalase, a peroxisomal protein, would appear to support an extranuclear role for ATM in peroxisomes (Watters *et al.*, unpublished data). These vesicles may not only be peroxisomes, since there is evidence that ATM interacts with β-adaptin proteins that are localized to secretory vesicles and the plasma membrane (205).

What then is its role outside the nucleus? Good evidence now exists for the in-

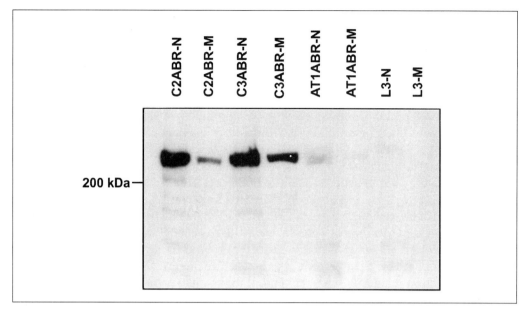

**Fig. 7.3** Immunoblotting to detect ATM protein in subfractionated extracts from control and A–T cells. C2ABR and C3ABR are controls and N and M refer to nuclear and microsomal extracts. AT1ABR (7636 del 9) and L3 (homozygous mutation at nucleotide 120 predicted to cause truncation) are A–T cell lines. The antibodies used here were ATM1.8 (Dr Tim Yen, Fox Chase, Philadelphia) that recognize poorly the near full-length ATM for AT1ABR extracts. L3 has no ATM protein.

volvement of extranuclear pathways for stress-induced signalling (38–40). For example, the epidermal growth factor receptor (EGFR) acts as a receptor for the transmission of signals caused by both UV light and ionizing radiation exposure, and it is likely that ROI are an important part of these events (38–40). Therefore it is possible that the extranuclear form of ATM is responding to the presence of ROIs. This response might occur by a reversible SH-group oxidation, as observed for the membrane-bound protein tyrosine phosphatase in the case of signalling through EGFR (41). As with the phosphatase this would activate ATM to initiate a response in the cytoplasm, which could lead to changes in gene expression by the activation of molecules such as c-jun and NF-*k*B. Alternatively, ATM may function as an iron–sulfur cluster protein serving as a biosensor of oxidants (101). The eukaryotic regulatory RNA-binding protein, IRP1 (102–104), and the bacterial transcription factors FNR (105, 106) and SoxR (107) are iron–sulfur proteins that respond to super-oxide, nitric oxide, or $H_2O_2$. In the latter case ROIs destabilize protein structure by removing iron from the cluster and, in the case of IRP1, this leads to activation of its RNA-binding activity. ATM has the appropriate SH-groups to accommodate either of these models and its purification may allow their discrimination. Based on its localization in cytoplasmic vesicles the ATM protein may have functions in common with the yeast vacuolar proteins TOR2 and Vps34. TOR2 is a target for the FKBP12–rapamycin complex that inhibits signal transduction events required for $G_1$-

to S-phase progression in yeast and mammalian cells. TOR2, which is located on the vacuolar surface, modulates vacuolar morphology and segregation as part of its role in signal transduction. The location of ATM in cytoplasmic vesicles, and what we know about the broader phenotype of A–T, could mean that it plays a similar role to a TOR-like protein but, unlike the TOR proteins, it is not a target for rapamycin (108).

## 5. *ATM* gene family

The ATM protein is a member of a growing family of large proteins in various organisms that share the PI3 kinase-like domain. Most of these proteins are involved in cellular responses to DNA damage and/or cell-cycle control (109, 110 and Table 7.2). Interestingly, the protein with the highest similarity to ATM in this group is Tel1p, which appears to be involved in maintaining telomere length in budding yeast (111, 112).

Telomere shortening was recently suggested to be another feature of the A–T cellular phenotype (113), and may be associated with the premature senescence observed in these cells. Other proteins of this family (Table 7.2) are involved in the control of cell-cycle checkpoints responding to DNA damage: Mec1p in *Saccharomyces cerevisiae* (114–117), Rad3p in *Schizosaccharomyces pombe* (118–121) and Mei-41 in *Drosophila melanogaster* (122–124). It has been suggested that Tel1p is functionally related to Mec1p in *S. cerevisiae* and that both proteins may play a role in

**Table 7.2** Members of the phosphoinositide 3-kinase family

| Protein | Origin | Role | Reference |
|---|---|---|---|
| p110 | Bovine | Mitogenic response | Kapeller and Cantley 1994 (204) |
| VPS34 | *S. cerevisiae* | Vacuole transport | |
| FRAP | Mammalian | $G_1/S$ progression | Brown *et al.* 1994 (129) |
| TOR1 | *S. cerevisiae* | $G_1/S$ progression | Heitman *et al.* 1991 (127) |
| TOR2 | *S. cerevisiae* | $G_1/S$ progression | Kunz *et al.* (128) |
| MEC1 | *S. cerevisiae* | DNA damage response checkpoint control | Weinert *et al.* 1994 (116) |
| RAD3 | *S. pombe* | DNA damage response checkpoint control | Jimenez *et al.* 1992 (120) |
| TEL1 | *S. cerevisiae* | Telomere control | Greenwell *et al.* 1995 (111) |
| MEI-41 | *D. melanogaster* | DNA damage response; cell-cycle control | Hari *et al.* 1995 (124) |
| ATM | Mammalian | DNA damage response; cell-cycle control | Savitsky *et al.* 1995 (28) |
| ATR/FRP1 | Mammalian | DNA damage response? cell-cycle control | Bentley *et al.* 1994 (126) Cimprich *et al.* 1994 (125) |
| DNA-PKcs | Mammalian | DNA damage response | Hartley *et al.* 1995 (133) |

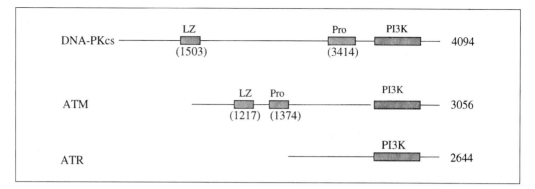

**Fig. 7.4** Subgroup of PI3-kinase family members. All three proteins are involved in DNA damage response, and in the case of ATM and ATR are also involved in cell-cycle control. They have been classified together because of the PI3-kinase domain, and DNA-PKcs and ATM also have leucine zipper (LZ) and proline-rich (Pro) regions which give them the capacity to interact with other proteins in intracellular signalling. Amino acid positions for these regions are indicated in parentheses and total amino acids appear at the right-hand side.

redundant checkpoint pathways. The similarity between the *ATM* gene product and the Mec1p, Rad3p, and *mei-41* proteins underscores the predicted role of this protein in the cell-cycle response to DNA damage, and highlights the link between the radiation sensitivity and chromosomal instability of A–T and cell-cycle regulation. It should be noted, however, that the *ATM* gene is not necessarily the closest or the only human homologue of any of these genes. For example, the recently identified FRPI/ATR protein (125, 126, Table 7.2) shows a higher degree of similarity to Rad3p, Mec1p, and *mei-41* than does the ATM protein. The TOR proteins (127–132, Table 7.2) are the targets of the immunosuppressant rapamycin, which blocks the progression from $G_l$ to the S phase of the cell cycle in yeast and mammalian cells. These proteins are probably not involved in cellular responses to DNA damage, but their role in the transduction of mitogenic signals and in cell-cycle progression highlights the involvement of the family of proteins with the PI3 kinase-like domain in cellular growth. The three human gene members of the PI3-kinase family, ATM, DNA-PKcs, and ATR are depicted in Fig. 7.4. They share a kinase domain and have leucine zippers and proline-rich domains that point to their interaction with other proteins.

## 5.1 DNA-dependent protein kinase

Of particular interest is the similarity of the ATM protein to the catalytic subunit of DNA-dependent protein kinase (DNA-PKcs) (133). This large protein has serine–threonine kinase activity, and is part of a heterotrimer that also contains the two subunits of the Ku autoantigen. DNA-PK is activated *in vitro* by DNA containing double-strand breaks and phosphorylates a variety of DNA binding regulatory proteins including several transcription factors. Co-localization of DNA-PK and its substrates on the same DNA molecule probably enhances the rate of phosphory-lation (18, 134). The Ku antigen subunits are thought to interact with the damaged DNA and activate the large catalytic subunit. Mutations inactivating DNA-PKcs in the mouse lead to the SCID phenotype, which shares several features with A–T such as radiosensitivity, chromosomal instability, and immune deficiency (24, 27). An attractive model for DNA-PK action is that following the detection of broken DNA ends, this enzyme complex recruits and/or activates the DNA repair machinery while inhibiting the action of several components of the transcription machinery which might interfere with DNA repair (18, 135). DNA-PK is thus emerging as a model system signalling the presence of DNA damage to numerous cellular regulatory systems via an enzymatic activity common to many signal transduction systems—protein phosphorylation. The similarity of DNA-PK to the ATM protein in size together with the presence of a carboxy-terminal PI3-kinase domain (Fig. 7.4), and the common features of A–T and the SCID phenotype, make DNA-PK a natural paradigm for the study of ATM's mode of action.

## 5.2   Atr (*a*taxia–*t*elangiectasia and *r*ad3-related)

The ATM-related proteins Rad3 (*Schizosaccharomyces pombe*), Mec1p (*S. cerevisiae*), and Mei-41 (*Drosophila*) monitor the integrity of genomic DNA and activate cell-cycle checkpoints to prevent the progression of cells through the cell cycle (115, 116, 119, 124). The human counterpart of these genes *ATR/FRP1* is the closest mammalian homologue to *ATM* (125, 126). The product of this gene can heteromultimerize with Rad3 *in vivo* and complements for mec1 UV sensitivity, suggesting that it too plays an important role in checkpoint control. Indeed, forced expression of ATR in the differentiation-competent murine myoblast cell line C2C12 led to loss of the $G_1/S$ cell-cycle checkpoint, aneuploidy, centrosome amplification, and inhibition of both MyoD-dependent transactivation and muscle cell differentiation (136).

Immunoblotting shows that ATR is expressed as a 300-kDa protein in human cell lines and in mouse testes (90). The highest level of ATR expression was found in pachytene spermatocytes (cells undergoing meiosis 1), it decreased with the advance of meiosis, and was not detectable in mature sperm. In early meiotic cells ATR is localized to the nucleus, which is compatible with a possible role in checkpoint control in these cells (90). The exact localization of ATR in meiotic chromosomes has been investigated using surface spreads of mouse spermatocytes. In order to sub-stage the binding of ATR, Cor1 staining was employed. Cor1 appears prior to the pairing of the axial elements to form the synaptonemal complex (137). As the homologues initiate synapsis in zygotene, ATR is detected at pairing forks as discrete foci along the asynapsed axes being lost as the homologues synapse (90). As discussed above ATR is a member of a family of protein kinases (125, 126), and the partial purification and immunoprecipitation of ATR revealed that it has an associated kinase activity (90). This would equip it well to participate in a signal transduction pathway for checkpoint control (138).

# 6. Mouse models

The mouse homologue of the human *ATM* gene, *Atm*, has been cloned and shown to contain an ORF of 9.198 kb, coding for 3066 amino acids, 10 more than for the human gene (86). The mouse protein has 84% overall identity with the human protein (which rises to 94% in the PI3-kinase domain) and the gene maps to chromosome 9, band C, a region that is syntenic with the distal end of the human 11q22–23 homology unit. The mouse mutant previously described with closest phenotype to A–T was *scid*, which is characterized by radiosensitivity, chromosomal instability, and defective double-strand break repair (139). However, human chromosome 8 complements this defect (140), and it has been demonstrated that *scid* is due to a nonsense mutation at Tyr-4046 in DNA-PKcs (23). The only mutant that maps close to *Atm* is *luxoid* which displays a different phenotype.

These observations suggested that the Atm null phenotype in mice might be embryonic lethal. However, these fears were unfounded when Barlow *et al.* (141) successfully inactivated the *Atm* gene by introducing a truncation mutation between nucleotides 5705 and 5882, a region that corresponds to a number of known human mutants (80). Mice homozygous for the disruption of *Atm* were growth retarded, no mature gametes were found in the gonads, T-lymphocyte differentiation was abnormal, animals succumbed more readily to radiation exposure, and they developed thymic lymphoblastic lymphomas between 2 and 4 months of age (141). Fibroblasts from phenotypically atm$^{-/-}$ mice grew poorly and were radiosensitive, characteristic of that seen in A–T patients. Targeted disruption of *Atm* was also described by Xu *et al.* (142) and Elson *et al.* (143). These mice also displayed a phenotype similar to that for A–T patients. While there was general consensus between the different mouse models for the degree of radiosensitivity, some differences were observed in radiation-induced apoptosis. Xu and Baltimore (144) reported that immature thymocytes in atm$^{-/-}$ mice were more resistant to apoptosis, whereas Elson *et al.* (143) observed similar extents of apoptosis in normal and *Atm*-deficient thymocytes (preliminary data). A similar number of TUNEL-positive (apoptotic) cells were detected in both the lung and thymus of normal and *Atm*-deficient mice after exposure to 10 Gy of radiation (145). More recently, Herzog *et al.* (146) showed a lack of apoptotic response in most regions of the brain of irradiated atm$^{-/-}$ mice. Overall it appears that radiation-induced apoptosis occurs to a lesser or equal extent in atm$^{-/-}$ and normal mice. Further support for a non-essential role for *Atm* in apoptosis is provided by the suppression of apoptosis in irradiated thymus from phenotypical atm$^{-/-}$ p53$^{-/-}$ double-mutant mice (147).

The most debilitating aspect of the A–T phenotype is the progressive neuronal degeneration (46). This characteristic is the one where the mouse models diverge most from the human disease. Histological analysis of Atm phenotype mutant brains showed normal architecture and no evidence of neuronal degeneration (141). In the cerebellum, Purkinje cell bodies were normal, the granular cell and molecular layer were of normal thickness, and dendritic arborization was normal (141, 142). However, there was evidence of motor impairment in atm$^{-/-}$ mice (141). Furthermore, the use of electron microscopy has detailed neuronal abnormalities in these mice including degenerating granule cells (21%), abnormal Purkinje cells (33%), and degenerating neurones (47%) (148). Further work is required to determine whether the neuronal changes in atm$^{-/-}$ are the same as those in patients with A–T.

# 7. Role of ATM in cell-cycle control

In the four years since the *ATM* gene was cloned there has been remarkable progress in understanding how ATM functions. This has been achieved from studies on the nature of the mutations in A–T patients, expression of ATM cDNA and correction of aspects of the A–T cellular phenotype, identification of interacting proteins, the description of certain activities for ATM, and the use of mouse models. However, an extensive body of information built up over 20 years on radiosensitivity and cell-cycle control in A–T cells had already provided important pointers to the likely roles of the protein defective in these cells. As outlined above, a hallmark of A–T cells is their extreme sensitivity to ionizing radiation (53, 54). Evidence has been provided that at least part of that radiosensitivity can be explained by a defect in the repair of a subcategory of double-strand breaks in DNA (31). It seems likely that ATM has a related but separate function from DNA-PKcs in sensing or detecting specific forms of damage in DNA prior to its repair by other enzymes (135). These events do not occur in isolation, but are also co-ordinated with and coupled to the normal progression of cells through the different phases of the cell cycle. When the integrity of genomic DNA is compromised, cells activate DNA damage checkpoints that prevent the progression of cells through the cycle (116). Checkpoints exist at the $G_1$/S-phase transition, during the S phase and at the transition from $G_2$ to mitosis (117, 149). A number of yeast mutants have been described that fail to undergo arrest at either $G_1$/S or $G_2$/M checkpoints in response to radiation damage (116, 118, 120, 150, 151). Some of these mutants, including *mec1* and *rad3*, fail to maintain the dependence of mitosis upon completion of DNA synthesis, they lose chromosomes spontaneously and they are hypersensitive to ionizing radiation.

## 7.1 $G_1$/S-phase checkpoint

A–T cells are characterized by a failure to activate either the $G_1$/S or $G_2$/M checkpoints shortly after radiation exposure, and they exhibit radioresistant DNA synthesis (55, 152–156, and Fig. 7.5). The inability of A–T cells to activate these checkpoints efficiently ultimately leads to the accumulation of irradiated cells in $G_2$/M phase where they die. The molecular nature of the defect in cell-cycle control in A–T cells has been extensively studied and it is evident that a pathway mediated by the *ATM* gene functioning through p53 is defective in A–T cells (157–161). The defect is evident not only at the level of radiation-induced stabilization of p53 but also at the level of its downstream effectors WAF1, gadd 45, and MDM2 (162, 163). Increased WAF1, as a consequence of exposure of control cells to radiation, leads to an inhibition of cyclin-dependent kinase activity, cyclinE–cdk2 at the $G_1$/S transition

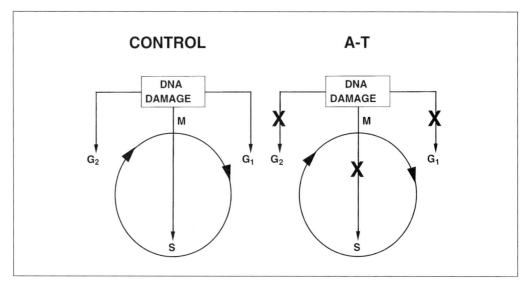

**Fig. 7.5** Cell-cycle checkpoint control in control and A–T cells in response to radiation damage. Checkpoints are activated at $G_1$/S, in S phase, and at $G_2$/M in control cells. A–T cells fail to activate $G_1$/S and $G_2$/M checkpoints shortly after irradiation and they display radioresistant DNA synthesis (S-phase checkpoint defect).

point (34). This inhibition is caused by a change in the stoichiometry of binding of WAF1 to a complex that includes cyclinE, cdk2, PCNA, and WAF1 (164).

In A–T cells WAF1 either does not increase in response to radiation or is delayed in its induction; consequently, inhibition of cyclin E–cdk2 is not observed and cells progress from $G_1$ into S phase without delay (163). Lack of inhibition of cyclin E–cdk2 in irradiated A–T cells appears to be due to little or no change in WAF1 association with the kinase (165). It is also of interest that the ratio of hyperphosphorylated to hypophosphorylated forms of retinoblastoma protein (RB) is high in A–T cells, which would favour unrestricted passage through the cell cycle (163).

As outlined in Fig. 7.6, ionizing radiation initiates a signalling pathway operating through ATM and p53 to activate the $G_1$/S checkpoint. It is unclear whether several steps are involved in this pathway between the two molecules, or whether they function in the pathway as a consequence of a direct interaction. Immuno-precipitation of cell lysates with anti-p53 antibody followed by immunoblotting with anti-ATM antibody revealed that ATM was associating with p53 (74). This association was either absent in A–T lysates, where a truncated protein was predicted, or occurred at low affinity in A–T cells producing mutated but near full-length ATM. It seems likely that the p53–ATM association is direct since an ATM cDNA fragment encoding the PI3-kinase domain activated the reporter gene β-galactosidase when co-transfected with a p53 cDNA in the yeast two-hybrid system (74). Binding of p53 to ATM was constitutive and was not influenced by exposure of the cells to ionizing radiation. Studies are currently underway to define the sites of interaction between ATM and p53. Recent data suggest that binding is complex, involving more than one

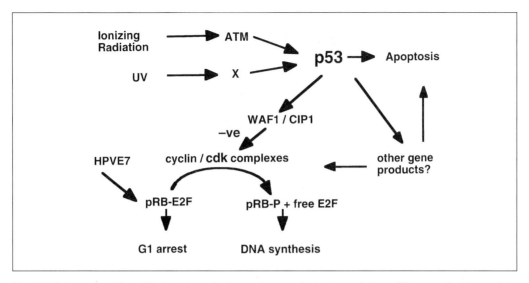

**Fig. 7.6** Schematic of the p53 signal transduction pathway activated by radiation. ATM is required to mediate the signalling from ionizing radiation but not UV light. Next, p53 transcriptionally activates p21/WAF1, which, in turn, binds to and inhibits the activity of cyclin-dependent kinase activity. When this enzyme is inhibited, substrates such as retinoblastoma protein (pRB) are not phosphorylated maintaining E2F inactive and thus preventing the passage of cells from $G_1$ to S phase.

site on ATM (166). The direct interaction between ATM and p53 suggests that p53 may be a substrate of ATM. In support of this is the observation that p53 is phosphorylated at ser15 after exposure of cells to radiation and that this leads to the reduced interaction of p53 with its negative regulator MDM2 (167). Furthermore, Siliciano *et al.* (168) have demonstrated that the loss of ATM in A–T cells leads to a reduced rate of ser15 phosphorylation on p53. More recently it has been demonstrated that ATM phosphorylates p53 on serine 15 in response to DNA damage *in vivo* and *in vitro* (166, 169, 170). ATM appears to be the protein kinase responsible for the rapid phosphorylation and activation of p53 in response to ionizing radiation but other modifications also occur to p53 some of which are ATM-dependent.

The defect at the $G_1$/S checkpoint has also provided further focus for identifying ATM interacting proteins. The finding that the product of the c-*Abl* gene, a non-receptor protein tyrosine kinase, is activated by certain DNA-damaging agents and that its overexpression causes arrest in $G_1$ phase (by a mechanism dependent on p53) suggested a potential role for c-Abl in growth arrest induced by DNA damage (171). Cells overexpressing dominant-negative, kinase-inactive c-Abl show a partial defect in ionizing radiation-induced $G_1$ arrest, and cells lacking c-Abl are impaired in their ability to downregulate cdk2 activity or undergo $G_1$ arrest (172). Since these results with c-Abl paralleled what was observed in A–T cells postirradiation, the possibility that ATM and c-Abl might interact in response to radiation was explored. Constitutive binding of c-Abl to ATM was observed in control cells but not in A–T

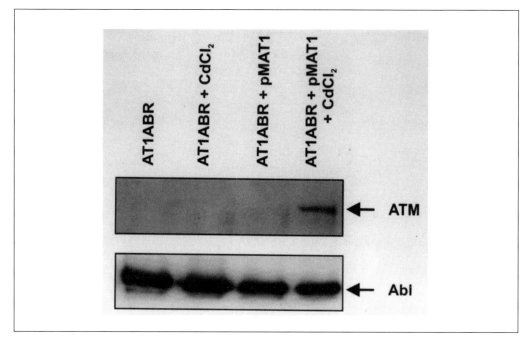

**Fig. 7.7** Restoration of ATM–Abl association after the ectopic expression of ATM cDNA. The cell line used for transfections was AT1ABR, which has an inframe deletion of 9 nucleotides (7636 del 9) and gives rise to a near full-length protein. In this experiment AT1ABR was transfected with pMAT1 (construct containing ATM cDNA) with or without induction with CdCl2 since it contains a metallothionein promoter. Extracts from the different cell types were immunoprecipitated with anti-c-Abl antibody and immunoblotted with antibody (ATM-3BA) against ATM. Loading was determined with c-Abl antibody.

cells (173). Ectopic expression of ATM cDNA in AT1ABR, an A–T cell line expressing defective protein which is incapable of binding to c-Abl (170), restored ATM–Abl binding in these cells (Fig. 7.7) thus underlying the physiological significance of this association. The association between ATM and c-Abl was direct and the SH3 domain of c-Abl was sufficient to bind a proline-rich sequence in ATM. Activation of c-Abl in response to ionizing radiation requires the presence of functional ATM protein: c-Abl was phosphorylated in response to ionizing radiation in normal human cells but not in A–T cells. In addition, embryonic fibroblasts from *ATM* knock-out mice were also defective in c-Abl activation following ionizing radiation exposure (174). The ATM-kinase domain alone was sufficient to restore c-Abl activity by phosphorylating serine 465. Of note, c-Abl also binds to DNA-PK, a protein kinase involved in DNA recombination and repair (175). The relationship of DNA-PK/c-Abl complexes with ATM and p53 remains unclear.

The involvement of ATM with c-Abl may be in a pathway separate to the p53 pathway. Recent data (Khanna *et al.*, unpublished) suggest that c-Abl phosphorylation by ATM is not required for the increased association of Abl with p53 after ionizing radiation exposure, since *ATM* null lines show a normal induction of

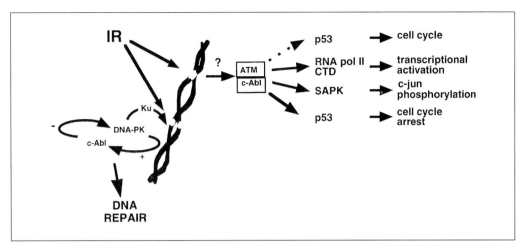

**Fig. 7.8** Damage response after exposure of cells to ionizing radiation. Ku recruits DNA-PKcs to double-strand breaks in DNA to produce an active DNA-PK complex, which is capable of phosphorylating a number of proteins involved in DNA repair and transcription. DNA-PK also phosphorylates c-Abl to activate stress signalling pathways through c-jun, and a feedback loop exists where c-Abl in turn phosphorylates DNA-PK and inactivates it. Since ATM also activates c-Abl in response to ionizing radiation (activating lesion not yet described) it might be expected that there would be some co-ordination between the two recognition processes. As indicated, the interaction between ATM and c-Abl may be responsible for activating several pathways for the regulation of transcription and cell-cycle control. The solid arrows refer to pathways involving ATM and c-Abl, while the broken arrow refers to ATM alone.

p53/c–Abl complexes after ionizing radiation. The involvement of ATM with c-Abl may be in a pathway operating through stress-activated protein kinase (SAPK), leading to the induction of stress-response genes via c-jun. c-*Abl* has been shown to be upstream of SAPK in a stress-response pathway (176) and SAPK is not activated in response to ionizing radiation in A–T cells (173, 177). In A–T cells, the defect in SAPK activation appears to be confined to the damaging agent to which these cells are hypersensitive (ionizing radiation), since SAPK activity is comparable in A–T and control cells after UV treatment or on exposure to the protein synthesis inhibitor anisomycin (176). Alternatively, the ATM/c-Abl pathway may modulate the expression of target genes by phosphorylating the C-terminal repeat domain (CTD) of RNA polymerase II, since ionizing radiation-induced CTD tyrosine phosphorylation by c-Abl is dependent on functional ATM (174). The proposed pathways, activated by ATM-Abl, are shown in Fig. 7.8.

## 7.2 S-phase and $G_2$/M checkpoints

A defect in ATM leading to an inadequate p53 response can explain the defective $G_1$/S checkpoint in A–T cells. However, since the S-phase and $G_2$/M checkpoints are also defective in A–T one might expect some overlap, given the likely involvement of ATM in control of the other checkpoints. In support of such an overlap, Aloni-

Grinstein *et al.* (178) have shown that p53 plays a role in γ-radiation-induced $G_2$ delay in pre-B cells, and evidence of a role for p53 at both the $G_2/M$ and $G_1$ checkpoints has also been presented (179). The resistance of several different cyclin-dependent kinase activities, that control the different checkpoints, to radiation-induced inhibition in A–T cells points to a wider role for the p53 pathway in radiation signal transduction (165). In this study, radiation caused a 5–20-fold increase in cdk-associated WAF1 in control cells synchronized at the $G_1/S$ transition, in S phase, and in $G_2$ phase. On the other hand, no increase in WAF1 associated with either cyclin E–cdk2, cyclin A–cdk2, or cyclin B–cdk2 complexes was observed in A–T cells, and these cyclin-dependent kinases were not inhibited (165). The increased association of WAF1 with cyclin A–cdk2 (cdc2) probably accounts for some of the inhibition of DNA synthesis observed in response to radiation damage. Given the complexity of the replication process, it is likely that more that a single pathway exists for radiation-induced inhibition. Phosphorylation of replication protein A (RPA), a trimeric, single-stranded DNA-binding protein complex, is observed when cells are exposed to ionizing radiation (180). This γ-radiation-induced phosphorylation of the p34 subunit of RPA is delayed in A–T cells. Since phosphorylation reduces the binding affinity of RPA for DNA it has been suggested that this represents a means of inhibiting DNA replication (181). Delayed phosphorylation of RPA could account for the radioresistant DNA synthesis observed in A–T cells. However, since rather high doses of radiation (10–50 Gy) were employed to expose this defect it is possible that other aspects of DNA replication control are also influenced by the presence of the ATM protein.

Postirradiation, cell-cycle abnormalities are also observed in *Atm* mutant mice (141, 142). The p53 response to radiation damage is defective in phenotypically atm$^{-/-}$-mouse embryonic fibroblasts and, presumably, reduced $G_1$ arrest also occurred as a consequence (144). Radioresistant DNA synthesis (RDS), which is also characteristic of A–T cells (55, 154), was evident in irradiated atm$^{-/-}$-mouse embryonic fibroblasts (141). It was recently reported that overexpression of wtATR complements the RDS phenotype of an A–T cell line, suggesting overlapping ATR and ATM functions.

Recent studies in yeast and human cells have also implicated chk1 protein in $G_2/M$-checkpoint control (182–184). The chk1 kinase becomes phosphorylated in response to DNA damage during $G_2$, and when overproduced causes mitotic delay in *S. pombe* (185, 186). After DNA damage, chk1 is modified in a Rad3-dependent manner and prevents cdc2 activation by inhibiting cdc25 (184, 187). The fact that ATM-related kinases act upstream of chk1 in *S. pombe* suggests that this entire checkpoint pathway may be conserved in all eukaryotes. Finally, a recent study (188) has established a connection between ATM and chk1 by demonstrating the ATM-dependent presence of chk1 protein in mouse testis. In this study, chk1 protein was present in a testis extract from atm$^{+/+}$ mice but was absent in atm$^{-/-}$ mice, suggesting that the synthesis or stability of chk1 depends on the ATM protein. However, more recent data from our laboratory suggest that in somatic cells there is no difference in the expression of chk1 protein in ATM-null or ATM-positive cells, but that, on the other hand, phosphorylation of the chk1 protein in response to DNA damage is dependent on the presence of functional ATM protein (Khanna *et al.* unpublished).

## 7.3  Cell cycle and radiosensitivity

The majority of earlier studies on A–T cells failed to identify a defect in the repair of strand breaks in DNA or the excision of other radiation-induced lesions in DNA (29). However, as pointed out above, up to 10% of the double-strand breaks induced by ionizing radiation remained unrepaired at 72-h postirradiation (31). This, in itself, could account for the increased sensitivity to radiation in A–T cells. Since the defect in A–T cells does not appear to be in the repair of breaks *per se* but rather in the recognition and processing of DNA damage, the unrepaired lesions could arise as a consequence of a failure to activate cell-cycle checkpoints. Shortly after irradiation, all cycle checkpoints are defective in A–T cells. The failure of A–T cells to mount a p53 response that is effective in delaying the progression of cells from $G_1$ to S phase is unlikely to provide an explanation for the enhanced radiosensitivity in these cells. Others have not observed a correlation between the absence of p53 or mutated (non-functional) p53 and a propensity to radiation sensitivity (189–192). It is more likely that the defective p53 response in $G_1$-phase cells leads to increased chromosomal instability as a consequence of these cells entering S phase prior to repairing DNA damage. Indeed, chromosomal instability and a predisposition to develop leukaemias, lymphomas, and, to a lesser extent, solid tumours are very characteristic of A–T (48–50). This is also the case in $atm^{-/-}$ mice, and a dramatic acceleration of tumour formation is observed in double-mutant $atm^{-/-}p53^{-/-}$ mice (193). In the absence of atm, mouse thymocytes showed a normal or partial resistance to radiation-induced apoptosis, but thymocytes were completely resistant in the double mutants (145, 193). Unexpectedly, ATM and p53 did not interact to protect against acute radiation toxicity since the double-mutant mice were killed to the same extent by radiation exposure (194). This toxicity is due to an increased sensitivity in intestinal and skin cells which remains the same in $atm^{-/-}$ $p53^{-/-}$ mice (Table 7.3). In essence, the loss of Atm in mouse thymocytes renders these cells partially resistant to apoptosis, but other cell types become extremely sensitive to radiation and the loss of p53 has no influence. It should be pointed out that the apoptosis data for thymocytes was determined over a 24-h period and it is possible that $atm^{-/-}$ thymocytes may be more sensitive to radiation-induced killing when determined at longer times postirradiation.

Some evidence has been provided that p53 may also play a role in other check-points in S phase and $G_2/M$ (165, 178, 179). Flow cytometric analysis has shown that when A–T cells are irradiated in $G_1$ or S phase and allowed to proceed through the cycle essentially they accumulate irreversibly in the following $G_2/M$ where they die (152). These results suggest that by ignoring the cell-cycle checkpoints, A–T cells ultimately carry chromosome damage into $G_2$ phase where they are incapable of normal chromosomal segregation and cell division. Treatment of irradiated A–T cells with caffeine allows them to bypass $G_2$ delay but leads to massive chromosomal

**Table 7.3** Radiation sensitivity and toxicity in atm/p53 mutant mice

| Phenotype | Radiation-induced apoptosis (killing) | Radiation toxicity | | |
|---|---|---|---|---|
| | Thymocytes | Lymphoid | Intestinal | Skin |
| Wild type | Sensitive | Some depletion | Resistant | Resistant |
| atm$^{-/-}$ | Sensitive/intermediate sensitivity | Some depletion | Some sensitivity | Oedematous |
| p53$^{-/-}$ | Resistant | Resistant | Resistant | Resistant |
| atm$^{-/-}$ p53$^{-/-}$ | Resistant | — | Severe sensitivity | Oedematous |

fragmentation and cell death; supporting the hypothesis that A–T chromosomes accumulate damage and, as a consequence, are incapable of proceeding to mitosis (194). Ignoring the G$_2$ checkpoint at short times postirradiation may also contribute to radiosensitivity in A–T cells, since overexpression of the *S. pombe chk1* gene in A–T cells has been shown to correct the radiosensitivity and aspects of cell-cycle control (195). Further elucidation of the role of the ATM protein will provide a greater understanding of the basis for radiosensitivity in A–T.

# 8. Role of ATM in meiosis

To ensure that the correct complement of chromosomes is distributed to each haploid nucleus during meiosis, specific structures have evolved to facilitate synapsis and homologous chromosome recombination (196). These structures include the synaptonemal complex (SC), with a central synapsed region and unsynapsed axial arms. Early meiotic nodules (MN) are found associated with the axial unsynapsed elements and newly formed SC, while recombination nodules (RN) are only found on the SC (197). Since the ATM protein is localized along the synapsed axes of SC it seemed likely that it might play a role in meiosis (90). In addition to this observation, data from A–T patients and atm$^{-/-}$ mice support such a role. Hypogonadism has been observed in male A–T patients but to a lesser extent than in female patients (44, 45). There are reports of puberty delay in male patients, histological evidence of testicular abnormalities, and incomplete spermatogenesis (198, 199). While the effects on germ-cell development are marginal in A–T patients, in contrast they are much more dramatic in atm$^{-/-}$ mice. These animals are infertile, have extremely small gonads, and spermatocytes are observed at various stages of degeneration (141–143). The development of spermatocytes is arrested between the zygotene and pachytene stages of meiotic prophase (142). When meiotic events were monitored with an antibody to Cor1, a protein component of axial elements and fully synapsed regions, it is evident that atm$^{-/-}$ zygotene spermatocytes have extensive unpaired axial elements. Nevertheless, these nuclei progress into pachynema where they experience increasing chromosome damage and cell death (142). Since high levels of p53, p21, and bax were only observed in the testes of atm$^{-/-}$ mice it was proposed that these proteins contribute to the severe meiotic phenotype in these animals (147). The use of double mutants, atm$^{-/-}$ p53$^{-/-}$ and atm$^{-/-}$ p21$^{-/-}$, led to an improvement in spermatocyte morphology compared to atm$^{-/-}$ and a decrease in apoptosis. In contrast to that for atm$^{-/-}$ mice these double mutants displayed nearly normal pachytene SC (147).

As mentioned above, two of the structures involved in synapsis and recombination are meiotic nodules (MN) and recombination nodules (RN). These are nucleoprotein complexes that ensure homology pairing and the subsequent strand exchange, and have been shown to contain proteins that are implicated in DNA repair: RAD51 (rec A homology), replication protein A (RPA), MLH1, (mismatch repair protein), ATM, and ATR. Initial studies showed that ATR was localized along unpaired axial elements, whereas ATM was associated with axial elements as they came into contact and was also found on fully synapsed bivalents (90). RAD51 is distributed on asynapsed axes in a pattern similar to that for ATR and BRCA1 (90, 200), but it also appears on synapsed SC in early pachynema (201). On the other hand, RPA appears in a focal pattern along the SC of synapsing chromosomes and co-localizes with RAD51 up to late pachynema, after which time RPA-only foci are observed. Co-localization of RPA and ATM has also been demonstrated along synapsed meiotic chromosomes and at sites where interactions between ectopic homologous chromosome regions appear to initiate (202). These results do not necessarily mean that ATM and RPA interact directly, but since phosphorylation of RPA is defective in mitotically dividing A–T cells, postirradiation, it suggests a functional relationship between the proteins. The mismatch repair protein MLH1 also co-localizes with RPA (201). Figure 7.9 depicts the series of protein composition changes that occur in meiotic nodules as determined by immunolocalization techniques (201).

In this model ATM becomes associated with meiotic nodules containing RPA and RAD51 on newly synapsed axes of SC. The latter proteins appear to be critical to homologous chromosome pairing. RAD51 is lost from the complex during progress through pachytene, leaving RPA and ATM to monitor the structure and facilitate the process of recombination between homologous pairs of chromosomes. A recent report showing that chromosomal fragmentation occurs preferentially at RPA sites along the SC in atm$^{-/-}$ spermatocytes adds further credence to 'sensor' and perhaps maintenance roles for ATM on the SC (202). This monitoring function of ATM can be compared to its involvement in the detection of damage-induced breaks in mitotic chromosomes.

# 9. Integrated view of the role of ATM

Based on the above description of the involvement of ATM in mitotic cells and in those undergoing meiosis it is reasonable to assume that this protein responds to breaks in DNA as part of the process of DNA repair or in DNA recombination. The

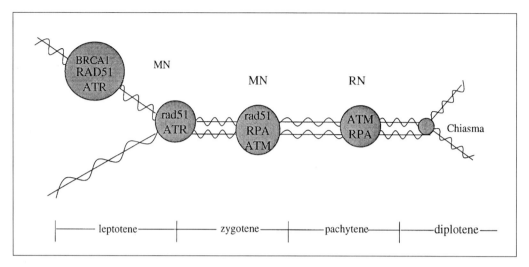

**Fig. 7.9** Change in protein composition of meiotic nodules (MN) and recombinant nodules (RN) during formation of the synaptonemal complex.

extranuclear location of ATM together with a complex phenotype associated with the loss of this protein suggests that it may have additional roles in signal transduction. Thus ATM plays a major role in DNA damage recognition in response to agents such as ionizing radiation that introduce double-strand breaks and other lesions into DNA. It would seem likely that the lesion involved is a double-strand break, since A–T cells are characterized by translocation involving hot spots for recombination (immunoglobulin and T-cell receptor genes) and by the presence of residual breaks at long times after DNA damage. In mouse spermatocytes in the absence of Atm, fragmentation of DNA occurs and a protein, RPA, that co-localizes with ATM on meiotic chromosomes, is found associated with a high percentage of these free ends, suggesting that ATM is responsible for ensuring that such breaks do not persist.

Unlike DNA-PKcs which is also a sensor of DNA damage, ATM transmits the DNA damage signal to the cell-cycle control machinery (Fig. 7.10). It does so at all three major checkpoints $G_1/S$, S phase, and $G_2/M$ (165). The series of events from p53 downstream in $G_1/S$-checkpoint control is reasonably well described, but there are subtleties to this pathway that require further investigation. For example, it has been shown recently that ATM interacts directly with WAF1 (Kedar *et al.*, unpublished). This adds an extra dimension of complexity to a pathway where ATM activates p53 that, in turn, acts to induce WAF1 to inhibit cyclin-dependent kinase activity, and it is now evident that WAF1 also interacts with ATM. It is possible that this represents some form of feedback loop similar to that where p53 induces MDM2, which, in turn, destabilizes p53 (167).

It remains unclear as to how ATM activates the pathway and how it is that ATM itself is responsive to radiation damage in DNA. One possibility is that ATM binds to DNA directly and responds to conformational change as a result of specific forms of

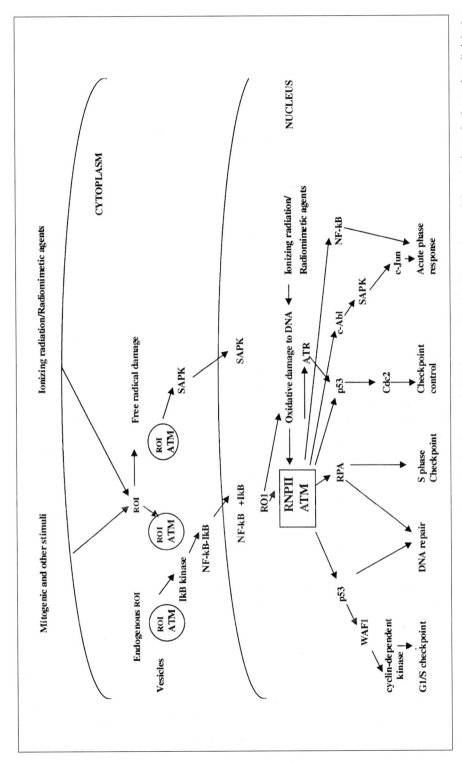

**Fig. 7.10** Role of ATM in radiation signal transduction and cell-cycle control. ATM is present both in the nucleus and the cytoplasm. In the nucleus it detects DNA damage (lesion unknown). As a consequence, a number of pathways are activated involving p53, c-Abl, and NF-*k*B, which, in turn, activate cell-cycle checkpoints and control transcription of stress-response genes such as *c-jun*. It is not clear how ATM functions in the cytoplasm, but it may also play a role in signalling in response to ROI.

damage. There is some evidence for ATM binding DNA as part of a purification procedure, but this appear to be rather weak (Boulton *et al.*, unpublished). In addition, there is good evidence that ATM is part of a nucleoprotein complex in meiosis, being associated with RPA and RAD51 (201). Its association with RPA would allow it to monitor this nucleoprotein complex and respond to DNA damage by phosphorylating RPA directly or through another kinase. Failure of RPA to be phosphorylated in A–T cells in response to radiation damage could account for radioresistant DNA synthesis, which is a universal characteristic of A–T (203). Such a monitoring role in $G_1$ phase could also apply where RPA was recruited to single-strand regions in DNA resulting from DNA damage. ATM might then associate with this complex and, since p53 is constitutively bound to ATM, it may represent the trigger for the activation of p53 by phosphorylation by ATM or another associated kinase.

How does c-Abl contribute to the $G_1$/S checkpoint? The interaction of c-Abl with ATM adds another dimension to the involvement of ATM in cell-cycle control. A role for c-Abl in the $G_1$/S checkpoint remains somewhat controversial, in that the data with c-abl$^{-/-}$ cells both support and provide evidence to the contrary for the involvement of c-Abl at the level of this control. Clearly the binding to c-Abl of ATM, and its activation by ionizing radiation as a protein kinase (only in cells with functional ATM) strongly suggest that c-Abl is a downstream effector of ATM. As referred to above, the ATM/c-Abl pathway may play a role in cell-cycle control or, perhaps more likely, this pathway is responsible for transcriptional control (Fig. 7.10). We have also provided evidence as to how ATM might be implicated in $G_2$/M-checkpoint control through its interaction with ATR and chk1 protein. *In toto*, these data suggest a complex signalling cascade downstream of ATM that includes c-Abl, p53, RPA, RAD51, ATR, and chk1 (Fig. 7.10). Data from functional knock-outs or overexpression of these proteins together with a definition of the complexes involved will provide a greater insight into the processes of DNA damage recognition and cell-cycle control.

# References

1. Vuillaume, M. (1987) Reduced oxygen species, mutation, induction and cancer initiation. *Mutat Res.*, **186**, 43.
2. Halliwell, B. and Gutteridge, J. M. C. (1990) Role of free radicals and catalytic metal ions in human disease: an overview. In *Methods in enzymology*, Vol. 186 (ed. L. Packer and A. N. Glazer), p. 1–85.
3. Chaudhri, G., Clark, I. A., Hunt, N. H., Cowden, W. B., and Ceredig, R. (1986) Effect of antioxidants on primary alloantigen-induced T cell activation and proliferation. *J. Immunol.*, **137**, 2646.
4. Schulze-Ostoff, K., Bauer, M. K., Vogt, M., and Wesselborg, S. (1997) Oxidative stress and signal transduction. *Int. J. Vitam. Nutr.*, **67**, 336.
5. Cerutti, P. A. and Remsen, J. F. (1978) Formation of repair of DNA damage induced by oxygen radical species in human cells. In *DNA processes* (ed. W. W. Nichols), p. 147. Symposium Specialists, Miami, Florida.

6. Acker, H. (1996) PO2 affinities, heme proteins, and reactive oxygen intermediates involved in intracellular signal cascades for sensing oxygen. *Adv. Exp. Mol. Biol.*, **410**, 59.

7. Blok, J. and Lohman, H. (1973) The effects of γ-radiation in DNA. *Curr. Topics Radiat. Res.*, **9**, 165.

8. Ward, J. F. (1985) Biochemistry of DNA lesions. *Radiat. Res.*, **104**, 103.

9. Van der Schans, G. P., Centen, H. B., and Lohman, P. H. M. (1992) The induction and repair of double-strand DNA breaks in normal and ataxia–telangiectasia cells exposed to $^{60}$Co-γ-radiation, 4-nitroquinoline-1-oxide or bleomycin. In *Ataxia–telangiectasia—A cellular and molecular link between cancer, neuropathology and immune deficiency* (ed. B. A. Bridges and D. G. Harnden), pp. 291. Wiley, Chichester.

10. Téoule, R. (1987) Radiation-induced DNA damage and its repair. *Int. J. Radiat. Biol.*, **51**, 573.

11. Radford, I. R. (1986) Evidence for a general relationship between the induced level of DNA double-strand breakage and cell killing after X-irradiation of mammalian cells. *Int. J. Radiat. Biol.*, **49**, 611.

12. Radford, I. R. (1986) Effect of radiomodifying agents on the ratios of X-ray induced lesions in cellular DNA: use in lethal lesion determination. *Int. J. Radiat. Biol.*, **49**, 621.

13. Bryant, P. E. (1985) Enzymatic restriction of mammalian cell DNA: evidence for double-strand breaks as potentially lethal lesions. *Int. J. Radiat. Biol.*, **48**, 55.

14. Badie, C., Iliakis, G., Foray, N., Alsbeih, G., Pantellias, G. E., Okayasu, R., *et al.* (1995) Defective repair of DNA double-strand breaks and chromosome damage in fibroblasts from a radiosensitive leukemia patient. *Cancer Res.*, **55**, 1232.

15. Lees-Miller, S. P., Godbout, R., Chan, D. W., Weinfeld, M., Day, R. S. III., Barron, G. M., *et al.* (1995) Absence of p350 subunit of DNA-activated protein kinase from a radiosensitive human cell line. *Science*, **267**, 1183.

16. Thompson, L. H. and Jeggo, P. A. (1995) Nomenclature of human genes involved in ionizing radiation sensitivity. *Mutat. Res.*, **337**, 131.

17. Zdzienicka, M. Z. (1995) Mammalian mutants defective in the response to ionizing radiation-induced DNA damage. *Mutat. Res.*, **336**, 203.

18. Gottlieb, T. M. and Jackson S. P. (1994) Protein kinases and DNA damage. *TIBS*, **19**, 500.

19. Kirchgessner, C. U., Patil, C. K., Evans, J. W., Cuomo, C. A., Fried, L. M., Carter, T., *et al.* (1995) DNA-dependent kinase (p350) as a candidate gene for the murine SCID defect. *Science*, **267**, 1178.

20. Taccioli, G. E., Rathbun, G., Oltz, G., Stamato, T., Jeggo, P., and Alt, F. W. (1993) Impairment of V(D)J recombination in double-strand break repair mutants *Science*, **260**, 207.

21. Blunt, T., Finnie, N. J., Taccioli, G., Smith, G. C. M., Demengeot, J., Gottlieb, T., *et al.* (1995) Defective DNA-dependent protein kinase activity is linked to V(D)J recombination and DNA repair defects associated with the murine scid mutation. *Cell*, **80**, 813.

22. Boubnov, N. V. and Weaver, D. T. (1995) Scid cells are deficient in Ku and replication protein A phosphorylation by the DNA-dependent protein kinase. *Mol. Cell. Biol.*, **15**, 5700.

23. Araki, R., Fujimori, A., Hamatani, K., Mita, K., Saito, T., Mori, M., *et al.* (1997) Nonsense mutation at Tyr-4046 in the DNA-dependent protein kinase catalytic subunit of severe combined immune deficiency mice. *Proc. Natl Acad. Sci. USA*, **94**, 2438.

24. Biedermann, K. A., Sun, J., Giaccia, A. J., Tosto, L. M., and Brown, J. M. (1991) Scid mutation in mice confers hypersensitivity to ionizing radiation and a deficiency in DNA double-strand break repair. *Proc. Natl Acad. Sci. USA*, **88**, 1394.

25. Bosma, C. C., Custer, R. R., and Bosma, M. J. (1983) A severe combined immunodeficiency mutation in the mouse. *Nature*, **301**, 527.

26. Hendrickson, E. A., Schatz, D. G., and Weaver, D. T. (1988) The scid gene encodes a trans-acting factor that mediates the rejoining event of Ig gene rearrangement *Genes Dev.*, **2**, 817.

27. Lieber, M, R., Hessem J. T., Lewis, S., Bosma, G. C., Rosenberg, N., Mizuuchi, K., *et al.* (1988) The defect in murine severe combined immune deficiency; joining of signal sequences but not coding segments in V(D)J recombination. *Cell*, **55**, 7.

28. Savitsky, K., Bar-Shira, A., Gilad, S., Rotman, G., Ziv, Y., Vanagaite, L., *et al.* (1995) A single ataxia–telangiectasia gene with a product similar to PI-3 kinase. *Science*, **268**, 1749.

29. Lavin, M. F. and Shiloh, Y. (1997) The genetic defect in ataxia–telangiectasia. *Annu. Rev. Immunol.*, **15**, 177.

30. Cornforth, M, W. and Bedford, J. S. (1985) On the nature of a defect in cells from individuals with ataxia–telangiectasia. *Science*, **227**, 1589.

31. Foray, N., Priestley, A., Alsbeih, G., Badie, C., Capulas, E. P., Arlett, C. F., *et al.* (1997) Hypersensitivity of ataxia–telangiectasia fibroblasts to ionizing radiation is associated with a repair deficiency of DNA double-strand breaks. *Int. J. Radiat. Biol.*, **72**, 271.

32. Hartwell, L. M. and Weinert, T. A. (1989) Checkpoints: controls that ensure the order of cell cycle events. *Science*, **246**, 629.

33. Kastan, M. B., Oneykwere, O., Sidransky, D., Vogelstein, B., and Craig, R. W. (1991) Participation of p53 protein in the cellular response to DNA damage. *Cancer Res.*, **51**, 6304.

34. Dulic, V., Kaufmann, W. K., Wilson, S. J., Tisty, T. D., Lees, E., Wade Harper, J., *et al.* (1994) p53-dependent inhibition of cyclin-dependent kinase activities in human fibroblasts during radiation-induced $G_1$ arrest. *Cell*, **76**, 1013.

35. Devary, Y., Rosette, C., Didonato, J. A., and Karin, M. (1993) NF-kB activation by ultraviolet light not dependent on a nuclear signal. *Science*, **261**, 1442.

36. Karin, M. (1995) The regulation of AP-1 activity by mitogen-activated protein kinase. *J. Biol. Chem.*, **270**, 16483.

37. Karin, M. and Hunter, T. (1995) Transcriptional control by protein phosphorylation: signal transmission from cell surface to the nucleus. *Curr. Biol.*, **5**, 747.

38. Coffer, P. J., Burgering, B. M., Peppelenbosch, M. P., Bos, J. L., and Kruijer, W. (1995) UV activation of RTK activity. *Oncogene*, **11**, 561.

39. Sachsenmaier, C., Radler-Pohl, A., Zinck, R., Nordheim A., Herrlich, P., and Rahmsdorf, H. J. (1994) Involvement of growth factor receptors in the mammalian UVC response. *Cell*, **78**, 963.

40. Schieven, G. L., Mittler, R. S., Nadler, S. G., Kirihara, J. M., Bolen, J. B., Kanner, S. B., *et al.* (1994) ZAP-70 tyrosine kinase, CD45 T-cell receptor involvement in UV and H2O2 induced signal transduction. *J. Biol. Chem.*, **260**, 20718.

41. Knebel, A., Rahmsdorf, H. J., Ullrich, A., and Herrlich, P. (1996) Dephosphorylation of receptor tyrosine kinases as targets of regulation by radiation, oxidants or alkylating agents. *EMBO J.*, **15**, 5314.

42. Syllaba, K. and Henner, K. (1926) Contribution a l'independance de l'athetose double idiopathique et congenitale. Atteinte familiale, syndrome dystrophique, signe de reseau vasculaire conjonctival, integrite psychique. *Rev. Neurol.*, **1**, 541.

43. Louis-Bar, D. (1941) Sur un syndrome progressif comprenant des telengiectasies capillaires cutanees et conjonctivales symetriques, a disposition naevode et de troubles cerebelleux. Confin. *Neurol. (Basel)*, **4**, 32.

44. Boder, E. and Sedgwick, R. P. (1957) Ataxia–telangiectasia. A familial syndrome of progressive cerebellar ataxia, oculocutaneous telangiectasia and frequent pulmonary

infection. A preliminary report on 7 children, an autopsy a case history. *Univ. S. Calif. Med. Bull.*, **9**, 15.

45. Boder, E. (1985) Ataxia–telangiectasia: an overview. In *Ataxia–telangiectasia* (ed. R. A. Gatti and M. Swift), pp. 1–63. Kroc Found. Ser. 19. Alan R. Liss, New York.

46. Sedgwick, R. P. and Boder, E. (1991) Ataxia–telangiectasia (208900; 208910; 208920). In *Hereditary neuropathies and spinocerebellar atrophies* (ed. J. M. B. Vianney De Jong), p. 347. Elsevier Science, Amsterdam.

47. Bowden, D. H., Danis P. G., and Sommers S. C. (1963) Ataxia–telangiectasia. A case with lesions of ovaries and adenohypophysis. *J. Neuropathol. Exp. Neurol.*, **22**, 549.

48. Boder, E. and Sedgwick, R. P. (1963) Ataxia–telangiectasia. A review of 101 cases. In *Little club clinics in developmental medicine*, No. 8 (ed. G. Walsh), p. 110. Heinemann Medical, London.

49. Spector, B. D., Filipovich, A. H., Perry, G. S. lll., and Kersey, J. H. (1982) Epidemiology of cancer in ataxia–telangiectasia. In *Ataxia–telangiectasia—a cellular and molecular link between cancer. Neuropathology and immune deficiency* (ed. B. A. Bridges and D. G. Harnden), pp. 103–138. Wiley, New York.

50. Hecht, F. and Hecht, B. K. (1990) Cancer in ataxia–telangiectasia patients. *Cancer Genet. Cytogenet.*, **46**, 9.

51. Gotoff, S. P., Amirmokri, E., and Liebner, E. J. (1967) Ataxia–telangiectasia. Neoplasia, untoward response to X-irradiation tuberous sclerosis. *Am. J. Dis. Child.*, **114**, 617.

52. Morgan, J. L., Holcomb, T. M., and Morrissey, R. W. (1968) Radiation reaction in ataxia–telangiectasia. *Am. J. Dis. Child.*, **116**, 557.

53. Taylor, A. M., Harnden, D. G., Arlett, C. F., Harcourt, S. A., Lehmann, A. R., Stevens, S., *et al.* (1975) Ataxia–telangiectasia: a human mutation with abnormal radiation sensitivity. *Nature*, **258**, 427.

54. Chen, P. C., Lavin, M. F., Kidson, C., and Moss D. (1978) Identification of ataxia telangiectasia heterozygotes, a cancer prone population. *Nature*, **274**, 484.

55. Painter, R. B. and Young, B. R. (1980) Radiosensitivity in ataxia–telangiectasia: a new explanation. *Proc. Natl Acad. Sci. USA*, **77**, 7315.

56. Coquerelle, T. M., Weibezahn, K. F., and Lucke-Huhle, C. (1987) Rejoining of double-strand breaks in normal human and ataxia–telangiectasia fibroblasts after exposure to $^{60}$Co γ-rays, $^{241}$Am α-particles or bleomycin. *Int. J. Radiat. Biol.*, **51**, 209.

57. Lavin, M. F. and Shiloh Y. (1996) Ataxia–telangiectasia: a multifacet genetic disorder associated with defective signal transduction. *Curr. Opin. Immunol.*, **8**, 459.

58. Shiloh, Y. (1995) Ataxia–telangiectasia: closer to unraveling the mystery. *Eur. J. Hum. Genet.*, **3**, 116.

59. Meyn, M. S., Lu-Kuo, J. M., and Herzing, L. B. K. (1993) Expression cloning of multiple human cDNAs that complement the phenotypic defects of ataxia-telangiectasia group D fibroblasts. *Am. J. Hum. Genet.*, **53**, 1206.

60. Ziv, Y., Bar-Shira, A., Jorgensen, T. J., Russell, P. S., Sartiel, A., Shows, T. B., Eddy, R. L., Buchwald, M., Legerski, R., Schimke, R. T., and Shiloh, Y. (1995) Human cDNA clones that complement the radiomimetic sensitivity of ataxia-telangiectasia (group a) cell. *Somat Cell Mol. Genet.*, **21**, 99.

61. Chen, P., Girjes, A. A., Hobson, K., Beamish, H., Khanna, K. K., Farrell, A., *et al.* (1996) Genetic complementation of radiation responses by 3' untranslated regions (UTR) of RNA. *Int J. Radiat. Biol.*, **69**, 385.

62. Rastinejad, F. and Blau, H. M. (1993) Genetic complementation reveals a novel regulatory role for 3' untranslated regions in growth and differentiation. *Cell*, **72**, 903.

63. Gatti, R. A., Berkel, I., Boder, E., Braedt, G., Charmley, P., Concannon, P., *et al.* (1988) Localization of an ataxia–telangiectasia gene to chromosome 11q22–23. *Nature*, **336**, 577.

64. Foroud, T., Sobel, E., Ziv, Y., Goradia, T., Wei, S., Charmley, P., *et al.* (1991) Localization of the AT locus to an 8 cM interval defined by STMY and S132. *Am. J. Hum. Genet.*, **49**, 1263.

65. Ziv, Y., Rotman, G., Frydman, M., Foroud, T., Gatti, R. A., and Shiloh, Y. (1991) The ATC (ataxia–telangiectasia complementation group) locus localizes to 11q22-q23. *Genomics*, **9**, 373.

66. McConville, C. M., Byrd, P. J., Ambrose, H. J., Stankovic, T., Ziv, Y., Bar-Shira, A., *et al.* (1993) Paired STSs amplified from radiation hybrids from associated YACs, identify highly polymorphic loci flanking the ataxia–telangiectasia locus on chromosome 11q22–23. *Hum. Mol. Genet.*, **2**, 969.

67. McConville, C. M., Byrd, P. J., Ambrose, H. J., and Taylor, A. M. R. (1994) Genetic and physical mapping of the ataxia–telangiectasia locus on chromosome 11q22–23. *Int. J. Radiat. Biol.*, **66**, S45.

68. Rotman, G., Savitsky, K., Vanagaite, L., Bar-Shira, A., Ziv, Y., Gilad, S., *et al.* (1994) Physical and genetic mapping at the ATA/ATC locus on chromosome 11q22–23. *Int. J. Radiat. Biol.*, **66**, S63.

69. Vanagaite, L., James, M. R., Rotman, G., Savitsky, K., Bar-Shira, A., Gilad, S., *et al.* (1995) A high-density microsatellite map of the ataxia–telangiectasia locus. *Hum. Genet.*, **95**, 451.

70. Lange, E., Borreson, A-L., Chen, X., Chessa, L., Chiplunkar, S., Concannon, P., *et al.* (1995) Localization of an ataxia–telangiectasia gene to a 850 kb interval on chromosome 11q23.1 by linkage analysis of 176 families in an international consortium. *Am. J. Hum. Genet.*, **57**, 112.

71. Byrd, P. J., McConville, C. M., Cooper, P., Parkhill, J., Stankovic, T., McGuire, G. M., *et al.* (1996) Mutations revealed by sequencing the 5' half of the gene for ataxia–telangiectasia. *Hum. Mol. Genet.*, **5**, 145.

72. Gilad, S., Khosravi, R., Shkedy, D., Uziel, T., Ziv, Y., Savitsky, K., *et al.* (1996) Predominance of null mutations in ataxia–telangiectasia. *Hum. Mol. Genet.*, **5**, 433.

73. Telatar, M., Wang, Z., Udar, W., Liang, T., Concannon, P., Bernatowska-Matuscklewicz, E., *et al.* (1996) Ataxia–telangiectasia: mutations in cDNA detected by protein truncation screening. *Am. J. Hum. Genet.*, **59**, 40.

74. Watters, D., Khanna, K. K., Beamish, H., Birrell, G., Spring, K., Kedar, P., *et al.* (1997) Cellular localisation of the ataxia–telangiectasia (ATM) gene proteins and discrimination between mutated and normal forms. *Oncogene*, **14**, 1911.

75. Wright, J., Teraoka, S., Onengut, S., Tolun, A., Gatti, R. A., Ochs, H. D., *et al.* (1996) A high frequency of distinct ATM gene mutations in ataxia–telangiectasia. *Am. J. Hum.Genet.*, **59**, 839.

76. Concannon, P. and Gatti, R. A. (1997) Diversity of ATM gene mutations detected in patients with ataxia–telangiectasia. *Hum. Mut.*, **10**, 100.

77. Gilad, S., Bar-Shira, A., Harnik, R., Shkedy, D., Ziv, Y., Khosravi, R., *et al.* (1996) Ataxia–telangiectasia: founder effect among North African Jews. *Hum. Mol. Genet.*, **5**, 2033.

78. Fukao, T., Song, X-X., Yoshida, T., Tashita, H., Kaneko, H., Teramoto, T., *et al.* (1998) Ataxia–telangiectasia (A–T) in the Japanese population: identification of R1917X, W2491R, R2909G, IVS33 (+2) gt to ga 7883 del 5, the latter two being relatively common mutations. *Hum. Mutat. Res.*, **12**, 338.

79. Telatar, M., Teraoka, S., Wang, Z., Chun, H. H., Liang, T., Castellvi-Bel, S., *et al.* (1998) Ataxia–telangiectasia: identification and detection of founder-effect mutations in the ATM gene in ethnic populations. *Am. J. Hum. Genet.*, **62**, 86.

80. Savitsky, K., Sfez, S., Tagle, D., Ziv, Y., Sartiel, A., Collins, F. S., *et al.* (1995) The complete sequence of the coding region of the ATM gene reveals similarity to cell cycle regulators in different species. *Hum. Mol. Genet.*, **4**, 2025.

81. Zhang, N., Chen, P., Khanna, K K., Scott, S., Gatei, M., Kozlov, S., *et al.* (1997) Isolation of full-length ATM cDNA and correction of the ataxia–telangiectasia cellular phenotype. *Proc. Natl Acad. Sci. USA*, **94**, 8021.

82. Ziv, Y., Bar-Shira, A., Pecker, I., Russell, P., Jorgensen, T. J., Tsarfati, I., *et al.* (1997) Recombinant ATM protein complements the cellular A–T phenotype. *Oncogene*, **15**, 159.

83. Scott, S. P., Zhang, N., Khanna, K. K., Khromykh, A., Hobson, K., Watters, D., *et al.* (1998) Cloning and expression of the ataxia–telangiectasia gene in baculovirus. *Biochem. Biophys. Res. Comm.*, **245**, 144.

84. Uziel, T., Savitsky, K., Platzer, M., Ziv, Y., Helbitz, T., Nehls, M., *et al.* (1996) Genomic organization of the ATM gene. *Genomics*, **33**, 317.

85. Ambrose, C. M., Duyao, M. P., Barnes, G., Bates, G. P., Lin, C. S., Srinidhi, J., *et al.* (1994) Structure and expression of the Huntington's disease gene: evidence against simple inactivation due to an expanded CAG repeat. *Somat. Cell. Mol. Genet.*, **20**, 27.

86. Pecker, I., Avraham, K. B., Gilbert, D. J., Savitsky, K., Rotman, G., Harnik, R., *et al.* (1996) Identification and chromosomal localization of Atm, the mouse homolog of ataxia–telangiectasia gene. *Genomics*, **35**, 39.

87. Brown, K. D., Ziv, Y., Sadanandan S. N., Chessa, L., Collins, F. S., Shiloh, Y., *et al.* (1997) The ataxia–telangiectasia gene product, a constitutively expressed nuclear protein that is not up-regulated following genome damage. *Proc. Natl Acad. Sci. USA*, **94**, 1840.

88. Chen, G. and Lee, E. Y-H. P. (1996) The product of the ATM gene is a 370-kDa nuclear phosphoprotein. *J. Biol. Chem.*, **271**, 33693.

89. Jung, M., Kondratyev, A., Lee, S., Dimtchev, A., and Dritschilo, A. (1997) ATM gene product phosphorylates IkB. *Cancer Res.*, **57**, 24.

90. Keegan, K. S., Holtzman, D. A., Plug, A. W., Christenson, E. R., Brainerd, E. E., Flaggs, G., *et al.* (1996) The Atr and Atm protein kinases associate with different sites along meiotically pairing chromosomes. *Genes Dev.*, **10**, 2423.

91. Lakin, N. D., Weber, P., Stankovic, T., Rottinghaus, S. T., Malcolm A., Taylor, R., *et al.* (1996) Analysis of the ATM protein in wild-type and ataxia telangiectasia cells. *Oncogene*, **13**, 2707.

92. Gilad, S., Chessa, L., Khosravi, R., Russell, P., Galenty, Y., Piane, M., *et al.* (1998) Genotype–phenotype relationships in ataxia–telangiectasia and variants. *Am. J. Hum. Genet.*, **62**, 551.

93. Paterson, M., Cerson, A. K., Smith, B. P., and Smith, P. J. (1979) Enhanced radiosensitivity of cultured fibroblasts from ataxia–telangiectasia heterozygotes manifested by defective colony-forming ability and reduced DNA repair replication after hypoxic gamma-irradiation. *Cancer Res.*, **39**, 3725.

94. Waghray, M., Al-Sedairy, S., Ozand, P. T., and Hannan, M. A. (1990) Cytogenetic characterization of ataxia–telangiectasia (A–T) heterozygotes using lymphoblastoid cell lines and chronic γ-irradiation. *Hum. Genet.*, **84**, 532.

95. Rosin, M. P. and Ochs, H. D. (1986) In vivo chromosomal instability in ataxia–telangiectasia homozygotes and heterozygotes. *Hum. Genet.*, **74**, 335.

96. Rudolph, N. S., Nagasawa, H., Little, J. B., and Latt, S. A. (1989) Identification of ataxia–telangiectasia heterozygotes by flow cytometric analysis of X-ray damage. *Mutation Res.*, **211**, 19.

97. Lavin, M. F., Le Poidevin, P., and Bates, P. (1992) Enhanced levels of radiation-induced $G_2$ phase delay in ataxia–telangiectasia heterozygotes. *Cancer Genet. Cytogenet.*, **60**, 183.

98. Sanford, K. K. and Parshad, R. (1990) Detection of cancer-prone individuals using cytogenetic response to X-rays. In *Chromosomal aberrations: basic and applied aspects* (ed. G. Obe and A. T. Natarajan), p. 113. Springer-Verlag, Berlin.

99. Zhang, N., Chen, P., Gatei, M., Scott, S., Khanna, K. K., and Lavin, M. F. (1998) An antisense construct of full-length ATM cDNA imposes a radiosensitive phenotype on normal cells. *Oncogene*, **17**, 811.

100. Chessa, L., Petrinelli, P., Antonelli, A., Fiorelli, M., Elli, R., Marcucci, L., *et al.* (1992) Heterogeneity in ataxia–telangiectasia: classical phenotype associated with intermediate cellular radiosensitivity. *Am. J. Med. Genet.*, **42**, 741.

101. Hentze, M. W. (1996) Iron–sulfur clusters and oxidant stress responses. *TIBS*, **21**, 282.

102. Hentze, M. W. and Argos, P. (1991) Homology between IRE-BP, a regulatory RNA-binding protein, aconitase isopropylomatate isomerase. *Nucl. Acids Res.*, **19**, 1739.

103. Martins, E. A., Robalinho, R. L, and Meneghini, R. (1995) Oxidative stress induces activation of a cytosolic protein responsible for the control of iron intake. *Arch. Biochem. Biophys.*, **316**, 128.

104. Rouault, T. A., Stout, C. D., Kaptain, S., Harford, J. B., and Klansner, R. D. (1991) Structural relationships between an iron-regulated RNA binding protein (IRE-BP) and aconitase: functional implications. *Cell*, **64**, 881.

105. Khoroshilova, N., Beinert, H., and Kiley, P. J. (1995) Association of a polynuclear iron–sulfur center with a mutant FNR protein enhances DNA binding. *Proc. Natl Acad. Sci. USA*, **92**, 2499.

106. Lazazzera, B., Beinert, H., Lhoroshilova, N., Kennedy, M. C., and Kiley, P. J. (1996) DNA binding and dimerization of the Fe-S-containing FNR protein from *Escherichia coli* are regulated by oxygen. *J. Biol. Chem.*, **271**, 2762.

107. Hidalgo, E. and Demple, B. (1996) Activation of SoxR-dependent transcription *in vitro* by noncatalytic or NifS-mediated assembly of [2Fe-2S] clusters into Apo-SoxR. *J. Biol. Chem.*, **271**, 7269.

108. Beamish, H., Williams, R., Chen, P., Khanna, K. K., Hobson, K., Watters, D., *et al.* (1996) Rapamycin resistance in ataxia–telangiectasia. *Oncogene*, **13**, 963.

109. Keith, C. T. and Schreiber, S. L. (1995) PIK-related kinases: DNA repair, recombination cell cycle checkpoints. *Science*, **270**, 50.

110. Zakian, V. A. (1995) ATM-related genes: what do they tell us about functions of the human gene? *Cell*, **82**, 685.

111. Greenwell, P. W., Kronmal, S. L., Porter, S. E., Gassenhuber, J., Obermaier, B., and Petes, T. D. (1995) TELl, a gene involved in controlling telomere length in *Saccharomyces cerevisiae*, is homologous to the human ataxia telangiectasia (ATM) gene. *Cell*, **82**, 823.

112. Morrow, D. M., Tagle, D. A., Shiloh, Y., Collins, F. S., and Hieter, P. (1995) TEL1, a *Saccharomyces cerevisiae* homologue of the human gene mutated in ataxia–telangiectasia, is functionally related to the yeast checkpoint gene MEC1/ ESRl. *Cell* **82**, 831.

113. Pandita, T. K., Pathak, S., and Geard, C. (1995) Chromosome end associations, telomeres and telomerase activity in ataxia–telangiectasia cells. *Cytogenet. Cell Genet.*, **71**, 86.

114. Allen, J. B., Zhou, Z., Siede, W., Friedberg, E. C., and Elledge, S. J. (1994) The SADl/ RAD53 protein kinase controls multiple checkpoints and DNA damage-induced transcription in yeast. *Genes Dev.*, **8**, 2401.

115. Kato, R. and Ogawa, H. (1994) An essential gene, ESRl, is required for mitotic cell growth, DNA repair and meiotic recombination in *Saccharomyces cerevisiae*. *Nucl. Acids Res.*, **22**, 3104.

116. Weinert, T. A., Kiser, G. L., and Hartwell, L. H. (1994) Mitotic checkpoint genes in budding yeast and the dependence of mitosis on DNA replication and repair. *Genes Dev.*, **8**, 652.

117. Paulovich, A. G. and Hartwell, L. H. (1995) A checkpoint regulates the rate of progression through S phase in *S. cerevisiae* in response to DNA damage. *Cell*, **82**, 841.

118. Al-Khodairy, F. and Carr, A. M. (1992) DNA repair mutants defining $G_2$ checkpoint pathways in *Schizosaccharomyces pombe*. *EMBO J.*, **11**, 1343.

119. Al-Khodairy, F., Fotou, E., Sheldrick, K. S., Griffiths, D. J. F., Lehmann, A. R., and Carr, A. M. (1994) Identification and characterisation of new elements involved in checkpoints and feedback controls in fission yeast. *Mol. Biol. Cell.*, **5**, 147.

120. Jimenez, G., Yucel, J., Rowley, R., and Subramani, S. (1992) The rad3+ gene of *Schizosaccharomyces pombe* is involved in multiple checkpoint functions and in DNA repair. *Proc. Natl Acad. Sci. USA*, **89**, 4952.

121. Enoch, T., Carr, A. M., and Nurse, P. (1992) Fission yeast genes involved in coupling mitosis to completion of DNA replication. *Genes Dev.*, **6**, 2035.

122. Boyd, J. B., Golino, M. D., Nguyen, T. D., and Green, M. M. (1976) Isolation and characterization of X-linked mutants of *Drosophila melanogaster* which are sensitive to mutagens. *Genetics*, **84**, 485.

123. Banga, S. S., Shenkar, R., and Boyd, J. B. (1986) Hypersensitivity of Drosophila mei-41 mutants to hydroxyurea is associated with reduced mitotic chromosome stability. *Mutat. Res.* **163**, 157.

124. Hari, K. L., Santerre, A., Sekelsky, J. J., McKim, K. S., Boyd, J. B., and Hawley, R. S. (1995) The *mei-41* gene of *Drosophila melanogaster* is functionally homologous to the human ataxia telangiectasia gene. *Cell*, **82**, 815.

125. Cimprich, K. A., Shin, T. B., Keith, C. T., and Schreiber, S. L. (1996) cDNA cloning and gene mapping of a candidate human cell cycle checkpoint protein. *Proc. Natl Acad. Sci. USA*, **93**, 2850.

126. Bentley, N. J., Holtaman, D. A., Keegan, K. S., Flaggs, G., DeMagio, A. J., Ford, J. C., *et al.* (1996) The *Schizosaccharomyces pombe* rad3 checkpoint gene. *EMBO J.*, **15**, 6641.

127. Heitman, J., Movva, N. R., and Hall, M. N. (1991) Targets for cell cycle arrest by the immunosuppressant rapamycin in yeast. *Science*, **253**, 905.

128. Kunz, J., Henriquez, R., Schneider, U., Deuter-Reinhard, M., Movva, N. R., and Hall, M. N. (1993) Target of rapamycin in yeast, TOR2, is an essential phosphatidylinositol kinase homolog required for $G_1$ progression. *Cell*, **73**, 585.

129. Brown, E. J., Albers, M. W., Shin, T. B., Ichikawa, K., Keith, C. T., Lane, W. S., *et al.* (1994) A mammalian protein targeted by $G_1$- arresting rapamycin–receptor complex. *Nature*, **369**, 756.

130. Brown, E. J., Beal, P. A., Keith, C. T., Chen, J., Shin, T. B., and Schreiber, S. L. (1995) Control of p70 S6 kinase by kinase activity of FRAP *in vivo*. *Nature*, **377**, 441.

131. Sabatini, D. M., Erdjument-Bromage, H., Lui, M., Tempst, P., and Snyder, S. H. (1994) RAFT1: a mammalian protein that binds to FKBPI2 in a rapamycin-dependent fashion and is homologous to yeast TORs. *Cell*, **78**, 35.

132. Sabers, C. J., Martin, M. M., Brunn, G. J., Williams, J. M., Dumont, F. J., Wiederrecht, G., *et al.* (1995) Isolation of a protein target of the FKBP12–rapamycin complex in mammalian cells. *J. Biol. Chem.*, **270**, 815.

133. Hartley, K. O., Gell, D., Smith, G. C. M., Zhang, H., Divecha, N., Connelly, M. A., *et al.* (1995) DNA-dependent protein kinase catalytic subunit: a relative of phosphatidyl-inositol 3-kinase and the ataxia telangiectasia gene product. *Cell*, **82**, 849.

134. Anderson, C. W. (1993) DNA damage and the DNA-activated protein kinase. *Trends Biochem. Sci.*, **18**, 433.
135. Jackson, S. P. (1995) Ataxia–telangiectasia at the crossroads. *Curr. Biol.*, **5**, 1210.
136. Smith, L., Liu, S-J., Goodrich, L., Jacobson, D., Degnin, C., Bentley, N., *et al.* (1998) Duplication of ATR by isochromosome formation inhibits MyoD, induces aneuploidy and eliminates the radiation induced $G_1$ arrest in rhabdomyosarcoma. *Nature Genet.*, **19**, 39.
137. Dobson, M. J., Pearlman, R. E., Karaiskakis, A., Sypropoulos, B., and Moens, P. B. (1994) Synaptonemal complex proteins, epitope mapping and chromosome disjunction. *J. Cell. Sci.*, **107**, 2749.
138. Carr, A. M. (1996) Checkpoints take the next step. *Science*, **271**, 314.
139. Bosma, M. J. and Carroll, A. M. (1991) The SCID mouse mutant: definition, characterization potential uses. *Rev. Immunol.*, **9**, 323.
140. Itoh, M., Hamatani, K., Komatsu, K., Araki, R., Takayama, K., and Abe, M. (1993) Human chromosome 8 (p12 q22) complements radiosensitivity in the severe combined immune deficiency (SCID) mouse. *Radiat Res.*, **134**, 364.
141. Barlow, C., Hirotsune, S., Paylor, R., Liyanage, M., Eckhaus, M., Collins, F., *et al.* (1996) Atm-deficient mice: a paradigm of ataxia–telangiectasia. *Cell*, **86**, 159.
142. Xu, Y., Ashley, T., Brainerd, E. E., Bronson, R. T., Meyn, S. M., and Baltimore, D. (1996) Targeted disruption of ATM leads to growth retardation, chromosomal fragmentation during meiosis, immune defects, thymic lymphoma. *Genes Dev.*, **10**, 2411.
143. Elson, A., Wang, Y., Daugherty, C. J., Morton, C. C., Zhou, F., Campos-Torres, J., *et al.* (1996) Pleiotropic defects in ataxia–telangiectasia protein-deficient mice. *Proc. Natl Acad. Sci. USA*, **93**, 13084.
144. Xu, Y. and Baltimore, D. (1996) Dual roles of ATM in the cellular response to radiation and in cell growth control. *Genes Dev.*, **10**, 2401.
145. Barlow, C., Brown, K. D., Deng, C. X., Tagle, D. A., and Wynshaw-Boris, A. (1997) Atm selectively regulates distinct p53-dependent cell-cycle checkpoint and apoptotic pathways. *Nat. Genet.*, **17**, 298.
146. Herzog, K. H., Chong, M. J., Kapsetaki, M., Morgan, J. I., and McKinnon, P. J. (1998) Requirement for Atm in ionizing radiation-induced cell death in the developing central nervous system. *Science*, **280**, 1098.
147. Barlow, C., Liyanage, M., Moens, P. B., Deng, C-X., Reid, T., and Wynshaw-Boris, A. (1997) Partial rescue of the prophase 1 defects of Atm-deficient mice by p53 and p21 null alleles. *Nature Genet.*, **17**, 462.
148. Kuljis, R. O., Xu. Y., Aguila, M. C., and Baltimore, D. (1997) Degeneration of neurons, synapses neuropil and glial activation in a murine Atm knockout model of ataxia–telangiectasia. *Proc. Natl Acad. Sci. USA*, **94**, 12699.
149. Murray, A. W. (1992) Creative blocks, cell-cycle checkpoints and feedback controls. *Nature*, **359**, 599.
150. Li, R. and Murray, A. W. (1991) Feedback control of mitosis in budding yeast. *Cell*, **66**, 519.
151. Weinert, T. A. and Hartwell, L. H. (1990) Characterization of RAD9 of *Saccharomyces cerevisiae* and evidence that its function acts post translationally in cell cycle arrest after DNA damage. *Mol. Cell. Biol.*, **10**, 6554.
152. Beamish, H. and Lavin. M. F. (1994) Radiosensitivity in ataxia–telangiectasia: anomalies in radiation-induced cell cycle delay. *Int. J. Radiat. Biol.*, **65**, 175.
153. De Wit, J., Jaspers, N. G. J., and Bootsma, D. (1981) The rate of DNA synthesis in normal and ataxia–telangiectasia cells after exposure to X-irradiation. *Mutat. Res.*, **80**, 221.

154. Houldsworth, J. and Lavin, M. F. (1980) Effect of ionizing radiation on DNA synthesis in ataxia telangiectasia cells. *Nucl. Acids Res.*, **8**, 3709.

155. Nagasawa, H. and Little, J. B. (1983) Comparison of kinetics of X-ray-induced cell killing in normal, ataxia–telangiectasia and hereditary retinoblastoma fibroblasts. *Mutat. Res.*, **109**, 297.

156. Scott, D. and Zampetti-Bosseler, F. (1982) Cell cycle dependence of mitotic delay in X-irradiated normal and ataxia–telangiectasia fibroblasts. *Int. J. Radiat. Biol.*, **42**, 679.

157. Artuso, M., Esteve, A., Bresil, H., Vuillaume, M., and Hall, J. (1995) The role of the ataxia–telangiectasia gene in the p53, WAF1/ClPl(p21) GADD45-mediated response to DNA damage produced by ionizing radiation. *Oncogene*, **8**, 1427.

158. Kastan, M. B., Zhan, O., El-Deiry, W. S., Carrier, F., Jacks, T., Walsh, W. V., *et al.* (1992) A mammalian cell cycle checkpoint pathway utilizing p53 and GADD45 is defective in ataxia–telangiectasia. *Cell*, **71**, 587.

159. Lu, X. and Lane, D. P. (1993) Differential induction of transcriptionally active p53 following UV or ionizing radiation: defects in chromosome instability syndromes? *Cell*, **75**, 765.

160. Mirzayans, R., Famulski, K. S., Enns, L., Fraser, M., and Paterson, M. C. (1995) Characterization of the signal transduction pathway mediating $\gamma$-ray-induced inhibition of DNA synthesis in human cells: indirect evidence for involvement of calmodulin but not protein kinase C nor p53. *Oncogene*, **8**, 1597.

161. Khanna, K. K. and Lavin, M. F. (1993) Ionizing radiation and UV induction of p53 protein by different pathways in ataxia–telangiectasia cells. *Oncogene*, **8**, 3307.

162. Canman, C. E., Wolff, A. C., Chen, C. Y., Fornace Jr, A. J., and Kastan, M. B. (1994) The p53-dependent $G_1$ cell cycle checkpoint pathway and ataxia–telangiectasia. *Cancer Res.*, **54**, 5054.

163. Khanna, K. K., Beamish, H., Yan, J., Hobson, K., Williams, R., Dunn, I., *et al.* (1995) Nature of $G_1/S$ cell cycle checkpoint defect in ataxia–telangiectasia. *Oncogene*, **11**, 609.

164. Xiong, Y., Zhang, H., and Beach, D. (1993) Subunit rearrangement of the cyclin-dependent kinases is associated with cellular transformation. *Genes Dev.*, **7**, 1572.

165. Beamish, H., Williams, R., Chen, P., Khanna, K. K., Hobson, K., Watters, D., *et al.* (1996) Defect in multiple cell cycle checkpoints in ataxia–telangiectasia. *J. Biol. Chem.*, **271**, 20486.

166. Khanna, K. K., Keating, K. E., Kozlov, S., Scott, S., Gatei, M., Hobson, K., Taya, Y., Gabrielli, B., Chan, D., Lees-Miller, S. P., and Lavin, M. F. (1998) ATM associates with and phosphorylates p53: mapping region of interaction. *Nat. Genet.*, **20**, 398.

167. Shieh, S-Y., Ikeda, M., Taya, I., and Prives, C. (1997) DNA damage-induced phosphorylation of p53 alleviates inhibition of MDM2. *Cell*, **91**, 325.

168. Siliciano, J. D., Canman, C. E., Taya, T., Sakaguchi, K., Appella, E, and Kastan, M. B. (1997) DNA damage induces phosphorylation of the amino terminus of p53. *Genes Dev.*, **11**, 3471.

169. Banin, S., Moyal, L., Shieh, S-Y., Taya, Y., Anderson, C. W., Chessa, L., Smorodinsky, N. I., Prives, C., Reiss, Y., Shiloh, Y., and Ziv, Y. (1998) Enhanced phosphorylation of p53 by ATM in response to DNA damage. *Science*, **281**, 1647.

170. Canman, C. E., Lim, D-S., Cimprich, K. A., Taka, Y., Tamai, K., Sakaguchi, K., Appella, E., Kastan, M. B., and Siliciano, J. D. (1998) Activation of the ATM kinase by ionizing radiation and phosphorylation of p53. *Science*, **281**, 1677.

171. Sawyers, C. L., McLaughlin, I., Goga, A., Havlik, M., and Witte, O. (1994) The nuclear tyrosine kinase cAbl negatively regulates cell growth. *Cell*, **77**, 121.

172. Yuan, A. M., Huang, Y., Whang, Y., Sawyers, C., Weichselbaum, R., Kharbanda, S., *et al.* (1996) Role for c-Abl tyrosine kinase in growth arrest response to DNA damage. *Nature*, **382**, 272.

173. Khanna, K. K., Shafman, T., Kedar, P., Spring, K., Kozlov, S., Yen, T., *et al.* (1997) Role of the ATM protein in stress response to DNA damage: evidence for interaction with c-Abl. *Nature*, **387**, 520.

174. Baskaran, R., Wood, L. D., Whitaker, L. L. Canman, C. E., Morgan, S. E., Zu, Y., *et al.* (1997) Ataxia–telangiectasia mutated gene product activates c-Abl tyrosine kinase in response to ionizing radiation. *Nature*, **387**, 516.

175. Kharbanda, S., Pandey, P., Jin, S., Inoue, S., Bharti, A., Yuan, Z-M., *et al.* (1997) Functional interaction between DNA-PK and c-Abl in response to DNA damage. *Nature*, **386**, 732.

176. Kharbanda, S., Ren, R., Pandey, P., Shafman, T. D., Feller, S. M., Weichselbaum, R. R., *et al.* (1995) Activation of the c-Abl tyrosine kinase in the stress response to DNA-damaging agents. *Nature*, **376**, 785.

177. Shafman, T. D., Saleem, A., Kyriakis, J., Weichselbaum, R., Kharbanda S., and Kufe, D. W. (1995) Defective induction of stress-activated protein kinase activity in ataxia–telangiectasia cells exposed to ionizing radiation. *Cancer Res.*, **55**, 3242.

178. Aloni-Grinstein, R., Schwartz, D., and Rotter, V. (1994) Accumulation of wild-type p53 protein upon γ-irradiation induces a $G_2$ arrest-dependent immunoglobulin **k** light chain gene expression. *EMBO J.*, **14**, 1392.

179. Agarwal, M. L., Agarwal, A., Taylor, W. R, and Stark, G. R. (1995) p53 controls both the $G_2/M$ and the $G_1$ cell cycle checkpoints and mediates reversible growth arrest in human fibroblasts. *Proc. Natl Acad. Sci. USA*, **92**, 8493.

180. Liu, V. F. and Weaver, D. T. (1993) The ionizing radiation-induced replication protein A phosphorylation response differs between ataxia–telangiectasia and normal human cells. *Mol. Cell. Biol.*, **13**, 7222.

181. Fried, L. M., Koumeris, C., Peterson, S. R., Green, S. L., Van Zijl, P., Allunis-Turner, J., *et al.* (1996) The DNA damage response in DNA-dependent protein kinase-deficient SCID mouse cells: replication protein A hyperphosphorylation and p53 induction. *Proc. Natl Acad. Sci. USA*, **93**, 825.

182. O'Connell, M. J., Raleigh, J. M., Verkade, H. M., and Nurse, P. (1997) Chk1 is a wee1 kinase in the $G_2$ DNA damage checkpoint inhibiting cdc2 by Y15 phosphorylation. *EMBO J.*, **16**, 545.

183. Rhind, N., Furnari, B., and Russell, P. (1997) Cdc2 tyrosine phosphorylation is required for the DNA damage checkpoint in fission yeast. *Genes Dev.*, **11**, 504.

184. Sanchez, Y, Wong, C., Thoma, R. S., Richman, R., Wu, Z., Piwnica-Worms, H., *et al.* (1997) Conservation of the Chk1 checkpoint pathway in mammals: linkage of DNA damage to Cdk regulation through Cdc25. *Science*, **277**, 1497.

185. Walworth, N., Davey, S., and Beach, D. (1993) Fission yeast chk1 protein kinase links the rad checkpoint pathway to cdc2. *Nature*, **363**, 368.

186. Walworth, N. and Bernards R. (1996) rad-dependent response of the chk1-encoded protein kinase at the DNA damage checkpoint. *Science*, **271**, 353.

187. Funari, B., Rhind, N., and Russell, P. (1997) Cdc25 mitotic inducer targeted by Chk1 DNA damage checkpoint kinase. *Science*, **277**, 1495.

188. Flaggs, G., Plug, A., Dunks, K. M., Mundt, K. E., Ford, J. C., Quiggle, M. R. E., *et al.* (1997) Atm-dependent interactions of a mammalian Chk1 homolog with meiotic chromosomes. *Curr. Biol.*, **7**, 977.

189. Clarke, A. R., Purdie, C. A., Harrison, D. J., Morris, R. G., Bird, C. C., Hooper, M. L, *et al.*

(1993) Thymocyte apoptosis induced by p53-dependent and independent pathways. *Nature*, **362**, 849.

190. Lee, J. H. and Bernstein, A. (1993) p53 mutation increases resistance to ionizing radiation. *Proc. Natl Acad. Sci. USA*, **90**, 5742.

191. Lowe, S., Schmitt, E., Smith, S., Osborne, B., and Jacks, T. (1993) p53 is required for radiation-induced apoptosis in mouse thymocytes. *Nature*, **362**, 847.

192. Slichenmyer, W. J., Nelson, W. G., Slebos, R. J., and Kastan, M. B. (1993) Loss of a p53 associated $G_1$ checkpoint does not decrease cell survival following DNA damage. *Cancer Res.*, **53**, 4164.

193. Westphal, C. H., Rowan, S., Schmaltz, C., Elson, A., Fisher, D. E., and Leder, P. (1997) ATM and p53 cooperate in apoptosis and suppression of tumourigenesis, but not in resistance to acute radiation toxicity. *Nature Genet.*, **16**, 397.

194. Bates, P. R., Imray, F. P., and Lavin, M. F. (1985) Effect of caffeine on γ-ray induced $G_2$ delay in ataxia telangiectasia. *Int. J. Radiat. Biol.*, **47**, 713.

195. Chen, P., Gatei, M., O'Connell, M. J., Khanna, K. K., Bugg, S. J., Hog, A., Scott, S. P., Hobson, K., and Lavin, M. F. (1999) Chk1 complements the G2/M checkpoint defect and radiosensitivity of ataxia-telangiectasia cells. *Oncogene*, **18**, 249.

196. Moses, M. J. (1968) Synaptonemal complex. *Annu. Rev. Genet.*, **2**, 363.

197. Carpenter, A. T. C. (1987) Gene conversion, recombinant nodules the initiation of meiotic synapsis. *BioEssays*, **6**, 232.

198. Strich, S. (1966) Pathological findings in 3 cases of ataxia–telangiectasia. *J. Neurol. Neurosurg. Psych.*, **29**, 489.

199. Aguilar, M. J., Kamoshita, S., Landing, B. H., Boder, E., and Sedgwick, R. P. (1968) Pathological observations in ataxia–telangiectasia. A report on 5 cases. *J. Neuropathol. Exp. Neurol.*, **27**, 659.

200. Scully, R., Chen, J., Plug, A. Ziao, Y., Weaver, D., Feunteun, J., *et al.* (1997) Association of BRCA1 with Rad51 in mitotic and meiotic cells. *Cell*, **88**, 265.

201. Plug, A. W., Peters, A. H. F. M., Keegan, K. S., Hoekstra, M. F., De Boer, P., and Ashley, T. (1998) Changes in protein composition of meiotic nodules during mammalian meiosis. *J. Cell Sci.*, **111**, 413.

202. Plug, A. W., Peters, A. H. F. M., Xu, Y., Keegan, K. S., Hoekstra, M. F., Baltimore, D., *et al.* (1997) Involvement of ATM and RPA in meiotic chromosome synapsis and recombination. *Nature Genet.*, **17**, 457.

203. Lavin, M. F. and Schroeder, A. L. (1988) Damage resistant DNA synthesis in eukaryotes. *Mutat. Res.*, **193**, 193.

204. Kapeller, R. and Cantley, L. C. (1994) Phosphatidylinositol 3-kinase. *BioEssays*, **16**, 565.

205. Lim, D. S., Kirsch, D. G., Canman, C. E., Ahn, J. H., Ziv, Y., Newman, L. S., Darnell, R. B., Shiloh, Y., and Kastan, M. B. (1998) ATM binds to β-adaptin in cytoplasmic vesicles. *Proc. Natl Acad. Sci.*, **95**, 10146.

# 8 | p53 and the integrated response to DNA damage

PAUL J. SMITH and CHRISTOPHER J. JONES

## 1. Introduction

In multicellular organisms the process of DNA repair is one facet of the cellular response to DNA damage and as such must be integrated with other pathways to effect physiologically and genetically satisfactory outcomes for the individual cell, affected tissue system, and organism. Integration has occurred through evolution, linking pathways for stress signalling, adaptive responses, recovery mechanisms, cell-cycle delay, transcription and replication, and not least the molecular guillotine of programmed cell death, apoptosis. The programmed senescence shown in many cell lineages appears to borrow signalling pathways used to herald the presence of DNA damage and change proliferative capacity. It is becoming increasingly clear that the fidelity of these integrated responses is affected by the accrual of genetic lesions which constitute the multistep progression typical of neoplasia. Currently, it is believed that malignant transformation requires several such mutations, co-operating in multistage pathways that differ from one tumour type to another. Some estimates suggest that as many as ten such mutations could be involved (1), whereas statistical analysis of radiation-induced malignancies is compatible with a smaller number of 'rate-limiting' events, typically about three (2).

This chapter will focus on the involvement of p53 (the product of the *TP53* tumour suppressor gene) in this integrated response, echoing themes developed in other chapters. The complex sequence of events leading to apoptosis and its upstream control through p53 will not be discussed. In particular, this overview touches upon the induction, processing, and repair of DNA damage, as well as the integration of these events with proliferation control and their impact upon cancer predisposition.

## 2. DNA damage: induction and processing

Historically, ionizing radiation and ultraviolet light have been studied intensively for their DNA damaging actions, both producing a wide spectrum of lesions that differ in their complexity and relevance to mutation induction and biological effects

(3). In part, the impetus to understanding the repair pathways involved in handling ionizing radiation- and ultraviolet light-induced DNA damage has been allied to anticancer therapy and environmental carcinogenesis, respectively.

Although radiation-induced membrane damage can occur, the biological effects of ionizing radiation largely result from DNA damage. Such damage can be caused directly by ionization within the DNA molecule or indirectly from the action of chemical radicals formed as a result of local ionization. DNA damage takes several general forms including single- and double-strand breaks, base damage, and DNA–protein cross-links. The cell may repair a high proportion of radiation-induced DNA damage with long-term biological consequences resulting from a proportion that is non-repairable or misrepaired. Resolution of DNA lesions can be effected through different discrete routes according to the nature of the lesion and the options available to the cell. Thus, repair depends on the operation of several pathways for which key genes are beginning to be identified including the *ATM* gene, (ref. 4 and Chapter 7), the *Ku* gene group thought to be involved in double-strand break repair (ref. 5 and Chapter 2), and the recently identified *XRCC2* gene (6). Repair may be incomplete if there is an underlying genetic defect, while a proportion of DNA lesions cannot be repaired successfully even in cells with competent repair genes. Some evidence suggests that clustered local damage in DNA (e.g. a double-strand break accompanied by additional breaks, base-damage, or DNA–protein cross-links) will be especially difficult for the cell to repair (7). However, at high lesion-induction rates even potentially repairable lesions may be misrepaired. Importantly, there may be a finite window of opportunity to repair a lesion before commitment to the division cycle or DNA synthesis, these processes being controlled, in part, by the fidelity of the p53 signalling system. Thus the tumour suppressor gene *TP53* (8–10) assumes a critical role in the cell-cycle arrest response to DNA damage and can act as a regulator of cell death. The p53 product is a complex protein and a series of domains have been identified that define its many functions in transcriptional activation, DNA binding, and oligomerization (see review ref. 11). Figure 8.1 illustrates the domain structure of p53 and indicates regions where it interacts with viral oncoproteins and proteins required for DNA repair and recombination, sites where modification by phosphorylation affect function, and those which are particularly prone to mutation in cancers. The interaction of p53 with Mdm2 mediates the destruction of the protein via proteasome-dependent degradation (12, 13). Mdm2 expression is controlled by p53 in an autoregulatory feedback loop.

Low rates of lesion induction may provide the best opportunity for repair but fail to trigger ancillary responses that maintain recovery potential and reduce the risk of genetic instability. Accordingly, attempts have been made to determine the extent to which DNA damage-sensing systems can detect low levels of critical lesions and trigger a biological response. Microinjection into cells of DNA substrates such as linearized plasmid DNA, circular DNA with a large gap, or single-stranded circular phage plasmid can induce p53-mediated $G_1$ arrest (14). Supercoiled and nicked plasmid DNA, and circular DNA with a small gap fail to induce a p53-mediated response. It appears that a single DNA molecule may be sufficient to induce the p53-

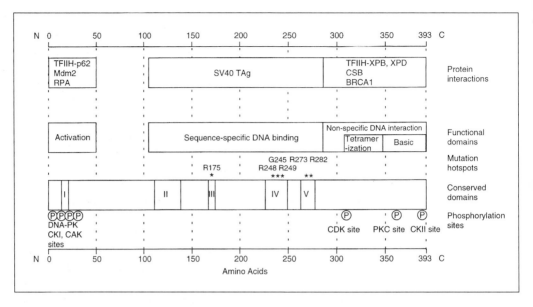

**Fig. 8.1** Map of the human p53 protein (derived, in part, from ref. 11). The p53 protein comprises 393 amino acids and significant areas of interest are highlighted. A large number of factors interact with p53 and those of relevance to this chapter are shown. The functional domains of the protein are illustrated and several mutation hotspots, apparent by their appearance in human cancer, are indicated. These hotspots tend to coincide with the regions I–V conserved during evolution. Finally, the many functions of p53 are mediated by phosphorylation and these sites and their respective kinases are indicated.

dependent $G_1$ arrest. This raises the question of whether p53 is actively involved in the recognition of DNA damage. A series of enzymatic activities have been attributed to p53. The p53 protein exhibits 3'-to-5' exonuclease activity (15), and there is evidence to suggest that p53 itself is modified by damaged DNA (16). Purified p53 from baculovirus cell lysates is sensitive to proteolysis in the presence of damaged DNA. A core domain, p35, is produced when p53 is incubated with single-stranded (ss) DNA or double-stranded (ds) DNA containing a mismatch; p35 itself acts as a protease which cleaves the N-terminus of p53. *In vitro*, p53 can be modified through both the C and N termini, which has implications for transcriptional activation, interaction with Mdm2, and the induction of apoptosis (11). The *in vitro* enzymatic activities of p53 raise the question of whether they are relevant in the intact cells.

The question also arises as to what access does p53 have to the region of DNA damage? It has been suggested that the incision intermediate may be a suitable site for p53 activation (16) and that a gapped DNA substrate of similar dimensions will activate $G_1$ arrest (14). However, p53 will have to compete with replication protein A (RPA)—the human single-stranded DNA binding protein—which is absolutely required for nucleotide excision repair (17), although it should be noted that p53/RPA associations have been demonstrated (18). Further discussions of the involvement of p53 in repair are found in Section 4.2.

# 3. DNA damage induction by anticancer agents

Cytotoxic anticancer drugs now present a formidable array of potentially genotoxic agents that are capable of damaging DNA. In many cases anticancer agents are xenobiotic molecules capable of producing unusual DNA lesions, providing probes for the complexity and cross-talk within repair systems. In some cases lesion processing may further compromise DNA integrity, with the cell becoming reliant upon signalling to stress response systems. The enormous expansion of our knowledge of the molecular processes of DNA repair has opened up the possibility that selective pathways could provide useful pharmacological targets for attack with novel inhibitors. Here, the therapeutic aim would be to modulate the responsiveness of tumour cells to anticancer drugs that damage DNA (see ref. 19 for a review).

Currently, the selection of therapy in the treatment of cancer depends upon perceived or empirical tumour sensitivity where responsiveness to a cytotoxic drug, and indeed radiation, is primarily a feature of the tumour rather than the class of agent (20). This is in keeping with our current understanding of neoplasia as a disease characterized by the sequential acquisition of genetic abnormalities (21). Together with the dysfunction or lack of integration of cellular pathways controlling proliferation and apoptosis (22), tumour cells may undergo changes that directly influence the efficacy of anticancer agents. Emerging from a complex picture is a consensus that, despite the presence of multiple factors in the control of transitions, there are a few dominant integration 'centres', namely the products of major tumour suppressor genes and proto-oncogenes (e.g. ras, myc, Rb, and p53). Furthermore, there is increasing evidence that some genes encoding cell-cycle control molecules can provide mutational targets for carcinogenesis in humans (23).

A major challenge in the effective deployment of cytotoxic drugs is the inherent or acquired resistance of many human tumours. Drug-resistance mechanisms include: mutation of target genes; amplification of target and mutated genes; and differences in repair capacity. There are several categories of cytotoxic agents, based upon their modes of action, and therefore category-specific mechanisms of drug resistance which need not involve DNA repair pathways. For example, differences in nucleoside and nucleobase salvage pathways can underpin resistance to antimetabolites (e.g. methotrexate (MTX) and 1-beta-D-arabinofuranosylcytosine (AraC); for a review see ref. 24). The following is a brief overview highlighting some of the features of the cytotoxic actions of three major classes of anticancer drugs capable of eliciting DNA damage responses.

## 3.1 Alkylating agents

Those cytotoxic agents referred to as 'alkylating drugs' remain among the most widely used in cancer chemotherapy. This class includes the nitrogen mustards (e.g. melphalan and cyclophosphamide), chloroethylnitrosoureas, triazenes (e.g. dacarbazine: dimethyl-triazenylimidazole carboxamide, DTIC), dimethanesulfonates (busulfan), aziridines (e.g. diaziquone), and the melamines (e.g. trimelamol). Early studies on the mutagenicity of mustard gas shifted attention to alkylating agents as carcinogens (25). Classically, they are thought to act through their ability to damage DNA and to interfere with cell replication. Unfortunately, this mode of action implicates these drugs as carcinogens (26). The hypothesis that DNA is the essential target of alkylating agents came from early studies providing evidence that the archetypal agent, mustard gas, could link guanine bases in DNA through their N-7 atoms (27). The extent of alkylation of tumours can be significantly higher than the average for leucocytes in the cases of ovarian and testicular tumours treated with cisplatin, and for plasma cell tumours treated with melphalan. However, in general, there is a lack of a favourable difference between the extent of alkylation in tumour DNA and normal cells, this being particularly apparent with the methylating drugs dacarbazine and procarbazine.

The transcription-coupled nucleotide excision repair pathway, activated by the arrest of RNA polymerase at a DNA lesion, results in preferential repair at transcriptionally active DNA sites. Interestingly, the mutagenic $O^6$-methylguanine ($O^6$-meGua) lesion, a model lesion for the genotoxic damage induced by alkylating agents, also appears to undergo preferential repair at transcriptionally active sites despite being bypassed by RNA polymerase (28). One possible explanation is that the repair enzyme $O^6$-methylguanine-DNA methyltransferase (MGMT) is normally concentrated at active transcription sites, thus providing an effective means of repairing alkylation damage (28). The ability of tumour cells to vary the transcriptional targeting of MGMT may be crucial in controlling chemotherapeutic responsiveness to alkylating agents. In human brain tumours, sensitivity to procarbazine, as measured by sensitivity in a xenograft tumour model (29), correlated inversely with the amounts of MGMT. It has been suggested that p53 mutations in brain tumours may contribute to procarbazine sensitivity by failing to induce arrest at the $G_1/S$ cell-cycle checkpoint, thereby preventing the repair of procarbazine-induced genetic alterations (30).

## 3.2   Antibiotics: parallels with ionizing radiation

The search for antitumour agents derived from microbial metabolites led to the discovery of bleomycin (BLM) in 1962 and aclacinomycin in 1975, and it is probable that this source will remain important for the future identification of new anticancer drugs (31). BLM is a cytotoxic drug, acting through the generation of DNA strand breaks, currently used in a restricted range of chemotherapy regimens. The antibiotic has an high intrinsic cytotoxicity once inside the cell, but its activity is limited because BLM it is unable to diffuse through the plasma membrane. The very low amounts of BLM that can reach the cell interior enter the cells by a mechanism that requires BLM interaction with a plasma membrane protein (32). BLM has been described as radiomimetic agent, principally because of the capacity of the glyco-peptide to generate free radical-mediated DNA strand breakage through a ferrous oxidase cycle. As a group, the enediyne antibiotics cause simultaneous site-specific free radical attack on sugar moieties in both strands of DNA, resulting in double-strand breaks as well as abasic sites with closely opposed strand breaks. In mammalian cells, abasic sites convert to form double-strand breaks. Genomic disruption consists primarily of small deletions, large deletions, and gene rearrangements, all of which probably result from errors in double-strand break repair by a non-homologous end-joining mechanism (33). Analysing BLM-induced, chromatid-type aberrations in $G_2$-phase fibroblasts derived from embryos from wild-type and p53 knock-out mice, the p53-deficient cells show an overdispersed distribution of BLM-induced chromatid aberrations accompanying a greater amount of overall genomic instability (34).

The hypersensitivity of several mammalian, double-strand break, repair-deficient mutants to these antibiotics confirms the role of these double-strand breaks in mediating cytotoxicity, and underlines the nature of their radiomimicity. However, BLM can have specific chromosome damaging effects, for example on chromosomes 4 and 5 (35). Specific-locus studies have shown chlorambucil (CHL) and BLM to be mutagenic in mouse oocytes. However, BLM does not appear to kill immature oocytes and thus differs markedly from radiation exposures equivalent for dominant-lethal induction. Therefore, the failure to recover specific-locus mutations cannot be ascribed to cell selection resulting from oocyte killing, as has sometimes been proposed for radiation. Russell and co-workers have estimated that the minimum proportion of large DNA lesions induced in oocytes by chemicals (such as CHL and BLM) is 35.3%, significantly different from the corresponding figure (approximately 70%) for radiation (30). Such studies reveal that, although an agent may be regarded as radiomimetic in terms of one aspect of its mode of action, there may be significant differences in the quality, targeting, and consequences of the genotoxic damage.

## 3.3   DNA topoisomerase inhibitors

Identification of topoisomerases as critical targets for a range of anticancer drugs was achieved following observations that such agents induced unique DNA lesions in cancer cells *in vitro* (for reviews see refs 36, 37). The drugs that generate these DNA lesions interfere with the breakage–reunion reaction in the enzyme's catalytic cycle, trapping the protein as a reaction intermediate known as a 'cleavable complex', which sequesters a double- or single-strand break. It is the processing of these unusual lesions during DNA metabolism that generates overt damage recognizable by damage sensing pathways and capable of causing cell-cycle arrest and the induction of apoptosis. *In vitro* short-term testing of the genotoxicity of topoisomerase-interactive drugs reveals that they are mutagenic, induce sister-chromatid exchanges, chromosomal aberrations, with evidence that they induce large DNA rearrangements and deletions (38). In general, there are two types of topoisomerases (types I and II) with different functionalities and drug sensitivities (39).

VP-16 (etoposide) and VM-26 (teniposide) are semi-synthetic derivatives of podophyllotoxin capable of acting as specific poisons for DNA topoisomerase II, the major type II DNA topoisomerase. Anticancer anthracyclines (e.g. doxorubicin), bis(alkylamino)anthraquinones (e.g. mitoxantrone), and anilinoacridines (e.g. acridinylamino-methanesulfonyl-*m*-anisidine, *m*-AMSA) have the ability to bind to DNA and are also capable of trapping DNA topoisomerase II molecules as cleavable complexes. Interestingly, in bacteriophage-infected bacterial cells the T4-encoded type II DNA topoisomerase is a major target for the antitumour agent *m*-AMSA, and mutations in several phage genes that encode recombination proteins (uvsX, uvsY, 46 and 59) increase the sensitivity of phage T4 to *m*-AMSA (40). Apurinic and, to some extent, apyrimidinic sites, arising from spontaneous DNA damage, can also act as position-specific poisons of topoisomerase II and stimulate DNA scission (41). There has been little evidence of a direct role for the type II enzyme in DNA repair and the use of topoisomerase inhibitors to modulate repair has been problematic due to the induction of lesions at inhibitory concentrations. However, the involvement of topoisomerase II in the processing of DNA damage has been inferred from the effects of an inhibitor (ICRF-193) which does not accumulate cleavable topoisomerase–DNA complexes. It was found that ICRF-193 synergistically enhances the yield of UVB-induced chromatid-type aberrations (42). It has been postulated that a catenation-sensitive checkpoint in $G_2/M$ can be activated through ICRF-193 inhibition of topoisomerase function (43). It appears likely that the UVB effects relate to topoisomerase II-dependent checkpoint activation in late $G_2$.

The first well-characterized inhibitor of type I topoisomerases was camptothecin (CPT). The cytotoxic actions of this alkaloid were shown to be dependent on the action of topoisomerase I (44). Topoisomerase I inhibitors stabilize a covalent bond between a tyrosine residue on the protein and the 3′-phosphoryl end of the single-strand of DNA that it breaks (45). The formation of cleavable complexes is due to inhibition of the re-ligation phase of the breakage–reunion reaction rather than through the promotion of cleavage (46). CPT and related agents, such as topotecan, appear to act through the cleavable complex pathway and kill cells when the unusual ternary complex comprising drug, DNA, and topoisomerase I, collides with a traversing replication complex. During this collision, the non-lethal reversible single-stranded DNA break is converted into a double-strand break (47) which acts as a trigger for cell death (apoptosis). Thus CPT is most cytotoxic to S-phase cells. Li (48) and Goldwasser *et al.* (49) have proposed that misrepair of damaged replicons and/or alterations in DNA damage checkpoints can determine chemosensitivity to CPT in cells in which there are no apparent abnormalities in the kinetics of topo-isomerase I-mediated DNA breakage. CPT–stabilized topoisomerase I-cleavable complexes, located on the template strand within a transcribed region, also appear to be converted into strand breaks by an elongating RNA polymerase, with implications for transcription-coupled repair (50). Indeed, enhanced DNA repair, resulting in UV-cross resistance, may contribute to resistance to topoisomerase I inhibition by CPT (51).

It has been suggested (52) that chemical activation of the DNA-unwinding activity of topoisomerase I can inhibit the fast component of potentially lethal damage repair, and thus enhance the lethality of X-rays against human laryngeal epidermoid carcinoma cells. Using an antibody-based method Subramanian *et al.* (53) examined genomic DNA cleavage by endogenous topoisomerase I in living cells and found that UVB irradiation stimulates covalent complex formation. The effect appears to be topoisomerase I-specific with rapid recruitment of pre-existing enzyme to sites of DNA damage. Because repair-deficient cells are additionally compromised in their ability to recruit topoisomerase I, a direct role for the enzyme in the DNA excision repair process *in vivo* has been proposed (53).

## 4. Cellular responses to stress and the role of p53

Cellular DNA repair capacity alone does not define the responses of human cells to DNA damage, since the process acts within the context of other important molecular pathways including those controlling cell-cycle progression, DNA replication, gene transcription, and programmed cell death. The link between DNA damage-induced growth arrest and the induction of apoptosis is often not clear. However, it appears that wild-type p53 protein can simultaneously induce the genetic programmes of both growth arrest and apoptosis within the same cell type, in which the apoptotic programme can proceed in either arrested or cycling cells. Figure 8.2 shows a general scheme for the integration of stress signals at p53, including the Mdm2 and phosphorylation control of protein stabilization and activation. Within this scheme the *ATM* gene product acts to enhance p53 stability and activation, providing an interface between direct DNA damage and cell-cycle arrest (54). Dysfunction of p53

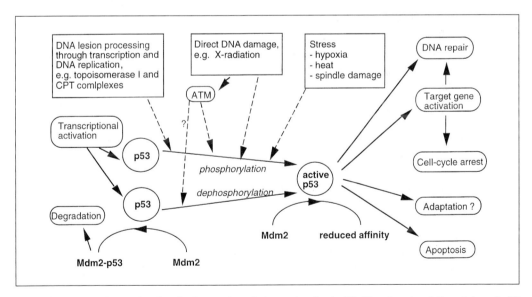

**Fig. 8.2** General scheme showing the integration of stress signals at p53. The phosphorylation status of p53 determines stability and availability to activate target genes or participate more directly in cell responses to stress. Here the term adaptation is used to describe the release from cell-cycle arrest, despite the presence of unrepaired damage discussed by Weinert (69). The dysfunction of ATM within this scheme for cells exposed to ionizing radiation is discussed by Nakamura (54).

acts to sustain genetic instability, and p53 mutations/abnormalities (see Fig. 8.1) are found in the majority of cancers. Importantly, cancers with p53 dysfunction tend to have a worse prognosis for responses to therapy than those showing wild-type alleles (55).

Recent evidence suggests that ionizing irradiation may also generate a state of genomic instability in the clonal descendants of surviving cells. This instability is seen as a long-term accumulation of random chromosome aberrations at an abnormally high rate (56). Heritable lesions that are not incompatible with cell division, and are carried by cells which escape the cell-death pathway, can include those which involve activation of oncogenes or inactivation of tumour suppressor genes. Thus the ability of a damaged cell to survive and divide is critical, while mutations which affect these processes can drive oncogenesis.

Understanding genomic stress response systems is increasingly seen as a prerequisite for the design of new therapies. The aim is to gain a therapeutic advantage based on differences between the responses of tumour and normal cells (57). For example, the finding that certain $G_2$ checkpoint reversing agents interact synergistically with DNA damaging agents in p53-defective cells suggests a new route for targeted destruction of tumour cells (58, 59). Central to this change in emphasis, from the importance of initial DNA damaging events, has been the new discoveries in cell-cycle control, programmed cell death, and the central roles played by damage-response genes such as *TP53*. There is a growing list of p53-activating factors ranging from genotoxic agents to non-genotoxic stresses revealing a complex signalling network (for a review see ref. 60).

## 4.1 p53 and cycle arrest

The main role for p53 is believed to be the restriction of neoplastic progression through the maintenance of genomic integrity (61). Inactivation of *TP53* in the germline of both experimental animals (62) and in the human Li–Fraumeni syndrome (63, 64) leads to a greatly increased predisposition to cancer in a range of tissues. Somatic mutations of the *p53* gene occur in over 50% of human tumours (65), and commonly lead to the expression of a stable, yet inactive, protein (66, 67). Mammalian cells have evolved multiple responses for dealing with DNA damage particularly the arrest of cells in the cell cycle, activation of p53, and DNA repair. Understanding the dynamics of these interlinked processes is complicated by the wide spectrum of lesions capable of eliciting cellular responses. For example, 3-methyladenine, a DNA lesion produced by endogenous cellular metabolites, is regarded as a relatively minor lesion induced by chemotherapeutic alkylating agents (see above) but can still cause S-phase arrest and the accumulation of p53 (68).

Transitions between different states of the cell cycle (e.g. $G_0$ exit, checkpoint traverse, or entry into mitosis) present control situations in which the cell interlaces elements of signal transduction, transcriptional regulation, and checks on metabolic status and genome integrity. Cell-cycle checkpoints are the surveillance mechanisms that enforce the proper order of cell-cycle events. Even in the most genetically accessible eukaryotic systems of fission and budding yeast there is little evidence of a direct role for checkpoint genes in the major DNA repair pathways. Although, that is not to say that under some circumstances specific repair of replication mutants do not show checkpoint defects. Current models for interfacing stalled replication and the de-repression of repair genes has been discussed by Weinert (69).

In mammalian cells, p53 has assumed a promiscuous role in the interface between damage signalling/DNA repair and cell-cycle arrest. Through this interface mammalian cells effect proliferation control in response to the environment, differentiation signals, and genomic-stress induced by drugs. Importantly, not only do the modes of action differ for the classes of cytotoxic agents but also do their effects on cell-cycle progression. For example, as described above, it is apparent that the trapping of DNA topoisomerase I causes replication-dependent cell death, and as such the critical events for lesion generation take place after the primary $G_1/S$ cell-cycle checkpoint guarded by p53. On the other hand, the direct induction of DNA damage in $G_1$ by ionizing radiation or BLM will cause lesions capable of involving early cell-cycle checkpoints. Thus, the differential expression of cell-cycle checkpoints and pathways for cell death will have different impacts according to the dynamics of damage induction and removal. There is no doubt that such complexities of mammalian cell responses to DNA damage have obscured the roles of proteins such as p53 in the perturbed cell cycle.

Most mammalian cells exhibit transient delays in the $G_1$ and $G_2$ phases of the cell cycle after treatment with radiation or radiomimetic compounds. Cell-cycle arrest is apparently an important stress response, and the role of p53 in $G_1/S$ arrest is now well established (reviewed in ref. 59). Normal, non-transformed cells express low levels of inactive p53 which can be activated by DNA damage, for example as a consequence of cytotoxic drug action, to acquire sequence-specific DNA binding (70) and transcriptional activation properties (71, 72). The activation of transcription of p53-inducible genes such as *WAF1* (73), *GADD45* (74), *mdm-2* (75), and *bax* (76) plays a major role in the induction of cell-cycle arrest and apoptosis by p53. This ability to activate transcription is lost by most tumour-derived *TP53* mutants (77). Upregulation of the p53 protein can induce cell-cycle arrest at the $G_1/S$ border and in some cases maintain arrest at the $G_2/M$ border (9, 10).

The p53-mediated arrest seen at the $G_1/S$ boundary appears to provide time for DNA repair to occur. Recently, a role for p53 in the $G_2/M$ transition has also been suggested. Arrest in $G_2$ arises from suppression of the activation of the cyclin B1/CDC2 kinase by DNA damage. Using p53-null, wild-type p53 and mutant p53-producer cell lines Schwartz *et al.* found that in $G_2$, p53 may also facilitate the repair of γ-irradiation induced DNA breaks (giving rise to micronuclei) and regulate the exit from the $G_2$ checkpoint (78). Cells lacking p53 or its downstream effector p21 (WAF1) fail to maintain a $G_2$ arrest following γ-irradiation (9, 10). Sustained $G_2$ arrest in cells expressing wild-type p53 is associated with nuclear localization of CDC2 rather than inactivation via inhibitory phosphorylation. Nuclear localization of CDC2 can be promoted in a *TP53*-null background by forced expression of p21 (10). Therefore the observation that the $G_1$ and $G_2$ checkpoints are interrelated and share initiators and effectors implies a control system that determines, depending on the extent of the damage, whether the cell needs to arrest cell-cycle progression at the subsequent checkpoint for further repair (79).

However, because of the frequent loss of p53 function in tumour cells, one would predict that such cells would become unusually reliant upon S-phase and p53-independent, late cell-cycle checkpoints (see refs 59, 80 for reviews). Thus the ability of cells to acutely downregulate DNA synthesis at the initiation step is potentially an important factor in defining the responses of tumour cells to DNA damaging agents. The loss of p53 in an embryonic stem-cell line does not alter the capacity to respond to DNA damage, with normal cessation of DNA synthesis after UV damage and a similar ultimate capacity to repair a transiently transfected reporter plasmid (81).

The consequences of p53 activation by DNA damaging agents for cell survival are even more complex and depend upon cell type and the nature of the lesion. In progressive neoplastic tissues, cellular resistance to DNA-damaging agents may be enhanced through a decrease in the ability to execute programmed cell death. Recent data suggest that aberrant function of the wild-type p53 protein may alter cellular survival following DNA damage through cellular pathways involving apoptosis and cell-cycle checkpoints. The central issue is whether the increased radiation survival shown by some mutant p53-expressing transformed fibroblasts and tumours is associated, in part, with an enhanced DNA and cellular repair capacity. The problem is to tease out the effects of changes in cycle arrest and apoptosis that qualitatively change the cell populations under investigation.

Bristow *et al.* (82) have reported that transformed rat embryo fibroblast clones expressing exogenous mutant p53 and p21ras proteins were generally radioresistant, irrespective of dose rate, compared either to transformed clones expressing p21ras protein alone or control cell lines expressing baseline endogenous levels of p21ras and wild type p53 protein. Radioresistance was not associated with a decreased radiation-induced $G_1$ arrest response, although mutant p53-expressing clones were found to be more proficient at the rejoining of DNA double-strand breaks (DNA-dsb), compared to wild type p53-expressing clones (82). In a cell-free system using extracts from human lymphoblastoid cell lines expressing wild-type, mutated, or essentially no p53 protein, an increased end-rejoining activity has been observed for extracts from cell lines lacking p53 or expressing mutated p53 (83). Interestingly, p53 status did not appear to influence the ratio of misrepair to correct repair, or the type of misrepair events.

A general view of cell-cycle arrest through p53-dependent pathways and DNA damage is shown in Fig. 8.3. In this scheme there is a putative DNA catenation-sensitive checkpoint in $G_2$ at which cells can be delayed by topoisomerase inhibition as discussed above (43). The breaching of the catenation-sensitive checkpoint can result in cells undergoing a 'normal' mitosis or progressing without cytokinesis to a polyploid cycle. Whether or not cells exit this checkpoint in a p53-dependent manner or retain p53-damaging signalling potential is not known, but they could be crucial issues in understanding the generation of abnormal ploidy patterns in genotoxically damaged cells.

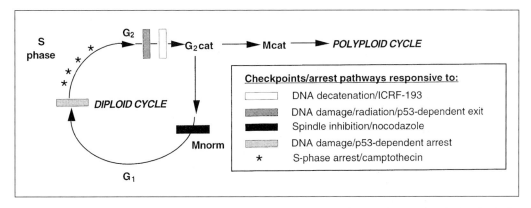

**Fig. 8.3** Diagram of the major cell-cycle arrest points in mammalian cells. $G_2$ cat refers to a late $G_2$ checkpoint for the putative completion of DNA decatenation (see text for details), after which cycle progression may be resumed. Progression may be through a normal mitosis (Mnorm) or without cytokinesis through a constrained mitosis (Mcat) to a polyploid cycle.

## 4.2   p53 and nucleotide excision repair

It is evident that the *TP53* gene is of considerable interest as a target for DNA damage, quite distinct from the role of its protein product in damage signalling and the control of stress responses. Denissenko *et al.* (84) have found that both preferential benzo(*a*)pyrene diol epoxide-adduct formation and slow repair contribute to hotspots for mutations at codons 157, 248, and 273, concluding that the strand bias of bulky adduct repair is primarily responsible for the strand bias of G-to-T trans-version mutations. Indeed, the frequency and diversity of mutations in *TP53* have been used to provide indirect evidence implicating environmental mutagens in human carcinogenesis (see ref. 85 for a review). The spatial formation and selective repair of DNA lesions in the *TP53* gene have been used to model the impact of such effects on this critical gene for limiting DNA damage-induced genetic instability. The human *TP53* gene is repaired in UV (254 nm)-irradiated xeroderma pigmentosum group C (XPC) cells as part of a large genomic region that is about twice the size of the gene (86). XPC mutant mouse embryo fibroblasts have been found to be specifically defective in the removal of pyrimidine(6–4)pyrimidone photoproducts from the non-transcribed strand of the transcriptionally active *TP53* gene (87). The repair pathway for lesion removal, nucleotide excision repair (NER), is consequently a fundamental process required for maintaining the integrity of critical genes, such as *TP53*, and the wider genome in cells exposed to environmental DNA damage, a process which is modulated by p53 activity *per se*.

The trait of cancer-proneness shown by patients with the Li–Fraumeni syndrome, has been used to reinforce the general concept that loss of p53 function may lead to greater genomic instability by reducing the efficiency of DNA repair. The critical and possibly rate-limiting step for the initiation of DNA repair pathways is recognition of the DNA lesion (for a review see ref. 88). The formation of multienzyme complexes of recognition proteins and active repair proteins is most clearly demonstrated in the case of nucleotide excision repair (NER). Ford and Hanawalt (89, 90) have produced compelling evidence for the involvement of the wild-type *TP53* gene product in nucleotide excision repair activity in UV-irradiated human cells. Li–Fraumeni syndrome fibroblasts, homozygous for *TP53* mutations, were shown to be deficient in the removal of UV-induced cyclobutane pyrimidine dimers from genomic DNA, but still proficient in the transcription-coupled repair pathway. In a UV-irradiated *TP53* homozygous mutant cell line, with tetracycline-regulated expression of a wild-type *TP53* gene, expression of wild-type p53 gave the expected result of cell-cycle checkpoint activation and UV-damage-induced apoptosis (90). The regulated expression of wild-type p53 resulting in the recovery of normal levels of repair did not alter the transcription-coupled repair of cyclobutane pyrimidine dimers (90). The disruption of wild-type p53 function in primary human fibroblasts, by human papillomavirus 16 E6 gene expression, results in enhanced UV sensitivity (91). This increased sensitivity, compared with normal controls, was attributed to the selective loss of p53-dependent global genomic NER, rather than UV-induced apoptosis (91).

The question arises as to how p53 activity can be coupled to DNA repair processes. However, p53 can act as a substrate for phosphorylation by cyclin-dependent kinase (cdk)-activating kinase (CAK). CAK comprises three subunits: cdk7, cyclin H, and p36MAT1 and the phosphorylation of p53 is enhanced by p36MAT1 (92). Since CAK is part of the transcription factor-IIH (TFIIH) multiprotein complex, required for RNA polymerase II transcription and nucleotide excision repair, p53 phosphorylation by this route may reveal how the regulation of p53 is coupled with DNA repair and the basal transcriptional machinery (93). Direct protein interaction may also regulate p53 activity. The replication protein A (RPA) is required for both DNA replication and nucleotide excision repair; however, the UV-irradiation of cells greatly reduces the ability of RPA to bind to p53 (94). This has led to the suggestion that RPA may participate in the co-ordination of DNA repair, normally acting through a release of p53 molecules and permitting the activation of downstream targets which operate p53-dependent checkpoint control (94).

## 5. p53 as a damage sensor in replicative senescence

One of the goals of research into the ageing process is to understand the controls regulating cellular lifespan and extrapolate these to the whole organism. It is becoming increasingly apparent that damage sensing, lesion removal, and damage reversal pathways are involved in determining the proliferative lifespan of cell lineages, with p53 again casting the shadow of its molecular involvement on the integrated manner by which cells execute senescence programmes. Here we highlight the roles of p53 in replicative senescence as a 'physiological model' for the role of p53 in guiding cellular responses to endogenous DNA damage.

## 5.1  Replicative senescence and telomeric clocks?

The cell's response to stress and DNA damage are integrated through the activity of p53. It also appears that p53 is involved in the ability of cells to arrest at or breach senescence barriers, with the proposal that endogenous DNA damage or pathways normally used for damage signalling activate these barriers. These concepts have arisen from the study of somatic cells in culture undergoing a state of viable growth arrest known as replicative senescence (95). One suggested trigger for replicative senescence is the excessive accumulation of DNA damage during prolonged periods in culture (96), although there is no substantial evidence to support this hypothesis (97). Recently, attention has been focused on telomere dynamics. It is now generally accepted that in a large number of cell types telomeres normally show progressive shortening. However cancer cells, which generally possess short telomeres, are able to proliferate indefinitely thus avoiding replicative senescence. Germ, stem, and cancer cells have reactivated an enzyme, telomerase, which modulates telomere length. It is also apparent that cancer cells with shorter tracts of telomeric DNA have further deregulated their response to shortened telomeres, which is consistent with established models of multistep carcinogenesis (98).

The hypothesis is that telomere shortening is a driving force for senescence. This is supported by the observation that the germline and stem cells maintain long telomeres thereby avoiding replicative senescence. However, there is considerable variation of telomere length in the presenescent cell and even within opposite arms of the same chromosome when analysed by quantitative fluorescent *in situ* hybridization (FISH) (99, 100). Thus it is not clear if telomere degradation signals, potentially used to trigger replicative senescence, arise from one critically degraded telomere or through an integrated signal representing total telomeric DNA status.

## 5.2  Telomeres: form and function

Telomeres are protein-bound DNA structures that define the ends of linear chromosomes. Telomeres are thought to act as a form of cellular clock counting off the number of cell divisions, serving as an indicator of the relative age of a culture and as a predictive marker for the remaining lifespan the population (101). In human cells they are characterized by the DNA repeat sequence TTAGGG, varying in length between 5 and 15 kb. The unique nature of the telomeric repeat sequence provides a substrate for two proteins. TRF1, purified as a telomere binding activity (102), has a role in telomere length maintenance (103, 104). TRF2 (105, 106) has a crucial role as it effects one aspect of telomere function, that of chromosome end protection (107). Expression of a dominant-negative form of TRF2, which inactivates the function of the endogenous protein, causes the fusion of chromosomes ends (107).

The number of TTAGGG repeats present in cells greatly exceeds that required to supply the end protective function. Telomeric DNA also offsets one of the detrimental affects of DNA replication. When considered in its most basic form, semiconservative replication of DNA requires the activity of an RNA primase followed by the action of DNA polymerases (108). Lagging strand synthesis is a discontinuous process involving the synthesis of RNA primers that are, in turn, removed by an RNase and replaced by DNA, the polymerase requiring a 3'-OH end to initiate synthesis. At the extreme terminus of a chromosome, however, synthesis is not possible following primer removal and this results in the production of an overhang of 50–200 bp. Unable to tolerate these structures the cell removes the overhang with the result that the chromosome is shortened. Telomere shortening, due to the end replication problem, was predicted by Olovnikov (109) and is detectable by Southern blotting (110). Telomeric DNA, resistant to enzymatic digestion, is detected by TTAGGG probes and is apparent as a smear representing a distribution of telomeric sequences of all the chromosomes from a population of cells. It is clear in somatic cells such as fibroblasts that telomeres shorten with an increasing number of population doublings (110), arriving at a threshold of about 5 kb in senescent cells (111).

## 5.3  Telomerase

Telomerase is the RNA-dependent DNA polymerase that synthesizes TTAGGG repeats *de novo* at chromosome ends. The enzyme's associated RNA molecule provides a template for the synthesis of TTAGGG repeats. The RNA component of telomerase, known as hTERC, was identified in 1995 (112). Expression of antisense hTERC in HeLa cultures abolishes telomerase activity and finally causes cell death (112). The mouse homologue of the RNA component mTERC was used as the target to produce a telomerase knock-out mouse (113, 114). Unfortunately, the RNA component is an unsuitable marker for telomerase, as hTERC expression does not always coincide with the enzyme's activity. (115, 116).

Telomerase can be detected in cell extracts by the telomeric repeat amplification protocol (TRAP assay), relying on the ability of telomerase to extend an artificial substrate *in vitro* (117). This extended substrate, an oligonucleotide of 18 bases, is then detected by the polymerase chain reaction (PCR) following the inclusion of a primer complementary to TTAGGG. Products are detected as ladders of 6-bp periodicity on acrylamide gels. This assay confirmed the prediction that most tumours and immortal cell lines possessed telomerase activity which was lacking in mortal cultures. Many confirmatory studies have since demonstrated telomerase activity in a wide range of tumour types (118). Infiltration by telomerase-positive inflammatory cells in some tumours can sometimes generate a false-positive result, and, conversely, false-negatives are apparent, as some cells possess inhibitors of *Taq* polymerase. In spite of these limitations the technique has shown that reactivation of the enzyme is a key contributor to carcinogenesis. Variations in telomerase activity after γ-irradiation have been reported for haematopoietic cell lines. Changes in activity were observed after several hours for doses within the 3 Gy range that could not be related to cell-cycle redistribution. Whether or not activation reflects the involvement of telomerase in DNA repair remains unclear (119).

## 5.4  p53 and replicative senescence

Cells that have entered a state of replicative senescence have a $G_1$ DNA content, and growth factors fail to stimulate them to enter S phase (120). Senescence is a dominant state as somatic cell hybrids of old and young cells retain the senescent phenotype (120). Upregulation of the cyclin-dependent kinase inhibitors p16 and p21 is apparent. The expression of p21 in senescent cells suggests that there is a p53-mediated damage response present and critically shortened telomeres may invoke this $G_1$ arrest. Abrogation of p53 function by SV40 large T antigen prior to senescence confers a limited extension of lifespan in fibroblasts of approximately 20 population doublings (121). Microinjection of anti-p53 antibodies into senescent fibroblasts allows cells to re-enter the cell cycle and resume proliferation (122). SV40 T-induced lifespan extension is accompanied by further telomere shortening until the cells enter crisis, due in part to chromosome fusion events. Extension of lifespan by abrogation of p53 function highlights two characteristics of senescent cells. First, telomeres in senescent cells are of sufficient length to preserve their end protective function and provide further potential for cell division. Second, abrogation of p53 function encourages further cell division. Therefore removal of the signal that invokes $G_1$ arrest via p53 should have the same effect.

This second aspect has been addressed following identification of the catalytic subunit of telomerase known as hTERT (123–125). The *hTERT* gene encodes a protein of calculated molecular weight, 127 kDa, which contains reverse transcriptase domains and a 'T' motif unique to members of the telomerase family (126). Ectopic expression of hTERT is sufficient to restore telomerase activity in cells, because hTERC, the RNA component, is present at adequate levels (127). Fibroblasts expressing telomerase avoid senescence and are essentially immortal due to increased telomere length associated with hTERT activity (128). In addition, cultures expressing telomerase lack an enzymatic activity normally associated with senescent cells. Senescent cells in culture, and also in tissue sections, are detected by a histochemical assay that detects a β-galactosidase activity which is unmasked at low pH (129). Figure 8.4 illustrates the concept of replicative senescence in terms of telomere length, and describes interventional protocols that allow extension of lifespan in culture. Interestingly, expression of hTERT alone is not sufficient to immortalize mammary epithelial cells, which require another event that knocks out the function of the Rb/p16 pathway (130). As a general phenomenon, lifespan extension in epithelial cells requires the abrogation of other pathways. For example, knocking out p53 function in cells derived from thyroid epithelium fails to induce an increase in proliferation (131).

## 5.5 Nature of the p53-activating signal at senescence and the role of DNA repair pathways?

If a critically shortened telomere is the signal that induces senescence, how does this structure induce a p53-mediated DNA damage response? Viewed in another way, telomeres are unrepaired double-strand DNA breaks and cells are well equipped to process these breaks (see Chapters 2 and 7). It is likely that telomere-associated proteins are involved in activating p53. One telomere binding protein, TRF1, has a

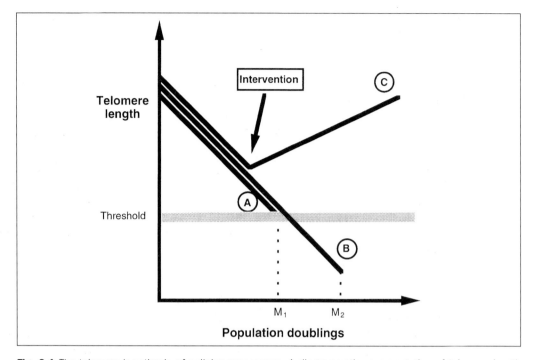

**Fig. 8.4** The telomere hypothesis of cellular senescence. A diagrammatic representation of telomere length versus population doublings in human diploid fibroblasts. Telomeres, comprising the DNA repeat TTAGGG and associated proteins, define the ends of linear chromosomes and shorten with every round of cell division (see Section 5.2). Normal human somatic cells have a finite proliferative capacity and cease dividing after a number of population doublings dependent on the cell type. For example, with no intervention (A) human diploid fibroblasts in culture will eventually arrest due to telomere shortening. It is not yet known whether the threshold is determined by either the total complement of telomeric DNA or a single critically shortened telomere. Replicative senescence or mortality stage 1 (M1) (173) can be overcome by abrogation of p53 function (B). This suggests that at M1 p53 is activated in a similar manner to that observed when DNA damage induces growth arrest at the $G_1$/S checkpoint. However, in (B) lifespan extension is limited as further telomere shortening is apparent, with concomitant gross chromosomal instability, causing growth arrest at mortality stage 2 (M2 or crisis). Reactivation of telomerase by ectopic expression of the catalytic subunit hTERT (C) effectively immortalizes cells by maintaining telomeres at lengths which are incapable of activating the p53 checkpoint function.

limited signalling function in as much as it regulates telomere length (103). Overexpression of TRF1 in telomerase-positive cells causes telomere shortening, while a dominant-negative form of the protein has the opposite effect.

An obvious suggestion is that proteins involved in the repair of double-strand breaks also act as a signal of the presence of shortened telomeres. There is compelling evidence in *Saccharomyces cerevisiae* that the proteins involved in dsb repair and non-homologous end -joining (NHEJ) have activities that are important for telomere function. Yeast telomeres consist of (C1–3 A) tandem repeats. Yeast strains defective in the homologues of Ku70 and Ku80 show significant loss of telomere repeats (132). Similarly, telomere shortening is a characteristic of strains with disrupted *RAD50*,

*XRS2*, or *MRE11* genes (133) which are involved in NHEJ. Loss of the Ku proteins also causes activation of yeast genes that are normally silenced due to their location at telomeres (134). This further implicates the role of DNA repair genes in telomere function.

Poly(ADP-ribose) polymerase (PARP) is activated in the presence of DNA strand breaks (135). Knock-out of PARP function in mice causes an elevation of recombination activity as evidenced by increased levels of sister chromatid exchange (136). In double mutants knock-out of PARP partially rescues the SCID phenotype, underlining the likelihood of involvement in similar pathways. PARP activity may play a role in the initiation of a p53-mediated $G_1$ arrest in senescent cells. The interaction of p53 protein with PARP has been shown, and inhibition of PARP activity abrogates p21 and Mdm2 expression after DNA damage (137). In the same report, hyperoxia, which activates PARP, was associated with p53 activation, telomere shortening, and premature senescence, while inhibition of PARP activity led to the extension of cellular lifespan. Again this links p53 to a response to telomere shortening, although it should be noted that PARP activation might just be an indicator of other cellular events. In addition, the PARP inhibitors used (1,5-dihydroxyisoquinoline and 3-aminobenzamide) may target other NAD-dependent enzymes. Furthermore, there is a telomere-associated PARP activity in cells. Tankyrase, a protein (138) identified by its ability to bind to TRF1, has significant homology to the catalytic domain of PARP. Recombinant tankyrase possesses PARP activity, which is inhibited by 3-aminobenzamide, and both TRF1 and tankyrase act as *in vitro* acceptors for adenosine diphosphate (ADP)-ribosylation. This modification decreases the affinity of TRF1 for telomeric DNA and suggests a method of regulating telomere function. Inhibition of this PARP activity may have mediated the response demonstrated by Vaziri *et al.* (137). The release of tankyrase from telomeres may occur when there are insufficient copies of its partner protein TRF1 present (i.e. when telomeres shorten).

Although end-joining functions are conserved between yeast and man, there is as yet no evidence to support the idea that the same proteins are involved in mammalian telomere regulation. Human cells contain binding activities (TRF1 and TRF2) not present in yeast. Differences in the nature of the tandem repeat sequence of telomeres may provide a simple explanation as to the lack conservation between species. However, it is clear that proteins involved in the repair of DNA strand breaks have activities which recruit p53.

Both ATM (139, 140) and DNA-PK (141) have demonstrable kinase activities directed towards p53 and on the same amino acid residue, namely serine 15. In the case of DNA-PK, cells derived from SCID mice and a human glioma cell line lacking DNA-PK activity, both exhibited no p53 sequence specific binding activity in response to DNA damage. Whether or not the enzymatic activities of DNA-PK and ATM activate p53 protein directly is contentious (142). For example, fibroblasts derived from SCID mice possess intact p53-mediated checkpoints (143). DNA-PK and ATM modify the amino-terminal region, which controls the interaction of p53 with the transcriptional apparatus and also Mdm2 (a protein that regulates p53 but targeting it for degradation by the proteasome) (12, 13).

# 6. Human cancer-prone disorders

It is clear that host factors can play a predominant role in allowing cells to survive genotoxic insults and provide a population of cells within which neoplastic progression can occur. The importance of DNA repair/protection in determining the balance of events that lead to neoplasia continues to be highlighted by studies on *de-novo* or established germline mutations in humans.

## 6.1 Perspective

The degree of DNA protection may differ from one individual to another. Genetic variation is observable in the human population for the expression of genes involved in the metabolism of organic chemical carcinogens (144) and oxidative chemical radicals (145). Thus, loss of function mutations in such genes may enhance cancer proneness. An alternative route for cancer predisposition is through the loss of function mutations in genes involved in the execution or regulation of DNA repair, or in genes maintaining genomic integrity (146, 147). There is also a widening group of genes which are now recognized to be either tumour suppressor genes, which normally act as negative regulators of various cellular functions, or proto-oncogenes, acting through gain of function mutations to deregulate cell-cycle control or cellular differentiation (21, 148). In general, a loss of function mutation in a tumour suppressor gene, such as the loss of ability to co-ordinate programmed cell death (apoptosis) with events imposing genomic stress, can enhance tumour formation (148–150).

The majority of germline mutations leading to disorders in DNA processing are autosomal recessive and tend to be rare. The sunlight-sensitivity and skin cancer-predisposing disorder xeroderma pigmentosum (XP) is now well understood (151), presenting a mutation in one of the many genes involved in the processing of UV-radiation induced DNA lesions (152). Cellular sensitivity to certain DNA damaging agents has also revealed DNA processing dysfunction in Bloom syndrome and Fanconi anaemia (153). Ionizing radiation has been used extensively as a probe for processing dysfunction, and this approach was effective in establishing cellular sensitivity in the rare disorders ataxia telangiectasia (A–T) and Nijmegen Breakage syndrome (NBS). A–T presents as a multisystem recessive disease characterized clinically by cerebellar ataxia, oculocutaneous telangiectasia, in which cellular sensitivity extends to radiomimetic agents. This pleiotropic disorder is caused by mutations in the *ATM* (mutated in A–T) gene, which is located in the human chromosomal region 11q22–q23. The *ATM* gene is discussed in more detail in Chapter 7. The *ATM* gene product is a member of a family of large proteins implicated in the regulation of the cell cycle and response to DNA damage. In both A–T and NBS there is predisposition to lymphohaemopoietic tumours, with each showing a characteristic but different spectrum (154, 155).

There are two well-characterized autosomal-dominant disorders in which mutations arise in DNA processing genes. Hereditary non-polyposis colon cancer

(HNPCC) and Li–Fraumeni syndrome (LFS) excess cancer is clearly seen within a familial context. HNPCC is an inherited disorder showing an excess of colon cancer and accounting for a few per cent of total colon cancer in the general population. Here the DNA processing abnormality is due to errors in mismatch correction (147). The rare disorder of LFS is often characterized by mutation in one copy of the *TP53* gene. The *TP53* gene product appears to be part of the cell's DNA damage signalling system and acts to co-ordinate cycle arrest and programmed death in damaged cells (156). Functionally LFS and HNPCC could be regarded as tumour suppressor gene disorders.

Genetically determined abnormalities in stress response systems, including DNA repair, can substantially modify the risk of neoplasia. For example, over 5% of the cancer patient population may be radiation-sensitive due to genetic status, and there may be overrepresentation of such individuals within the group undergoing anticancer treatment. These individuals include not only rare ataxia–telangiectasia (157) homozygotes with an up to threefold normal radiation sensitivity, but also far more numerous patients with slightly enhanced radiosensitivity (158). Dominant effects of the *ATM* gene, in A–T heterozygotes, may contribute to cancer risk in the general population, although recent estimates suggest that this only applies to a few per cent of breast cancer cases (159).

The existence of occult high-risk groups of patients with underlying DNA repair defects could effectively reduce the maximum tolerable radiation and drug exposures considered appropriate for cancer therapy in the general population. One option would be to identify high-risk individuals and modify the intensity of treatment with the expectation that the tumour cells would retain any genetically determined defect, and therefore any reduction in radiation/drug dose would not result in undertreatment of the tumour (160). An accompanying strategy of more aggressive therapy of apparently normal-risk individuals would have the attendant problem of increased risk of secondary tumours. It is clear that the impact of genetically determined variations in DNA repair capacity on normal and tumour cell responses to anticancer therapies must be evaluated before treatment strategies can accommodate the existence of occult high-risk groups.

## 6.2 Germline disorders involving dominant proto-oncogenes and dominant tumour suppressor genes

Germline mutations in the RET proto-oncogene appear to be involved in the familial thyroid cancer predisposition in multiple endocrine neoplasia (MEN) 2A/2B and in familial medullary thyroid cancer (FMTC) (153). There are also strong pointers to an underlying defect in the *p16* proto-oncogene (161), involved in the regulation of cyclin-dependent kinase activity, in certain kindreds afflicted with familial suscept-ibility to melanoma. However, it is the expanding group of germline tumour suppressor gene abnormalities which is generating intense interest, not least because of the degree of tissue specificity for excess cancer and the options for screening for individuals at risk. Indeed, the highest prevalence in the population appears to be for disorders associated with excess breast, ovarian, and prostate cancer. For example, the major contributory genes for breast and ovarian cancer are *BRCA1* and *BRCA2*, both genes showing high penetrance (153). Both BRCA1 and BRCA2 produce embryonic lethality in null mouse embryos. This phenotype is partially rescued by abrogation of p53 function (162), suggesting a role for familial breast cancer genes in DNA damage repair—particularly the repair of DNA double-strand breaks (163, 164). Studies with BRCA1 have also revealed the existence of p53 interactions with proteins involved with recombination and repair. BRCA1 interacts with hRAD51, the human homologue of RecA (165), demonstrated by co-immunoprecipitation and co-localization at synaptonemal complexes in meiotic cells. BRCA1 also physically associates with p53 and stimulates its transcriptional activity (166). In addition to its p53-related functions, BRCA1 is the founder of a protein family displaying BRCT motifs. These motifs are present in a wide range of DNA repair proteins. For example, a BRCT domain in DNA ligase III-alpha mediates its interaction with XRCC1 (167), although the same domains in DNA ligase IV are not essential to bind to XRCC4 (168).

The paradigm for excess cancer risk accompanied by early onset of malignancy has been the retinoblastoma gene (*RB1*) (149). Another high penetrance, but even lower prevalence, disorder is the nevoid basal cell carcinoma syndrome (NBCCS). Molecular studies, particularly on *RB1*, affirm the hypothesis of Knudson (169) that the loss of one germline copy from all somatic cells imposes a dominant pre-disposition to cancer because of the likelihood of losing the remaining wild-type copy. One would predict that such a risk would reflect the fidelity of pathways that act to limit the consequences of DNA damage and the genotoxic load experienced by various tissue compartments. Importantly, RB and NBCCS individuals subjected to conventional radiotherapy appear to develop an excess of secondary neoplasia (170–172).

# 7. Conclusions

It is clear that cells possess complex mechanisms which act to preserve the integrity of the genetic material, and many of them are described in the preceding chapters. The detrimental consequences of the misfunction of these processes are apparent when considering the high frequency of cancer incidence in human populations. It is of no surprise that the appearance of mutations in key genes often underlie initiating events in tumorigenesis. These mutations can arise as a consequence of DNA repair mechanisms working at less than 100 per cent efficiency, or as a result of defects in the lesion processing machinery itself. Integration of stress activated mechanisms, in terms of detection, repair, and determining whether a successful response has been mounted, is vital in order for the cell to resume or maintain its normal function. The p53 protein is an attractive candidate for assuming the role of command and control at the heart of the cell's response to stress.

# References

1. Renan, M. J. (1993) How many mutations are required for tumorigenesis? Implications from human cancer data. *Mol. Carcinog.*, **7**, 139–46.

2. Little, M. P. (1995) Are two mutations sufficient to cause cancer? Some generalizations of the two-mutation model of carcinogenesis of Moolgavkar, Venzon, and Knudson, and of the multistage model of Armitage and Doll. *Biometrics*, **51**, 1278–91.

3. Goodhead, D. T., Thacker, J., and Cox, R. (1993) Weiss Lecture. Effects of radiations of different qualities on cells: molecular mechanisms of damage and repair. *Int. J. Radiat. Biol.*, **63**, 543–56.

4. Savitsky, K., Bar-Shira, A., Gilad, S., Rotman, G., Ziv, Y., Vanagaite, L., *et al.* (1995) A single ataxia telangiectasia gene with a product similar to PI-3 kinase. *Science*, **268**, 1749–53.

5. Ross, G. M., Eady, J. J., Mithal, N. P., Bush, C., Steel, G. G., Jeggo, P. A., *et al.* (1995) DNA strand break rejoining defect in xrs-6 is complemented by transfection with the human Ku80 gene. *Cancer Res.*, **55**, 1235–8.

6. Tambini, C. E., George, A. M., Rommens, J. M., Tsui, L. C., Scherer, S. W., and Thacker, J. (1997) The XRCC2 DNA repair gene: identification of a positional candidate. *Genomics*, **41**, 84–92.

7. Goodhead, D. T. (1994) Initial events in the cellular effects of ionizing radiations: clustered damage in DNA. *Int. J. Radiat. Biol.*, **65**, 7–17.

8. Lane, D. P. (1994) The regulation of p53 function: Steiner Award Lecture. *Int. J. Cancer*, **57**, 623–7.

9. Bunz, F., Dutriaux, A., Lengauer, C., Waldman, T., Zhou, S., Brown, J. P., *et al.* (1998) Requirement for p53 and p21 to sustain $G_2$ arrest after DNA damage. *Science*, **282**, 1497–501.

10. Winters, Z. E., Ongkeko, W. M., Harris, A. L., and Norbury, C. J. (1998) p53 regulates Cdc2 independently of inhibitory phosphorylation to reinforce radiation-induced $G_2$ arrest in human cells. *Oncogene*, **17**, 673–84.

11. Ko, L. J. and Prives, C. (1996) p53: puzzle and paradigm. *Genes Dev.*, **10**, 1054–72.

12. Kubbutat, M. H., Jones, S. N., and Vousden, K. H. (1997) Regulation of p53 stability by Mdm2. *Nature*, **387**, 299–303.

13. Haupt, Y., Maya, R., Kazaz, A., and Oren, M. (1997) Mdm2 promotes the rapid degradation of p53. *Nature*, **387**, 296–9.

14. Huang, L. C., Clarkin, K. C., and Wahl, G. M. (1996) Sensitivity and selectivity of the DNA damage sensor responsible for activating p53-dependent $G_1$ arrest. *Proc. Natl Acad. Sci. USA*, **93**, 4827–32.

15. Mummenbrauer, T., Janus, F., Muller, B., Wiesmuller, L., Deppert, W., and Grosse, F. (1996) p53 Protein exhibits 3'-to-5' exonuclease activity. *Cell*, **85**, 1089–99.

16. Okorokov, A. L., Ponchel, F., and Milner, J. (1997) Induced N- and C-terminal cleavage of p53: a core fragment of p53, generated by interaction with damaged DNA, promotes cleavage of the N-terminus of full-length p53, whereas ssDNA induces C-terminal cleavage of p53. *EMBO J.*, **16**, 6008–17.

17. Coverley, D., Kenny, M. K., Munn, M., Rupp, W. D., Lane, D. P., and Wood, R. D. (1991) Requirement for the replication protein SSB in human DNA excision repair. *Nature*, **349**, 538–41.

18. Dutta, A., Ruppert, J. M., Aster, J. C., and Winchester, E. (1993) Inhibition of DNA replication factor RPA by p53. *Nature*, **365**, 79–82.

19. Barret, J. M. and Hill, B. T. (1998) DNA repair mechanisms associated with cellular resistance to antitumor drugs: potential novel targets. *Anticancer Drugs*, **9**, 105–23.

20. Souhami, R. (1995) Cancer therapy, cell cycle control and death. In *Apoptosis and cell cycle control in cancer* (ed. N. S. B. Thomas), pp. 149–160. Bios Scientific, Oxford.

21. Bishop, J. M. (1991) Molecular themes in oncogenesis. *Cell*, **64**, 235–48.

22. Thomas, N. S. B. (1995) Introduction to life and death. In *Apoptosis and cell cycle control in cancer* (ed. N. S. B. Thomas), pp. 1–16. Bios Scientific, Oxford.

23. Rolfe, M. (1995) Novel cell cycle targets in cancer. In *Apoptosis and cell cycle control in cancer* (ed. N. S. B. Thomas), pp. 191–203. Bios Scientific, Oxford.

24. Kinsella, A. R., Smith, D., and Pickard, M. (1997) Resistance to chemotherapeutic anti-metabolites: a function of salvage pathway involvement and cellular response to DNA damage. *Br. J. Cancer*, **75**, 935–45.

25. Lawley, P. D. (1994) From fluorescence spectra to mutational spectra, a historical overview of DNA-reactive compounds. *IARC Sci. Publ.*, **125**, 3–22.

26. Lawley, P. D. (1995) Alkylation of DNA and its aftermath. *Bioessays*, **17**, 561–8.

27. Lawley, P. D. and Phillips, D. H. (1996) DNA adducts from chemotherapeutic agents. *Mutat. Res.*, **355**, 13–40.

28. Ali, R. B., Teo, A. K., Oh, H. K., Chuang, L. S., Ayi, T. C., and Li, B. F. (1998) Implication of localization of human DNA repair enzyme O6-methylguanine-DNA methyltransferase at active transcription sites in transcription-repair coupling of the mutagenic O6-methylguanine lesion. *Mol. Cell. Biol.*, **18**, 1660–9.

29. Russell, S. J., Ye, Y. W., Waber, P. G., Shuford, M., Schold, S. C. Jr, and Nisen, P. D. (1995) p53 mutations, O6-alkylguanine DNA alkyltransferase activity, and sensitivity to procarbazine in human brain tumors. *Cancer*, **75**, 1339–42.

30. Russell, L. B., Hunsicker, P. R., and Shelby, M. D. (1996) Chlorambucil and bleomycin induce mutations in the specific-locus test in female mice. *Mutat. Res.*, **358**, 25–35.

31. Nicolaou, K. C., Smith, A. L., and Yue, E. W. (1993) Chemistry and biology of natural and designed enediynes. *Proc. Natl Acad. Sci. USA*, **90**, 5881–8.

32. Mir, L. M., Tounekti, O., and Orlowski, S. (1996) Bleomycin: revival of an old drug. *Gen. Pharmacol.*, **27**, 745–8.

33. Povirk, L. F. (1996) DNA damage and mutagenesis by radiomimetic DNA-cleaving agents: bleomycin, neocarzinostatin and other enediynes. *Mutat. Res.*, **355**, 71–89.

34. Donner, E. M. and Preston, R. J. (1996) The relationship between p53 status, DNA repair and chromatid aberration induction in $G_2$ mouse embryo fibroblast cells treated with bleomycin. *Carcinogenesis*, **17**, 1161–5.

35. Wu, X. F., Spitz, M. R., Delclos, G. L., Connor, T. H., Zhao, Y., Siciliano, M. J., *et al.* (1996) Survival of cells with bleomycin-induced chromosomal lesions in the cultured lymphocytes of lung cancer patients. *Cancer, Epidemiol. Biomarkers Prev.*, **5**, 527–32.

36. Cummings, J. and Smyth, J. F. (1993) DNA topoisomerase I and II as targets for rational design of new anticancer drugs. *Ann. Oncol.*, **4**, 533–43.

37. Smith, P. J. and Soues, S. (1994) Multilevel therapeutic targeting by topoisomerase inhibitors. *Br. J. Cancer* (Suppl.), **23**, S47–51.

38. Anderson, R. D. and Berger, N. A. (1994) International Commission for Protection Against Environmental Mutagens and Carcinogens. Mutagenicity and carcinogenicity of topoisomerase-interactive agents. *Mutat. Res.*, **309**, 109–42.

39. Chen, A. Y. and Liu, L. F. (1994) DNA topoisomerases: essential enzymes and lethal targets. *Annu. Rev. Pharmacol. Toxicol.*, **34**, 191–218.

40. Neece, S. H., Carles-Kinch, K., Tomso, D. J., and Kreuzer, K. N. (1996) Role of

recombinational repair in sensitivity to an antitumour agent that inhibits bacteriophage T4 type II DNA topoisomerase. *Mol. Microbiol.*, **20**, 1145–54.

41. Kingma, P. S. and Osheroff, N. (1997) Spontaneous DNA damage stimulates topoisomerase II-mediated DNA cleavage. *J. Biol. Chem.*, **272**, 7488–93.

42. Ikushima, T., Shima, Y., and Ishii, Y. (1998) Effects of an inhibitor of topoisomerase II, ICRF-193 on the formation of ultraviolet-induced chromosomal aberrations. *Mutat. Res.*, **404**, 35–8.

43. Downes, C. S., Clarke, D. J., Mullinger, A. M., Gimenez-Abian, J. F., Creighton, A. M., and Johnson, R. T. (1994) A topoisomerase II-dependent $G_2$ cycle checkpoint in mammalian cells. *Nature*, **372**, 467–70.

44. Eng, W. K., Faucette, L., Johnson, R. K., and Sternglanz, R. (1988) Evidence that DNA topoisomerase I is necessary for the cytotoxic effects of camptothecin. *Mol. Pharmacol.*, **34**, 755–60.

45. Hsiang, Y. H., Hertzberg, R., Hecht, S., and Liu, L. F. (1985) Camptothecin induces protein-linked DNA breaks via mammalian DNA topoisomerase I. *J. Biol. Chem.*, **260**, 14873–8.

46. Robinson, M. J. and Osheroff, N. (1990) Stabilization of the topoisomerase II-DNA cleavage complex by antineoplastic drugs: inhibition of enzyme-mediated DNA religation by 4'-(9-acridinylamino)methanesulfon-*m*-anisidide. *Biochemistry*, **29**, 2511–15.

47. Zhang, H., D'Arpa, P., and Liu, L. F. (1990) A model for tumor cell killing by topoisomerase poisons. *Cancer Cells*, **2**, 23–7.

48. Li, L. H., Fraser, T. J., Olin, E. J., and Bhuyan, B. K. (1972) Action of camptothecin on mammalian cells in culture. *Cancer Res.*, **32**, 2643–50.

49. Goldwasser, F., Shimizu, T., Jackman, J., Hoki, Y., O'Connor, P. M., Kohn, K. W., *et al.* (1996) Correlations between S and $G_2$ arrest and the cytotoxicity of camptothecin in human colon carcinoma cells. *Cancer Res.*, **56**, 4430–7.

50. Wu, J. and Liu, L. F. (1997) Processing of topoisomerase I cleavable complexes into DNA damage by transcription. *Nucl. Acids Res.*, **25**, 4181–6.

51. Fujimori, A., Gupta, M., Hoki, Y., and Pommier, Y. (1996) Acquired camptothecin resistance of human breast cancer MCF-7/C4 cells with normal topoisomerase I and elevated DNA repair. *Mol. Pharmacol.*, **50**, 1472–8.

52. Boothman, D. A., Trask, D. K., and Pardee, A. B. (1989) Inhibition of potentially lethal DNA damage repair in human tumor cells by beta-lapachone, an activator of topoisomerase I. *Cancer Res.*, **49**, 605–12.

53. Subramanian, D., Rosenstein, B. S., and Muller, M. T. (1998) Ultraviolet-induced DNA damage stimulates topoisomerase I-DNA complex formation *in vivo*: possible relationship with DNA repair. *Cancer Res.*, **58**, 976–84.

54. Nakamura, Y. (1998) ATM: the p53 booster. *Nature Med.*, **4**, 1231–2.

55. Callahan, R. (1992) p53 mutations, another breast cancer prognostic factor. *J. Natl Cancer Inst.*, **84**, 826–7.

56. Morgan, W. F., Day, J. P., Kaplan, M. I., McGhee, E. M., and Limoli, C. L. (1996) Genomic instability induced by ionizing radiation. *Radiat Res.*, **146**, 247–58.

57. Coleman, C. N. (1996) Modulating the radiation response. *Stem Cells*, **14**, 10–15.

58. DeFrank, J. S., Tang, W., and Powell, S. N. (1996) p53-null cells are more sensitive to ultraviolet light only in the presence of caffeine. *Cancer Res.*, **56**, 5365–8.

59. O'Connor, P. M. (1997) Mammalian $G_1$ and $G_2$ phase checkpoints. *Cancer Surv.*, **29**, 151–82.

60. Wang, X. and Ohnishi, T. (1997) p53-dependent signal transduction induced by stress. *J. Radiat. Res.(Tokyo)*, **38**, 179–94.

61. Lane, D. P. (1992) p53, guardian of the genome. *Nature*, **358**, 15–16.

62. Donehower, L. A., Harvey, M., Slagle, B. L., McArthur, M. J., Montgomery, C. A., Jr., Butel, J. S., *et al.* (1992) Mice deficient for p53 are developmentally normal but susceptible to spontaneous tumours. *Nature*, **356**, 215–21.

63. Malkin, D., Li, F. P., Strong, L. C., Fraumeni, J. F., Jr., Nelson, C. E., Kim, D. H., *et al.* (1990) Germ line p53 mutations in a familial syndrome of breast cancer, sarcomas, and other neoplasms. *Science*, **250**, 1233–8.

64. Srivastava, S., Zou, Z. Q., Pirollo, K., Blattner, W., and Chang, E. H. (1990) Germ-line transmission of a mutated p53 gene in a cancer-prone family with Li–Fraumeni syndrome. *Nature*, **348**, 747–9.

65. Greenblatt, M. S., Bennett, W. P., Hollstein, M., and Harris, C. C. (1994) Mutations in the p53 tumor suppressor gene: clues to cancer etiology and molecular pathogenesis. *Cancer Res.*, **54**, 4855–78.

66. Barnes, D. M., Dublin, E. A., Fisher, C. J., Levison, D. A., and Millis, R. R. (1993) Immunohistochemical detection of p53 protein in mammary carcinoma: an important new independent indicator of prognosis? *Hum. Pathol.*, **24**, 469–76.

67. Dowell, S. P. and Hall, P. A. (1995) The p53 tumour suppressor gene and tumour prognosis: is there a relationship? *J. Pathol.*, **177**, 221–4.

68. Engelward, B. P., Allan, J. M., Dreslin, A. J., Kelly, J. D., Wu, M. M., Gold, B., *et al.* (1998) A chemical and genetic approach together define the biological consequences of 3-methyladenine lesions in the mammalian genome. *J. Biol. Chem.*, **273**, 5412–18.

69. Weinert, T. (1998) DNA damage and checkpoint pathways: molecular anatomy and interactions with repair. *Cell*, **94**, 555–8.

70. Tishler, R. B., Calderwood, S. K., Coleman, C. N., and Price, B. D. (1993) Increases in sequence specific DNA binding by p53 following treatment with chemotherapeutic and DNA damaging agents. *Cancer Res.*, **53**, 2212–16.

71. Lu, X. and Lane, D. P. (1993) Differential induction of transcriptionally active p53 following UV or ionizing radiation: defects in chromosome instability syndromes? *Cell*, **75**, 765–78.

72. Hupp, T. R., Sparks, A., and Lane, D. P. (1995) Small peptides activate the latent sequence-specific DNA binding function of p53. *Cell*, **83**, 237–45.

73. el-Deiry, W. S., Tokino, T., Velculescu, V. E., Levy, D. B., Parsons, R., Trent, J. M., *et al.* (1993) WAF1, a potential mediator of p53 tumor suppression. *Cell*, **75**, 817–25.

74. Kastan, M. B., Zhan, Q., el-Deiry, W. S., Carrier, F., Jacks, T., Walsh, W. V., *et al.* (1992) A mammalian cell cycle checkpoint pathway utilizing p53 and GADD45 is defective in ataxia-telangiectasia. *Cell*, **71**, 587–97.

75. Barak, Y., Juven, T., Haffner, R., and Oren, M. (1993) mdm2 expression is induced by wild type p53 activity. *EMBO J.*, **12**, 461–8.

76. Miyashita, T. and Reed, J. C. (1995) Tumor suppressor p53 is a direct transcriptional activator of the human bax gene. *Cell*, **80**, 293–9.

77. Kern, S. E., Pietenpol, J. A., Thiagalingam, S., Seymour, A., Kinzler, K. W., and Vogelstein, B. (1992) Oncogenic forms of p53 inhibit p53-regulated gene expression. *Science*, **256**, 827–30.

78. Schwartz, D., Almog, N., Peled, A., Goldfinger, N., and Rotter, V. (1997) Role of wild type p53 in the $G_2$ phase: regulation of the gamma-irradiation-induced delay and DNA repair. *Oncogene*, **15**, 2597–607.

79. Pellegata, N. S., Antoniono, R. J., Redpath, J. L., and Stanbridge, E. J. (1996) DNA damage and p53-mediated cell cycle arrest: a reevaluation. *Proc. Natl Acad. Sci. USA*, **93**, 15209–14.

80. Larner, J. M., Lee, H., and Hamlin, J. L. (1997) S phase damage sensing checkpoints in mammalian cells. *Cancer Surv.*, **29**, 25–45.

81. Prost, S., Bellamy, C. O., Clarke, A. R., Wyllie, A. H., and Harrison, D. J. (1998) p53-independent DNA repair and cell cycle arrest in embryonic stem cells. *FEBS Lett.*, **425**, 499–504.

82. Bristow, R. G., Hu, Q., Jang, A., Chung, S., Peacock, J., Benchimol, S., *et al.* (1998) Radioresistant MTp53-expressing rat embryo cell transformants exhibit increased DNA-dsb rejoining during exposure to ionizing radiation. *Oncogene*, **16**, 1789–802.

83. Bill, C. A., Yu, Y., Miselis, N. R., Little, J. B., and Nickoloff, J. A. (1997) A role for p53 in DNA end rejoining by human cell extracts. *Mutat. Res.*, **385**, 21–9.

84. Denissenko, M. F., Pao, A., Pfeifer, G. P., and Tang, M. (1998) Slow repair of bulky DNA adducts along the nontranscribed strand of the human p53 gene may explain the strand bias of transversion mutations in cancers. *Oncogene*, **16**, 1241–7.

85. Pfeifer, G. P. and Denissenko, M. F. (1998) Formation and repair of DNA lesions in the p53 gene: relation to cancer mutations? *Environ. Mol. Mutagen*, **31**, 197–205.

86. Tolbert, D. M. and Kantor, G. J. (1996) Definition of a DNA repair domain in the genomic region containing the human p53 gene. *Cancer Res.*, **56**, 3324–30.

87. Cheo, D. L., Ruven, H. J., Meira, L. B., Hammer, R. E., Burns, D. K., Tappe, N. J., *et al.* (1997) Characterization of defective nucleotide excision repair in XPC mutant mice. *Mutat. Res.*, **374**, 1–9.

88. Boulikas, T. (1996) DNA lesion-recognizing proteins and the p53 connection. *Anticancer Res.*, **16**, 225–42.

89. Ford, J. M. and Hanawalt, P. C. (1995) Li–Fraumeni syndrome fibroblasts homozygous for p53 mutations are deficient in global DNA repair but exhibit normal transcription-coupled repair and enhanced UV resistance. *Proc. Natl Acad. Sci. USA*, **92**, 8876–80.

90. Ford, J. M. and Hanawalt, P. C. (1997) Expression of wild-type p53 is required for efficient global genomic nucleotide excision repair in UV-irradiated human fibroblasts. *J. Biol. Chem.*, **272**, 28073–80.

91. Ford, J. M., Baron, E. L., and Hanawalt, P. C. (1998) Human fibroblasts expressing the human papillomavirus E6 gene are deficient in global genomic nucleotide excision repair and sensitive to ultraviolet irradiation. *Cancer Res.*, **58**, 599–603.

92. Ko, L. J., Shieh, S. Y., Chen, X., Jayaraman, L., Tamai, K., Taya, Y., *et al.* (1997) p53 is phosphorylated by CDK7-cyclin H in a p36MAT1-dependent manner. *Mol. Cell. Biol.*, **17**, 7220–9.

93. Jones, C. J. and Wynford-Thomas, D. (1995) Is TFIIH an activator of the p53-mediated $G_1/S$ checkpoint? *Trends Genet.*, **11**, 165–6.

94. Abramova, N. A., Russell, J., Botchan, M., and Li, R. (1997) Interaction between replication protein A and p53 is disrupted after UV damage in a DNA repair-dependent manner. *Proc. Natl Acad. Sci. USA*, **94**, 7186–91.

95. Hayflick, L. (1965) The limited *in vitro* lifetime of human diploid cell strains. *Exp. Cell Res.*, **37**, 614–636.

96. Rubin, H. (1998) Telomerase and cellular lifespan: ending the debate? *Nature Biotechnol.*, **16**, 396–7.

97. Faragher, R. G., Jones, C. J., and Kipling, D. (1998) Telomerase and cellular lifespan: ending the debate? *Nature Biotechnol.*, **16**, 701–2.

98. Lengauer, C., Kinzler, K. W., and Vogelstein, B. (1998) Genetic instabilities in human cancers. *Nature*, **396**, 643–9.

99. Lansdorp, P. M., Verwoerd, N. P., van de Rijke, F. M., Dragowska, V., Little, M. T., Dirks, R. W., *et al.* (1996) Heterogeneity in telomere length of human chromosomes. *Hum. Mol. Genet*, **5**, 685–91.

100. Martens, U. M., Zijlmans, J. M., Poon, S. S., Dragowska, W., Yui, J., Chavez, E. A., Ward, R. K., and Lansdorp, P. M. (1998) Short telomeres on human chromosome 17p. *Nature Genet.*, **18**, 76–80.

101. Allsopp, R. C., Vaziri, H., Patterson, C., Goldstein, S., Younglai, E. V., Futcher, A. B., *et al.* (1992) Telomere length predicts replicative capacity of human fibroblasts. *Proc. Natl. Acad. Sci. USA* **89**, 10114–18.

102. Chong, L., Vansteensel, B., Broccolli, D., Erdjument-Bromage, H., Hanish, J., Tempst, P., *et al.* (1995) A human telomeric protein. *Science*, **270**, 1663–7.

103. van Steensel, B. and de Lange, T. (1997) Control of telomere length by the human telomeric protein TRF1. *Nature*, **385**, 740–3.

104. Smith, S. and de Lange, T. (1997) TRF1, a mammalian telomeric protein. *Trends Genet.*, **13**, 21–26.

105. Bilaud, T., Brun, C., Ancelin, K., Koering, C. E., Laroche, T., and Gilson, E. (1997) Telomeric localization of TRF2, a novel human telobox protein. *Nature Genet.*, **17**, 236–9.

106. Broccoli, D., Smogorzewska, A., Chong, L., and de Lange, T. (1997) Human telomeres contain two distinct Myb-related proteins, TRF1 and TRF2. *Nature Genet.*, **17**, 231–5.

107. van Steensel, B., Smogorzewska, A., and de Lange, T. (1998) TRF2 protects human telomeres from end-to-end fusions. *Cell*, **92**, 401–13.

108. Waga, S. and Stillman, B. (1994) Anatomy of a DNA replication fork revealed by reconstitution of SV40 DNA replication *in vitro*. *Nature*, **369**, 207–12.

109. Olovnikov, A. M. (1973) A theory of marginotomy: the incomplete copying of template margin in enzymatic synthesis of polynucleotides and biological significance of the problem. *J. Theor. Biol.*, **41**, 181–190.

110. Harley, C. B., Futcher, A. B., and Greider, C. W. (1990) Telomeres shorten during ageing of human fibroblasts. *Nature*, **345**, 458–60.

111. Allsopp, R. C. and Harley, C. B. (1995) Evidence for a critical telomere length in senescent human fibroblasts. *Exp. Cell Res.*, **219**, 130–6.

112. Feng, J., Funk, W. D., Wang, S.-S., Weinrich, S. L., Avilion, A. A., Chiu, C.-P., *et al.* (1995) The RNA component of human telomerase. *Science*, **269**, 1236–41.

113. Lee, H. W., Blasco, M. A., Gottlieb, G. J., Horner, J. W. 2nd, Greider, C. W., and DePinho, R. A. (1998) Essential role of mouse telomerase in highly proliferative organs. *Nature*, **392**, 569–74.

114. Blasco, M. A., Lee, H. W., Hande, M. P., Samper, E., Lansdorp, P. M., DePinho, R. A., *et al.* (1997) Telomere shortening and tumor formation by mouse cells lacking telomerase RNA. *Cell*, **91**, 25–34.

115. Blasco, M. A., Rizen, M., Greider, C. W., and Hanahan, D. (1995) Differential regulation of telomerase activity and telomerase RNA during multistage carcinogenesis. *Nature, Genet.*, **12**, 200–4.

116. Avilion, A. A., Piatyszek, M. A., Gupta, J., Shay, J. W., Bacchetti, S., and Greider, C. W. (1996) Human telomerase RNA and telomerase activity in immortal cell lines and tumor tissues. *Cancer Res.*, **56**, 645–50.

117. Kim, N. W., Piatyszek, M. A., Prowse, K. R., Harley, C. B., West, M. D., Ho, P. L. C., *et al.* (1994) Specific association of human telomerase activity with immortal cells and cancer. *Science*, **266**, 2011–15.

118. Shay, J. W. and Bacchetti, S. (1997) A survey of telomerase activity in human cancer. *Eur. J. Cancer*, **33**, 787–91.

119. Leteurtre, F., Li, X., Gluckman, E., and Carosella, E. D. (1997) Telomerase activity during the cell cycle and in gamma-irradiated hematopoietic cells. *Leukemia*, **11**, 1681–9.

120. Campisi, J. (1997) The biology of replicative senescence. *Eur. J. Cancer*, **33**, 703–9.

121. Shay, J. W., Pereira-Smith, O. M., and Wright, W. E. (1991) A role for both RB and p53 in the regulation of human cellular senescence. *Exp. Cell Res.*, **196**, 33–9.

122. Gire, V. and Wynford-Thomas, D. (1998) Reinitiation of DNA synthesis and cell division in senescent human fibroblasts by microinjection of anti-p53 antibodies. *Mol. Cell. Biol.*, **18**, 1611–21.

123. Nakamura, T. M., Morin, G. B., Chapman, K. B., Weinrich, S. L., Andrews, W. H., Lingner, J., *et al.* (1997) Telomerase catalytic subunit homologs from fission yeast and human. *Science*, **277**, 955–9.

124. Meyerson, M., Counter, C. M., Eaton, E. N., Ellisen, L. W., Steiner, P., Caddle, S. D., *et al.* (1997) hEST2, the putative human telomerase catalytic subunit gene, is up-regulated in tumor cells and during immortalization. *Cell*, **90**, 785–95.

125. Kilian, A., Bowtell, D. D., Abud, H. E., Hime, G. R., Venter, D. J., Keese, P. K., *et al.* (1997) Isolation of a candidate human telomerase catalytic subunit gene, which reveals complex splicing patterns in different cell types. *Hum. Mol. Genet.*, **6**, 2011–19.

126. Kipling, D. (1997) Mammalian telomerase: catalytic subunit and knockout mice. *Hum. Mol. Genet.*, **6**, 1999–2004.

127. Weinrich, S. L., Pruzan, R., Ma, L., Ouellette, M., Tesmer, V. M., Holt, S. E., *et al.* (1997) Reconstitution of human telomerase with the template RNA component hTR and the catalytic protein subunit hTRT. *Nature Genet.*, **17**, 498–502.

128. Bodnar, A. G., Ouellette, M., Frolkis, M., Holt, S. E., Chiu, C. P., Morin, G. B., *et al.* (1998) Extension of life-span by introduction of telomerase into normal human cells. *Science*, **279**, 349–52.

129. Dimri, G. P., Lee, X. H., Basile, G., Acosta, M., Scott, C., Roskelley, C., *et al.* (1995) A biomarker that identifies senescent human-cells in culture and in aging skin in vivo. *Proc. Natl Acad. Sci. USA*, **92**, 9363–7.

130. Kiyono, T., Foster, S. A., Koop, J. I., McDougall, J. K., Galloway, D. A., and Klingelhutz, A. J. (1998) Both Rb/p16INK4a inactivation and telomerase activity are required to immortalize human epithelial cells. *Nature*, **396**, 84–8.

131. Wynford-Thomas, D. (1995) Molecular genetics of thyroid cancer. *Curr. Opin. Endocrinol. Diabetes*, **2**, 429–36.

132. Porter, S. E., Greenwell, P. W., Ritchie, K. B., and Petes, T. D. (1996) The DNA-binding protein Hdf1p (a putative Ku homologue) is required for maintaining normal telomere length in *Saccharomyces cerevisiae*. *Nucl. Acids Res.*, **24**, 582–5.

133. Critchlow, S. E. and Jackson, S. P. (1998) DNA end-joining: from yeast to man. *Trends Biochem. Sci.*, **23**, 394–8. [In Process citation.]

134. Weaver, D. T. (1998) DNA repair: bacteria to humans. A conference sponsored by the Genetics Society of America, April 16–19, 1998. *Biochim. Biophys. Acta*, **1378**, R1–9.

135. Satoh, M. S. and Lindahl, T. (1992) Role of poly(ADP-ribose) formation in DNA repair. *Nature*, **356**, 356–8.

136. Morrison, C., Smith, G. C., Stingl, L., Jackson, S. P., Wagner, E. F., and Wang, Z. Q. (1997) Genetic interaction between PARP and DNA-PK in V(D)J recombination and tumorigenesis. *Nature Genet.*, **17**, 479–82.

137. Vaziri, H., West, M. D., Allsopp, R. C., Davison, T. S., Wu, Y. S., Arrowsmith, C. H., *et al.* (1997) ATM-dependent telomere loss in aging human diploid fibroblasts and DNA damage lead to the post-translational activation of p53 protein involving poly(ADP-ribose) polymerase. *EMBO J.*, **16**, 6018–33.

138. Smith, S., Giriat, I., Schmitt, A., and de Lange, T. (1998) Tankyrase, a Poly(ADP-ribose) polymerase at human telomeres. *Science*, **282**, 1484–7.

139. Canman, C. E., Lim, D. S., Cimprich, K. A., Taya, Y., Tamai, K., Sakaguchi, K., *et al.* (1998) Activation of the ATM kinase by ionizing radiation and phosphorylation of p53. *Science*, **281**, 1677–9.

140. Banin, S., Moyal, L., Shieh, S., Taya, Y., Anderson, C. W., Chessa, L., *et al.* (1998) Enhanced phosphorylation of p53 by ATM in response to DNA damage. *Science*, **281**, 1674–7.

141. Woo, R. A., McLure, K. G., Lees-Miller, S. P., Rancourt, D. E., and Lee, P. W. (1998) DNA-dependent protein kinase acts upstream of p53 in response to DNA damage. *Nature*, **394**, 700–4.

142. Lane, D. (1998) Awakening angels. *Nature*, **394**, 616–17.

143. Huang, L. C., Clarkin, K. C., and Wahl, G. M. (1996) p53-dependent cell cycle arrests are preserved in DNA-activated protein kinase-deficient mouse fibroblasts. *Cancer Res.*, **56**, 2940–4.

144. Harris, C. C. (1991) Chemical and physical carcinogenesis: advances and perspectives for the 1990s. *Cancer Res.*, **51**, 5023s–5044s.

145. Ward, J. F. (1991) DNA damage and repair. In *Physical and chemical mechanisms in molecular radiation biology* (ed. W. A. Glass and M. N. Varma), pp. 403–21. Plenum, New York.

146. Murnane, J. P. and Kapp, L. N. (1993) A critical look at the association of human genetic syndromes with sensitivity to ionizing radiation. *Semin. Cancer Biol.*, **4**, 93–104.

147. Fishel, R. and Kolodner, R. D. (1995) Identification of mismatch repair genes and their role in the development of cancer. *Curr. Opin. Genet. Dev.*, **5**, 382–95.

148. Karp, J. E. and Broder, S. (1995) Molecular foundations of cancer: new targets for intervention. *Nature Med.*, **1**, 309–20.

149. Weinberg, R. A. (1991) Tumor suppressor genes. *Science*, **254**, 1138–46.

150. Skuse, G. R. and Ludlow, J. W. (1995) Tumour suppressor genes in disease and therapy. *Lancet*, **345**, 902–6.

151. Kraemer, K. H., Levy, D. D., Parris, C. N., Gozukara, E. M., Moriwaki, S., Adelberg, S., *et al.* (1994) Xeroderma pigmentosum and related disorders: examining the linkage between defective DNA repair and cancer. *J. Invest. Dermatol.*, **103**, 96S–101S.

152. Lehmann, A. R. (1998) Dual functions of DNA repair genes: molecular, cellular and clinical implications. *Bioessays*, **20**, 146–155.

153. Auerbach, A. D. and Verlander, P. C. (1997) Disorders of DNA replication and repair. *Current Opinion in Pediatrics*, **9**, 600–16.

154. Taylor, A. M. R., Scott, D., Arlett, C. F., and Cole, J. (1994) Ataxia–telangiectasia: the effect of a pleiotropic gene. Proceedings of the 6th Ataxia–telangiectasia Workshop. Birmingham, UK, 22–25 May 1994. *Int. J. Radiat. Biol.*, **66**, S1–201.

155. Weemaes, C. M., Smeets, D. F., and van der Burgt, C. J. (1994) Nijmegen Breakage syndrome: a progress report. *Int. J. Radiat. Biol.*, **66**, S185–8.

156. Kinzler, K. W. and Vogelstein, B. (1996) Lessons from hereditary colorectal cancer. *Cell*, **87**, 159–70.

157. Vorechovsky, I., Luo, L., Lindblom, A., Negrini, M., Webster, A. D., Croce, C. M., and Hammarstrom, L. (1996) ATM mutations in cancer families. *Cancer Res.*, **56**, 4130–3.

158. Bishop, D. T. and Hopper, J. (1997) AT-tributable risks? [news; comment]. *Nature Genet.*, **15**, 226.

159. FitzGerald, M. G., Bean, J. M., Hegde, S. R., Unsal, H., MacDonald, D. J., Harkin, D. P., *et al.* (1997) Heterozygous ATM mutations do not contribute to early onset of breast cancer. *Nature Genet.*, **15**, 307–10.

160. Busch, D. (1994) Genetic susceptibility to radiation and chemotherapy injury: diagnosis and management. *Int. J. Radiat. Oncol. Biol. Phys.*, **30**, 997–1002.

161. Wainwright, B. (1994) Familial melanoma and p16–a hung jury. *Nature Genet.*, **8**, 3–5.

162. Ludwig, T., Chapman, D. L., Papaioannou, V. E., and Efstratiadis, A. (1997) Targeted mutations of breast cancer susceptibility gene homologs in mice: lethal phenotypes of Brca1, Brca2, Brca1/Brca2, Brca1/p53, and Brca2/p53 nullizygous embryos. *Genes Dev.*, **11**, 1226–41.

163. Kinzler, K. W. and Vogelstein, B. (1997) Cancer-susceptibility genes. Gatekeepers and caretakers. *Nature*, **386**, 761–3.

164. Connor, F., Bertwistle, D., Mee, P. J., Ross, G. M., Swift, S., Grigorieva, E., *et al.* (1997) Tumorigenesis and a DNA repair defect in mice with a truncating Brca2 mutation. *Nature Genet.*, **17**, 423–30.

165. Scully, R., Chen, J., Plug, A., Xiao, Y., Weaver, D., Feunteun, J., *et al.* (1997) Association of BRCA1 with Rad51 in mitotic and meiotic cells. *Cell*, **88**, 265–75.

166. Zhang, H., Somasundaram, K., Peng, Y., Tian, H., Bi, D., Weber, B. L., *et al.* (1998) BRCA1 physically associates with p53 and stimulates its transcriptional activity. *Oncogene*, **16**, 1713–21.

167. Taylor, R. M., Wickstead, B., Cronin, S., and Caldecott, K. W. (1998) Role of a BRCT domain in the interaction of DNA ligase III-alpha with the DNA repair protein XRCC1. *Curr. Biol.*, **8**, 877–80.

168. Grawunder, U., Zimmer, D., and Leiber, M. R. (1998) DNA ligase IV binds to XRCC4 via a motif located between rather than within its BRCT domains *Curr. Biol.*, **8**, 873–6.

169. Knudson, A. G., Jr. (1986) Genetics of human cancer. *Annu. Rev. Genet.*, **20**, 231–51.

170. Wong, F. L., Boice, J. D., Jr., Abramson, D. H., Tarone, R. E., Kleinerman, R. A., Stovall, M., *et al.* (1997) Cancer, incidence after retinoblastoma. Radiation dose and sarcoma risk. *J. Am. Med. Assoc.*, **278**, 1262–7.

171. Strong, L. C. (1977) Genetic and environmental interactions. *Cancer*, **40**, 1861–6.

172. Heimler, A., Friedman, E., and Rosenthal, A. D. (1978) Naevoid basal cell carcinoma syndrome and Charcot–Marie–Tooth disease: two autosomal dominant disorders segregating in a family. *J. Med. Genet.*, **15**, 288–91.

173. Wright, W. E., Pereira-Smith, O. M., and Shay, J. W. (1989) Reversible cellular senescence: implications for immortalisation of normal human diploid fibroblasts. *Mol. Cell. Biol.* **9**, 3088–92.

# Index